Soil Science

Soil Science

Edited by **Brian Bechdal**

SYRAWOOD
PUBLISHING HOUSE

New York

Published by Syrawood Publishing House,
750 Third Avenue, 9th Floor,
New York, NY 10017, USA
www.syrawoodpublishinghouse.com

Soil Science
Edited by Brian Bechdal

International Standard Book Number: 978-1-68286-150-9 (Hardback)

Printed in the United States of America.

Contents

Permissions

List of Contributors

Preface

The main aim of this book is to educate learners and enhance their research focus by presenting diverse topics covering this vast field. This is an advanced book which compiles significant studies by distinguished experts. This book addresses successive solutions to the challenges arising in the area of application, along with it; the book provides scope for future developments.

The rapidly increasing world population has raised concerns regarding preservation of soil and arable land, soil degradation and increasing per capita food consumption. Soil is an essential natural resource for survival on the earth that is constantly being damaged due to human activities. In recent decades, soil science has emerged as a specialised discipline to study soil formation, soil classification and its properties. As an interdisciplinary field, soil science interests a wide range of academicians from different disciplines like geologists, archaeologists, microbiologists and many more. This book is a comprehensive account on soil formation and classification, fertility properties and management of soils. It is an excellent reference for students, researchers and professionals engaged in agriculture or related fields.

It was a great honour to edit this book, though there were challenges, as it involved a lot of communication and networking between me and the editorial team. However, the end result was this all-inclusive book covering diverse themes in the field.

Finally, it is important to acknowledge the efforts of the contributors for their excellent chapters, through which a wide variety of issues have been addressed. I would also like to thank my colleagues for their valuable feedback during the making of this book.

Editor

Effect of soil amendment with yeasts as bio-fertilizers on the growth and productivity of sugar beet

Ramadan AGAMY[1], Mohamed HASHEM[2,3]* and Saad ALAMRI[3]

[1]Department of Agricultural Botany, Faculty of Agriculture, Fayoum University, Fayoum, Egypt.
[2]Botany Department, Faculty of Science, Assiut University, Assiut, Egypt, 71516.
[3]Biology Department, Faculty of Science, King Khalid University, P. O. Box 10255, Abha 61321, Saudi Arabia.

The use of yeast as a bio-fertilizer in agriculture has received considerable attention because of their bioactivity and safety for human and the environment. This study evaluated the effect of soil amendment with three newly isolated yeast strains on the productivity and the external and internal structure of sugar beet to prove their application as bio-fertilizer. We conducted a two-year pot experiment to investigate the effects of *Kluyveromyces walti*, *Pachytrichospora transvaalensis* and *Sacharromycopsis cataegensis* on the growth and productivity of sugar beet. Soil was inoculated with three doses of each strain (0.0, 50.0 and 100.0 ml pot^{-1} with concentration of ~10^8 cfu ml^{-1}). Results showed that application of the yeasts significantly ($P < 0.05$) increased the photosynthetic pigments, soluble sugars, sucrose, and total soluble proteins of sugar beet. *K. walti* showed the best results among the three yeasts. It increased the sucrose content by about 43% of the control. Anatomy of the leaf and the root showed an increase in thickness of the blade, midvein, dimensions of the vascular bundles, and number and diameter of xylem vessels as the result of application of yeasts. Gas chromatography–mass spectrometry (GC-MS) analysis of the culture filtrates of the yeasts detected some beneficial secondary metabolites that could enhance the plant vigor and the physical and chemical properties of the soil. We assume that application of *K. walti*, *P. transvaalensis* and *S. cataegensis* as bio-fertilizers is a good alternative of the chemicals in the sustainable and organic farming and safe for human and environment.

Key words: Bio-fertilizer, sugar beet, yeast, anatomy, secondary metabolites.

INTRODUCTION

Organic farming strategy is growing rapidly all over the world to conserve human health and the environment, which became under risk because of the unbalance use of pesticides and chemical fertilizers. The dangerous effect is because the repeated use of chemical fertilizers destroys soil biota (Boraste et al., 2009). Organic farmingfarming is 'zero impact' on the environment (www.seedbuzz.com).

Bio-fertilizers are formulations of beneficial microorganisms, which upon application can increase the availability of nutrients by their biological activity and help to improve the soil health. Microbes involved in the formulation of bio-fertilizers not only mobilize N and P but is the process of producing crops and foods naturally. This method avoids the use of synthetic chemical fertilizers and genetically modified organisms to influence the growth of crops.

The main idea behind organic also secrete various plant growth and health promoting substances (Pandya and Saraf, 2010). Bio-fertilizers are low cost, effective and renewable source of plant nutrients to supplement chemical fertilizers (Boraste et al., 2009). In addition to their role in enhancing the growth of the plants, bio-fertilizers can act as biocontrol agents in the rhizosphere at the same time. This synergistic effect, when present,

*Corresponding author. E-mail: drmhashem69@yahoo.com.

increases the role of application of bio-fertilizers in the sustainable agriculture.

Many attempts were made to prepare a bio-fertilizer from wastes using effective microorganism including bacteria and yeasts. Yeasts synthesize antimicrobial and other useful substances required for plant growth from amino acids and sugars secreted by bacteria, organic matter and plant roots (Boraste et al., 2009). *Saccharomyces cerevisiae* is considered as a new promising plant growth promoting yeast for different crops. Recently, it became a positive alternative to chemical fertilizers safely used for human, animal and environment (Omran, 2000). A growing number of studies indicate that plant root growth may be directly or indirectly enhanced by yeasts in the rhizosphere (Nassar et al., 2005; El-Tarabily and Sivasithamparam, 2006; Cloete et al., 2009). A wide diversity of soil yeasts have been researched for their potential as bio-fertilizers (Gomaa and Mohamed, 2007; Eman et al., 2008). Representatives of *Candida*, *Geotrichum*, *Rhodotorula*, *Saccharomyces*, and *Williopsis* are able to nitrify ammonium to nitrate via nitrite *in vitro* (Al-Falih, 2006). Whereas the *ascomycetous* genera *Williopsis* and *Saccharomyces* were able to oxidize elemental sulfur *in vitro* to produce phosphate, tetrathionate, and sulfate (Al-Falih and Wainwright, 1995).

On the other hand, sugar beet (*Beta vulgaris* L.) is the second important sugar crop after sugar cane; produce about 30% of total world production. Few preliminary studies revealed the suitability of growing sugar beet in Saudi Arabia. Nowadays, a great attention has been focused on the possibility of using natural and safe agents for promoting growth of sugar beet. Little information is available about the effect of application of yeasts as bio-fertilizers on the productivity and growth enhancement of sugar beet. Yeasts are a poorly investigated group of microorganisms that represent an abundant and dependable source of bioactive and chemically novel compounds. Searching new yeasts as bio-fertilizers and studying their productivity of bioactive chemical compounds expand our knowledge about their approached mechanisms to enhance the plant growth and soil characteristics.

We assume that a good understanding of the role of soil yeasts in the rhizosphere holds a key to future sustainable agricultural practices. Therefore, the aim of this study was to investigate the impact of soil amendment with *Kluyveromyces walti*, *Sccaromycopsis cataegensis* and *Pachytichospora transvaalensis* as bio-fertilizers on the growth parameters and productivity of sugar beet.

MATERIALS AND METHODS

Plant materials and yeasts

Seeds of sugar beet (*Beta vulgaris* L.) cv Hind were obtained from the agricultural commercial market, Egypt and used in this study. We obtained the yeast strains; *Pachytichospora transvaalensis* UFOSY-1240, *Kluyveromyces walti* UFOSY-1175 and *Sccaromycopsis cataegensis* UFOSY-1365 from Professor J.L.F. Kock, University of the Free State, South Africa for research purpose.

Yeast culture preparation

Yeast cultures were prepared by growing the yeast strains in yeast extract malt extract broth (YMB) (yeast extract 3 gL^{-1}, malt extract 3 gL^{-1}, peptone 5 gL^{-1} and glucose, 10 gL^{-1}) at 25 $\pm$ 1 °C with shaking (150 rpm) for 48 to 72 h. The yeast cells were pelletized by centrifugation (5000 r.p.m) for 10 min and resuspended in sterilized tap water to the desired concentration (~10^8 cfu ml^{-1}).

Experimental design

The pot experiment was conducted during the two-growing seasons on 15th September, 2010 and 2011. Each season extended for 5 months. Five seeds of sugar beet were sown in plastic pots (30-cm in diameter). Each pot was filled with 12 kg loamy soil (pH = 7.66, EC = 1.42 dS m^{-1}, CaCO$_3$, 5.72% and organic matter, 1.52%). Different doses of the three yeasts (0.0, 50.0 and 100.0 ml pot^{-1} with concentration of ~10^8 cfu ml^{-1}) and the final volume was completed to two liters per pot using tap water. The selection of the doses was based on preliminary studies (data not shown). This application was regularly repeated three-week intervals during the season. Each treatment was set in five replicates. After complete germination, plants in each pot were thinned to one plant. Pots were arranged in a complete randomized design in the greenhouse at temperature range 20 to 30 °C, 12 h dark and 12 h light. Irrigation was done twice a week (Using two liters of tap water per pot in all treatments including control) and each pot was irrigated every two weeks with 50 ml pot^{-1} of Hoagland's nutrient solution.

Morphological measurements

Growth parameters including plant height, number of leaves per plant, root length and diameter, fresh and dry weight of the shoot as well as fresh and dry weight of roots were estimated in the treated plants after 150 days from sowing cultivation.

Photosynthetic pigments

Photosynthetic pigments (chlorophyll a, b and carotenoids) were determined in fresh leaf samples of 120-days old plants. Leaf samples (0.5 g) were homogenized in acetone (90% v/v), filtered and made up to a final volume of 50 ml. Pigment concentrations were calculated from the absorbance of extract at 663, 645 and 470 nm using the formulae of Lichtenthaler (1987) as given below:

Chlorophyll a (mg/g FW) = (11.75 × A663 - 2.35 × A645) × 50/500

Chlorophyll b (mg/g FW) = (18.61 × A645 - 3.96 × A663) × 50/500

Carotenoids (mg/g FW) = ((1000 × A470) − (2.27× Chl a)-(8.14×Ch b)/227) × 50)/500

Total soluble sugars

Total soluble sugars (TSS) were extracted by overnight submersion of fresh leaves in 10 ml of 80% (v/v) ethanol at 25 °C with periodic shaking, and centrifuged at 600 rpm. The supernatant was evaporated to complete dryness, and then dissolved in a known volume of distilled water to be ready for determination of soluble sugars (Homme et al., 1992). TSS was analyzed by reacting of 0.1ml of ethanolic extract with 3.0 ml freshly prepared anthrone (150 mg anthrone + 100 ml of 72% H_2SO_4) in boiling water bath for ten minutes and reading the cooled samples at 625 nm using Spekol Spectrocololourimeter VEB Carl Zeiss (Yemm and Willis, 1994). Sucrose was estimated in fresh roots of sugar beet root by using Saccharometer according to the method described by A.O.A.C. (1995).

Total soluble proteins

Total soluble proteins content of the fresh leaves and roots was determined according to the method described by Bradford (1976) with bovine serum albumin as a standard. An amount of 2 g of samples was ground in a mortar with 5 ml of phosphate buffer (pH 7.6) and was then transformed to the centrifuge tubes. The homogenate was centrifuged at 8000 rpm for 20 min. The supernatant of different samples was put in separate tubes. The volume of all of the samples in tubes was then made equal by adding a phosphate buffer solution and the extraction were stored in the refrigerator at 4°C for further analysis. After extraction, 30 µl of different samples were taken out in separate tubes and were mixed with 70 µl of distilled water. In all of these separate sample tubes, 2.9 ml of the Coosmassic Brilliant Blue solution was then added and mixed thoroughly. The total volume was 3 ml in each tube. All tubes were incubated for 5 min at room temperature and then, the absorbance was recorded at 600 nm against the Blank. A standard curve of absorbance (600 nm) versus concentration (µg) of total soluble proteins was calculated.

Anatomical study

Samples of 150-days old from the middle of the fifth leaf from apex and root from 2 cm from base of the main root were taken. Samples were killed and fixed in F.A.A. solution (50 ml 95% ethyl alcohol + 10 ml formalin + 5 ml glacial acetic acid + 35 ml distilled water) for 48 h. Thereafter, samples were washed in 50% ethyl alcohol, dehydrated and cleared in tertiary butyl alcohol series, embedded in paraffin wax of 54 to 56°C mp. Cross sections with 20 µ thick were cut with a rotary microtome, adhered by Haupt's adhesive and stained with the crystal violet-erythrosin combination (Sass, 1961), cleared in carbolxylene and mounted in Canada balsam. Measurements were done, using a micrometer eyepiece and an average of 10 readings were calculated.

Gas chromatography–mass spectrometry (GC–MS) analysis

The three yeasts were grown on YMB at 25 ± 1°C with shaking (150 rpm) for 72 h. The yeast suspension was centrifuged at 10000 rpm for 15 min under cooling and the supernatant was filtered through cellulose membrane filter (0.45 µm), and then was extracted by chloroform solution. An aliquot of one µl extract (chloroform extract) of cell free extract was injected into the GC–MS (6890 N/5975B).

The HP-5MS column was 30 m in length, 0.25 mm i.d., and 0.25 mm in thickness. The carrier gas was helium with average velocity 36 cm sec^{-1}, and flow 1 ml min^{-1}. The operating condition of GC oven temperature was maintained as follows: initial temperature 40°C for 9 min, 150°C for 8 min, at 15°C min^{-1} up to final tempe-rature 310°C with isotherm for 3 min at 25°C min^1. The injector and detector temperatures were set at 250 and 280°C, respectively, according to the standard method 8270 EPA (Cakir et al., 2004). Identification of the components of the prepared extract was assigned by comparison of their retention indices, relative to a series of n-alkane indices on the capillary column and GC–MS spectra from the Wiley 6.0 MS data.

Statistical analysis

Treatments were arranged in a completely randomized design. Analysis of variance was performed using the SPSS software package. Analysis of variance (ANOVA) was performed on the data to determine the least significant difference (LSD) among treatment at $P < 0.05$ and Duncan's multiple range tests were applied for comparing the means (Duncan, 1955).

RESULTS

Growth parameters and yield

Results showed that addition of any of the three yeasts (*P. transvaalensis*, *K. walti* and *S. cataegensis*) as bio-fertilizer to the soil cultivated with sugar beet significantly increased the yield and enhanced the growth of the plants (Table 1). The two doses (50 and 100 ml pot^{-1}) increased the plant height, number of leaves, root length, root diameter, fresh and dry weight of shoots and roots significantly as compared to the control in almost cases. However, the highest dose (100 ml pot^{-1}) showed better results than the lowest one. *K. walti* involved in the highest increase in all parameters significantly compared to the untreated and treated plants.

Our results revealed that application of the three yeasts induced the formation of photosynthetic pigments (chlorophyll a and b). However, *K. walti* (100 ml plant^{-1}) involved in the highest increase in the pigments' contents (0.86 and 0.22 mg g^{-1} fresh leaves, respectively). The other two yeasts increased the pigments' content significantly, compared to the untreated control. Carotenoids content was either did not change because of the application of the yeast (*K. walti*, 100 ml pot^{-1}), or decreased significantly in the rest of the treatments (Table 2). Consequently, the content of total sugars in leaves and total soluble proteins in both leaves and roots increased significantly because of the application of the yeasts except in one case (*P. transvaalensis*, 50 ml pot^1). The data indicate that the three yeasts induced sucrose formation in the beet roots significantly as compared with the control. The highest dose (100 ml pot^{-1}) was the best inducer among the all cases. However, *K. walti* (100 ml/plant) caused the highest increase in the sucrose content. It increased the sucrose content by 42.45% of the yield of the control.

Anatomical studies

Root

Data in Table 3 and Figure 1 show that treatment of *P. transvaalensis*, *K. walti* and *S. cataegensis* increased the

Table 1. Effect of *Kluyveromyces walti*, *Pachytichospora transvaalensis* and *Saccharomycopsis cartaegensis* on growth parameters and yield of sugar beet plants.

Treatment	Plant height (cm)	Number of leaves plant⁻¹	Root length (cm)	Root diameter (cm)	FW of shoots plant⁻¹ (g)	DW of shoots plant⁻¹ (g)	FW of roots plant⁻¹ (g)	DW of roots plant⁻¹ (g)
Control	27.33^d	20.33^c	12.50^c	6.20^d	55.57^f	6.61^d	58.47^d	10.50^b
K. walti (50 ml pot⁻¹)	34.00^b	23.00^b	14.00^{bc}	6.90^{bcd}	76.67^{cd}	8.62^c	62.10^d	11.15^b
K. walti (100 ml pot⁻¹)	36.00^a	26.67^a	16.83^a	8.23^a	96.19^a	11.37^a	87.26^a	15.78^a
P. transvaalensis (50 ml pot⁻¹)	27.67^d	24.33^{ab}	13.67^{bc}	6.70^{cd}	71.11^e	9.09^c	62.41^d	11.52^b
P. transvaalensis (100 ml pot⁻¹)	31.00^c	25.33^a	13.83^{bc}	7.17^{bc}	84.91^b	9.40^{bc}	78.66^b	14.95^a
S. cartaegensis (50 ml pot⁻¹)	30.33^c	20.67^c	13.67^{bc}	7.73^{ab}	82.03^{bc}	10.19^b	74.72^b	14.85^a
S. cartaegensis (100 ml pot⁻¹)	32.00^b	24.00^{ab}	16.00^a	8.23^a	78.34^{cd}	10.18^b	78.16^b	15.74^a
LSD (P < 0.05)	1.89	1.43	1.41	0.82	5.06	0.94	6.96	2.38

Values in the same column followed by the same letter(s) are not significantly different at LSD, P < 0.05.

Table 2. Effect of *K. waltii*, *P. transvaalensis* and *S. crataegensis* on photosynthetic pigment content, total soluble proteins content and total soluble sugars content of sugar beet plants.

Treatment	Photosynthetic pigments (mg g⁻¹ fresh leaves)			Proteins in leaves	Total soluble metabolites (mg g⁻¹ fresh matter)		
	Chl. a	Chl. b	Carotenoids		Proteins in roots	Sugars in leaves	Sucrose in roots
Control	0.61^e	0.14^c	0.47^a	7.38^b	2.47^d	15.39^d	115.44^g
K. walti (50 ml pot⁻¹)	0.66^d	0.17^b	0.37^b	7.31^a	3.08^{bc}	17.36^b	123.04^e
K. walti (100 ml pot⁻¹)	0.86^a	0.22^a	0.48^a	8.65^a	3.44^b	21.31^a	164.45^a
P. transvaalensis (50 ml pot⁻¹)	0.69^c	0.21^a	0.40^c	7.52^b	2.90^c	17.16^b	117.65^f
P. transvaalensis (100 ml pot⁻¹)	0.74^b	0.19^b	0.42^b	8.42^a	3.14^{bc}	20.85^a	141.13^c
S. cartaegensis (50 ml pot⁻¹)	0.67^d	0.14^c	0.38^d	7.42^b	3.46^b	16.45^c	128.22^d
S. cartaegensis (100 ml pot⁻¹)	0.69^c	0.18^b	0.38^d	7.80^b	4.09^a	20.77^a	150.45^b
LSD (P < 0.05)	0.01	0.02	0.01	0.50	0.33	0.55	1.23

Values in the same column followed by the same letter(s) are not significantly different at LSD, P < 0.05.

thickness of growth rings of sugar beet roots by increasing the average diameter of the cells. Similarly, average diameter of secondary xylem vessels was also increased as compared to control. The maximum growth of rings' thickness (833.33) and average diameter of the cells (29.17) was obtained by 100 ml pot⁻¹ of K. walti.

Leaf

Table 4 and Figure 2 show that inoculation of the soil cultivated with sugar beet plants with *P. transvaalensis*, *K. walti* and *S. cartaegensis* increased the thickness of the leaf blade and mid-vein by increasing length and width of the vascular bundles. The average diameter of the vessels and average number of vessels/bundle increased significantly as compared to the control. In consistence with the obtained results from morphological and physiological analysis, the maximum increase was obtained as the result of application of 100 ml pot⁻¹ of K. walti. It increased

Table 3. Effect of *K. walti*, *P. transvaalensis* and S. *cataegensis* on anatomical structure of root of sugar beet plants.

| Treatment | Average number of vessels bundle^{-1} | Growth rings | | Secondary growth | |
		Thickness (μ)	Diameter (μ)	Number of vessels row^{-1}	Diameter of vessel (μ)
Untreated plant	35.0	446.7	20.6	4	26.3
K. walti (50 ml pot^{-1})	52.7	583.3	22.9	4	27.5
K. walti (100 ml pot^{-1})	63.3	883.3	23.3	5	27.5
P. transvaalensis (50 ml pot^{-1})	47.33	550.0	20.2	4	27.5
P. transvaalensis (100 ml pot^{-1})	60.0	633.3	21.7	6	29.6
S. cataegens (50 ml pot^{-1})	36.0	583.3	22.9	4	29.5
S. cataegens (100 ml pot^{-1})	46.0	625.0	26.3	5	26.9

Table 4. Effect of *K. walti*, *P. transvaalensis* and S. *cataegensis* on anatomical structure of leaf blade of sugar beet plants.

| Treatment | Midvein thickness (μ) | Blade thickness (μ) | Dimensions of vascular bundles (μ) | | Average diameter of vascular bundles (μ) |
			length	width	
Control	1875.0	260	350.0	340.0	33.3
K. walti (50 ml pot^{-1})	2400.0	220	243.3	223.3	38.8
K. walti (100 ml pot^{-1})	2875.0	280	583.3	500.0	40.0
P. transvaalensis (50 ml pot^{-1})	2012.5	210	533.3	456.7	34.2
P. transvaalensis (100 ml pot^{-1})	2312.5	230	550.0	550.0	35.8
S. cataegens (50 ml pot^{-1})	2312.5	250	386.7	313.3	32.5
S. cataegens (100 ml pot^{-1})	2337.5	260	500.0	423.3	37.1

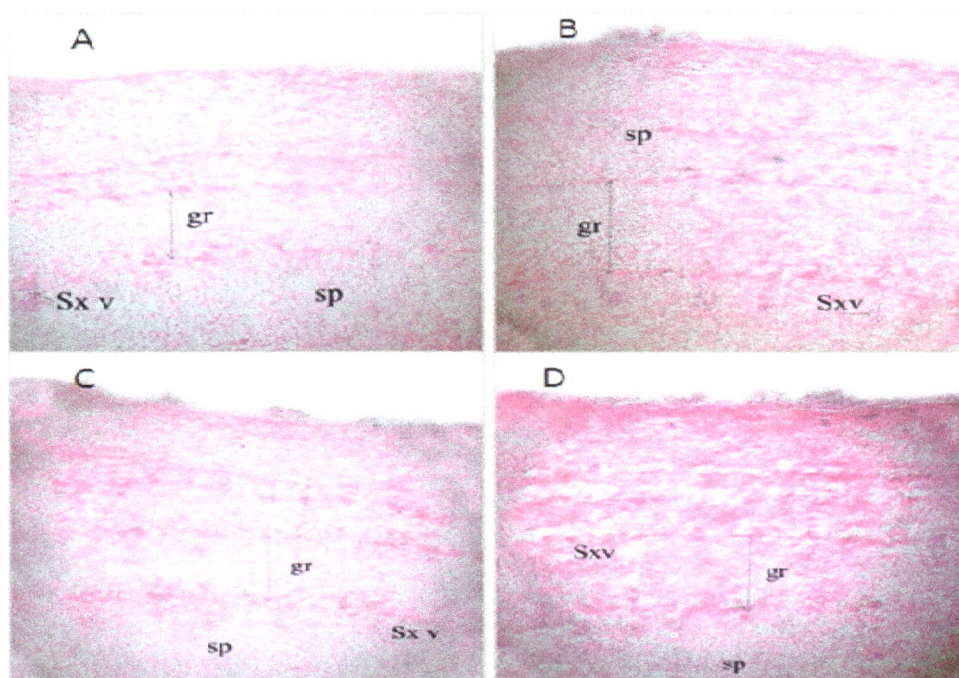

Figure 1. Transections of sugar beet root as affected by application of yeasts. A) untreated plant, B) *K.walti* (100 ml pot^{-1}), C) *P. transvaalensis* (100 ml pot^{-1}), D) *S. cataegensis* (100 ml pot^{-1}), gr; growth ring, Sx v; secondary xylem vessels and sp; storage parenchyma.

Table 5. Chemical composition chloroform extract of the culture filtrate of K. *walti*, P. *transvaalensis* and S. *cataegensis*.

Chemical compound	Abundance (%)	RT
K. *walti*		
Diisodecyl trimethyladipate	9.227	36.388
3-Methylundecane	9.117	36.114
1-(4'-Methoxy-6-methoxy-quinolin-2-yl)-3-methyl-pryrazol-5-ol	6.388	36.639
Dioctyl adipate	5.728	35.700
6-Ethyl-oct-3-ylheptatyloxalate	5.385	36.517
Didecyl sebacate	4.623	35.793
Decyl ether	4.038	36.197
Decane	3.407	36.882
Bis(6-ethyloct-3-yl)oxalate	3.261	35.536
2-(3-Methyl-2-butenyl)-4-nitrophenol	3.100	23.150
P. *transvaalensis*		
2-Hexyl-1-decanol	7. 345	21.332
2-(3-Methyl-2-butenyl)-4-nitrophenol	6 .872	23.158
Bacchotricuneatin c	5 .514	21. 441
2,4-Bis (1,1- dimethylethyl) phenol	5 .486	16.974
Methyl 14-methylpentadecanoate	3.983	23.485
2,2-Dimethoxy-1,2-diphenylethanone	3 .773	22.998
1,3 -Bis(1,1- dimethylethyl) benzene	3 .766	12.698
4-Methyl-2-undecene	3 .384	10.052
Dibutylcyanamide	3 .248	13.518
4-Isopropylphenyl methyl phathalate	3 .033	20.227
S. *cataegensis*		
3-methylundecane	27.392	35.746
Decane	7.96	36.699
Dihexyl hexanedioate	5.806	36.429
Diisodecyl trimethyladipate	5.461	35.347
1-(Tert-butoxycarbonyl)-2-methoxy-3-(5-phenyl-4-pentynyl)piperidine	4.316	37.007
Diisooctyl adipate	3.562	36.234
4-Methylundecane	2.746	36.633
4-Hydroxy-3-(3-methyl-2-butenyl)nitrobenzene	2.064	23.159
Capric ether	1.887	36.338
1-(6'-Methoxy-4'-methyl-2'-quinolyl)-3-methyl-1H-pryrazol-5-ol)	1.873	37.301

RT= Retention time.

the midvein thickness to 2875 μ and the average diameter of the vessels to 40 μ.

Analysis of the secondary metabolites

Table 5 and Figures 3 to 5 shows that the culture filtrate of the three yeasts contains many aliphatic and aromatic compounds. We considered only the top ten secondary metabolites from each yeast. Methyl undecane or its derivatives were common among the metabolites of the three yeasts.

Adipate compounds were detected in the filtrate of K. *walti* and S. *cataegensis* as dioctyl adipate and diisooctyl adipate, respectively. Didecyl sebacate was detected in

the culture filtrate of K. *walti* in considerable concentration (5.385%). Sebacate is an organic compound which is the diester of sebacic acid and 2-ethylhexanol.

DISCUSSION

Our results showed that all growth parameters of sugar beet plants were significantly enhanced as the result of application of yeasts, especially K. *walti*. Increase in fresh and dry weight of the root is a good indicator for enhancement of the yield.

Increase in vegetative growth of plants because of the application of bio-fertilizers was reported in previous works

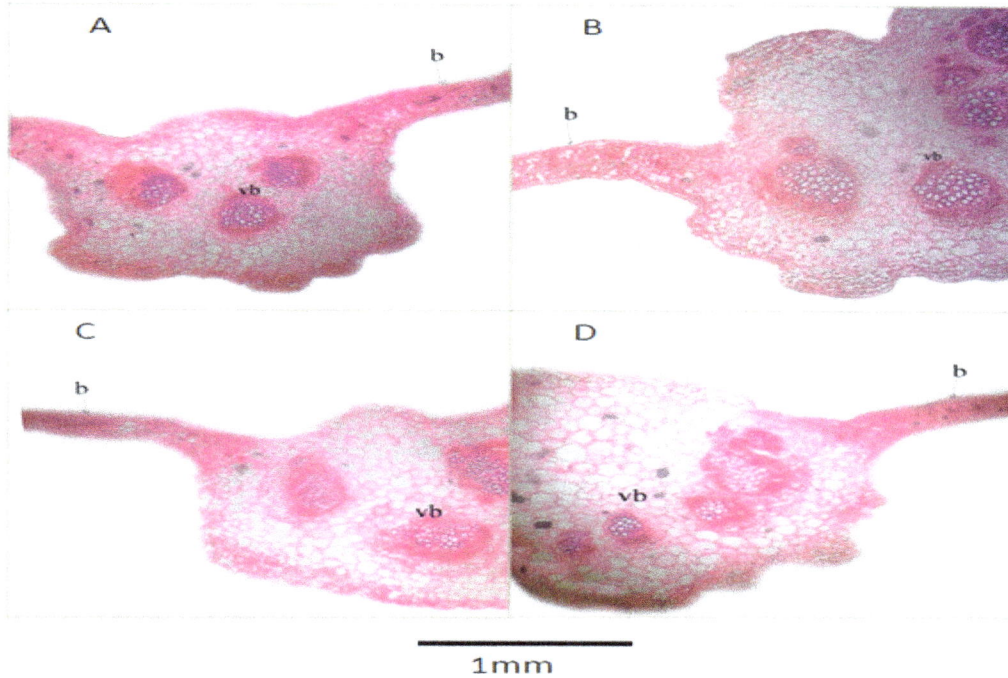

Figure 2. Transections of sugar beet leaf blade as affected by application of yeasts. (A) Untreated plant, (B) *K. walti* at 100 ml pot[-1], (C) *P. transvaalensis* 100 ml pot[-1], (D) *S. cataegensis* at 100 ml pot[-1]; b, blade; vb, vascular bundles.

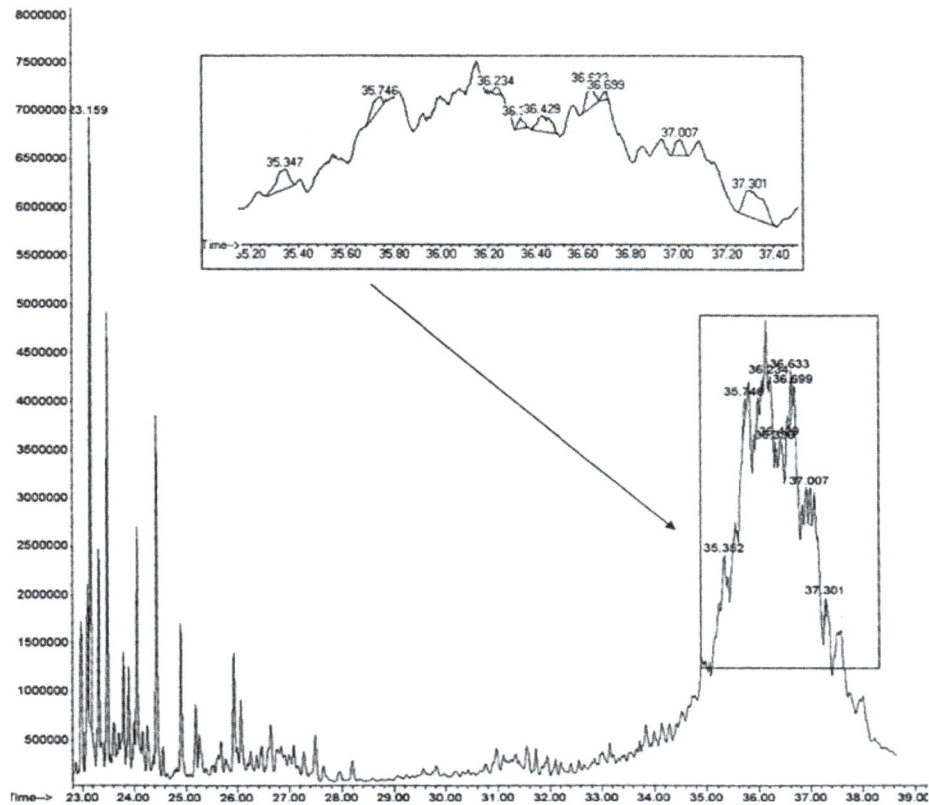

Figure 3. GC-MS chart of chloroform extract of the culture filtrate of *K. walti.*

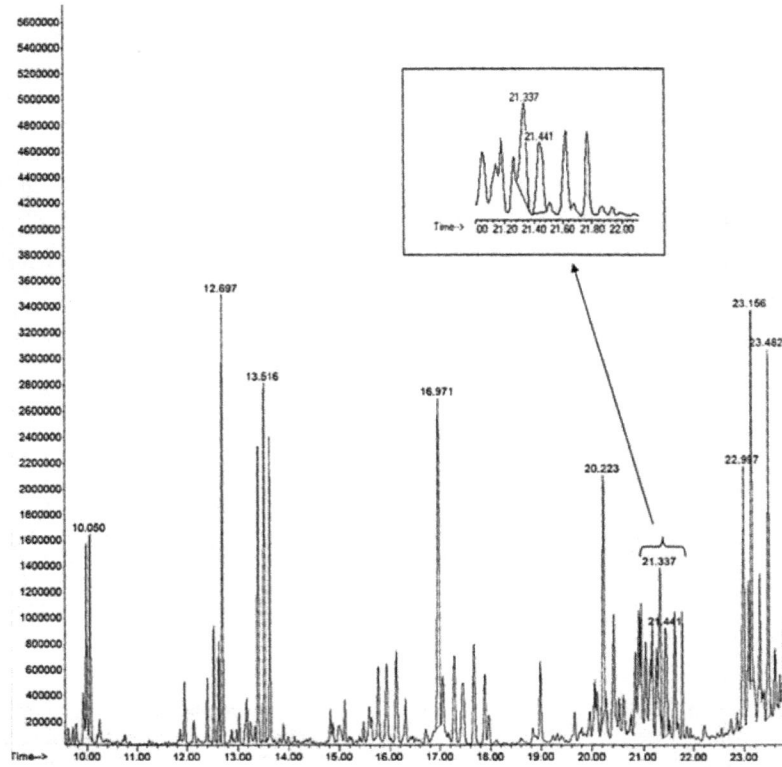

Figure 4. GC-MS chart of chloroform extract of the culture filtrate of *P. transvaalensis.*

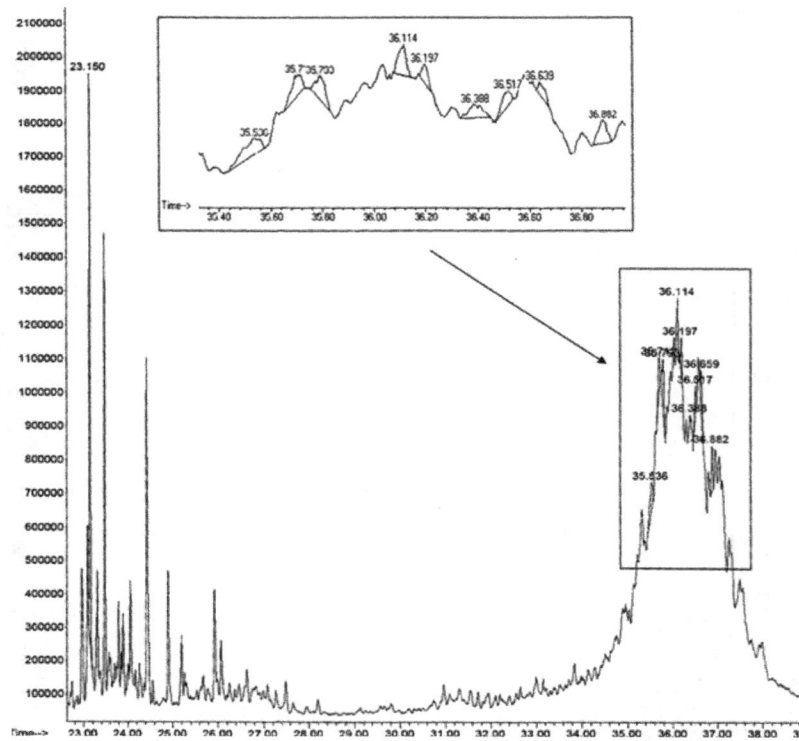

Figure 5. GC-MS chart of chloroform extract of the culture filtrate of *S. cataegensis.*

(Mahdi et al., 2010). Saber (1994) stated that this increment might be due to the availability of soil microorganisms to convert the unavailable forms of nutrient elements to available forms by generating of carbon dioxide from bio-fertilizers (Kurtzman and Fell, 2005). In agreement with our results, Wali Asal (2010) indicated that yeast has good efficiency on growth characters of wheat plants. Nakayan et al., (2009) reported that combination of yeast strain *Pichia* sp. CC1 and a half dose of chemical fertilizer (½CF) increased lettuce dry weight to 107%. The positive effect of yeast is supported by the findings of Mekki and Ahmed (2005). They stated that the increase in yield components because of yeast treatment is mainly attributed to the effect of yeast, which can play a very significant role in making available nutrient elements for plants. In addition, yeast content of macro and micronutrients, growth regulators and vitamins stimulate the plant to build up dry matters (Mirabal Alonso et al., 2008; Hesham and Mohamed, 2011).

The promoting effect of yeasts could be due to the biologically active substance produced by these bio-fertilizers such as auxins, gibberellins, cytokinins, amino acids and vitamins (Bahr and Gomaa, 2002). Afifi et al. (2003) obtained similar results where they found that inoculation of maize with *Rhodotorula* and *Azotobacter* in the presence of half the recommended doses of NPK induced growth parameters to match those of the recommended doses of NPK.

In this study, the application of the three yeasts induced the formation of photosynthetic pigments (chlorophyll a and b). However, *K. walti* (100 ml plant^{-1}) involved in the highest increase in the pigments' contents (0.86 and 0.22 mg g^{-1} fresh leaves, respectively). The difference among yeast strains efficiency to enhance the growth of plants was reported (Amprayn et al., 2012). For example, Nassar et al. (2005) mentioned that yeast isolates vary greatly in their efficiency for IAA production.

Data indicate that the three yeasts induced sucrose formation in the beet roots significantly as compared with the control. The positive effect of yeasts on chl. a and b is in consistence with the result obtained by Hayat (2007) and Stino et al. (2009), who stated that the increase in chl. a and b leads to a consequent increase in total carbohydrates, because the yeast application could enhance role in cell division, cell elongation producing more leaf area. Hussain et al. (2002) reported that *Saccharomyces* sp. is among the microorganisms, which improve crop growth and yield by increasing photosynthesis, producing bioactive substances, such as hormones and enzymes and controlling soil diseases.

The increase in the total soluble proteins content could be attributed to the growth hormones produced by yeast (Gaballah and Gomaa, 2004; Khalil and Ismael, 2010), direct stimulation of the synthesis of protein (Stino et al., 2009), providing plants with essential nutrient elements required for protein formation (Hayat, 2007). Previously.

Castelfranco and Beale (1983) stated that the increase in photosynthetic pigment formation could be attributed to the role of yeast cytokinins delaying the aging of leaves by reducing the degradation of chlorophyll and enhancing the protein and RNA synthesis.

Anatomical studies of the root showed that *P. transvaalensis*, *K. walti* and *S. cataegensis* increased the thickness of growth rings of sugar beet roots and average diameter of secondary xylem vessels. A growing number of studies indicate that plant root growth may be directly or indirectly enhanced by yeasts in the rhizosphere (El-Tarabily and Sivasithamparam 2006; Cloete et al., 2009). Warring and Philips (1973) stated that yeast is rich in tryptophan which consider precursor of IAA (indole acetic acid) which stimulate cell division and elongation.

Yeasts are grown rapidly on simple carbohydrates, often through fermentative as well as respiratory pathways (Botha, 2011). As a consequence of their nutritional preference, yeast populations are generally an order of magnitude higher in the rhizosphere as opposed to the bulk soil (Cloete et al., 2009; Botha, 2011). A diverse range of yeasts exhibit plant growth promoting characteristics, including pathogen inhibition (El-Tarabily and Sivasithamparam, 2006; Sansone et al., 2005); phytohormone production (Nassar et al., 2005); phosphate solubilisation (Mirabal Alonso et al., 2008); N and S oxidation (Al-Falih and Wainwright, 1995); siderophore production (Sansone et al., 2005) and stimulation of mycorrhizal-root colonization (Mirabal Alonso et al., 2008). Also, the anatomy of leaf showed that yeasts increased the thickness of the leaf blade and midvein by increasing length and width of the vascular bundles. We assume that the increase in the thickness of the leaf blade and the midvein is expected and consequent effect of the overall enhancement of the plant nutrition, production of the phytohormones, cell division and elongation (El-Tarabily and Sivasithamparam, 2006; Nassar et al., 2005; Mirabal Alonso et al., 2008). To the best of our knowledge, this is the first report about the effect of yeast application on the anatomical structure of the sugar beet leaves.

GC-MS analysis approves the presence of some important chemical compound like Methylundecane, Adipate, didecyl sebacate and Bacchotricuneatin c. Methylundecane is an aliphatic natural product belonged to semiochemicals, which are defined as chemicals that mediate communication between individual organisms. Although some semiochemicals are released purposefully (sex pheromones, the scent of flowers), others are released as a consequence of normal metabolism, but nevertheless still convey information (Am et al., 1992). It was reported that adipic acid may be released into the environment in various waste streams from its production and use in the manufacture of synthetic fibers, plasticizers, resins plastics, and as a food acidulant (Mitchell et al., 1982). Sebacate is an organic compound which is the diester of sebacic acid and 2-ethylhexanol. It is an oily colorless liquid and used as a plasticizer. We could assume that releasing such compounds in soil might improve the physical and chemical properties of soil that increase water holding capacity, prevent nutrient leaching and add more mineral nutrients to the soil, especially in the rhizosphere zone.

This assumption could be supported by finding of many authors; for example, Botha (2006), who reported that yeasts, such as cryptococci, may affect soil texture by producing extracellular substances that form connective bridges between soil particles or sand grains, thereby contributing to aggregate formation. Genera such as *Cryptococcus*, *Lipomyces*, and *Rhodotorula* are well known producers of extracellular polymeric substances (Vishniac, 1995; Cho et al., 2001).

These substances usually form a capsule enveloping the yeast cell (Kurtzman and Fell, 1998) and may contribute to biofilm formation (Joubert et al., 2003). The ability to resist desiccation and predation has been attributed to capsule formation (Steenbergen et al., 2003), Some soil yeasts are found to assimilate interme- diates of lignin degradation, that is, ferulic acid, gallic acid, 4-hydroxybenzoic acid, protocatechuic acid and vanillic acid (Sampaio, 1999; Botha, 2011). The diterpene compound "Bacchotricuneatin c" and its relatives were isolated from plant species and show biological activities (Simirgiotis et al., 2000). Fatty acid was detected in the filtrate of *S. cataegensis* in the form of capric ether. We could state that production of such chemical compounds by the yeasts could serve as precursors or intermediates of beneficial compounds for the plants like growth hormones, fungicides, soil particles aggregators, or plants could assimilate them into valuable compounds. Hence, these compounds could directly or indirectly enhance the growth and the productivity of the sugar beet plants. Because of lacking of the literatures dealing with the production of such compounds by yeasts, we could conclude that yeasts are a poorly investigated group of microorganisms that represent an abundant and dependable source of bioactive and chemically novel compounds. Study the production of such products and their bioactivities in this field holds exciting promise.

Conclusion

Our results are promising in the field of bio-fertilizers. Application of yeasts increased the sugar content sugar beet by about 43%. It significantly enhanced the overall growth of the treated plants. The mechanisms which could be involved include the bioavialbility of macro and micronutrients, production of growth hormones, and reduction of the phytopathogens' growth. In addition, they could improve the physical and chemical properties of soil that increase water holding capacity, prevent nutrient leaching and add more mineral nutrients to the soil. We assume that studying the production of yeasts' secondary metaboilites and their bioactivities in the rhizospher holds

exciting promise. We recommend further study dealing with the identification of the secondary metabolites of the yeasts and their bioactivities in the rhizosphere as well as their direct and indirect relationships with the plant growth and productivity.

ACKNOWLEDGMENT

The authors thank Professor J. L. F. Kock, University of the Free State, South Africa for kindly providing the three yeast strains tested in this research.

REFERENCES

A.O.A.C. (1995). Association of Official Analytical Chemists. Official methods of analysis, 16th edition, AOAC International, Washington, DC.

Afifi MH, Manal FM, Gomaa AM (2003). Efficiency of applying bio-fertilizers to maize crop under different levels of mineral fertilizers. Annal. Agric. Sci. Moshtohor 41(4):1411-1420.

Al-Falih AM (2006). Nitrogen transformation in vitro by some soil yeasts. Saudi. J. Biol. Sci. 13(2):135-140.

Al-Falih AM, Wainwright M (1995). Nitrification, S oxidation and P-solubilization by the soil yeast *Williopsis californica* and by *Saccharomyces cerevisiae*. Mycol. Res. 99:200-204.

Am H, Toth M, Priesner E (1992). List of Sex Pheromones of Lepidoptera and Related attractants, 2nd edn, OILB-SROP/IOBC-WRPS, Paris. pp. 1-179.

Amprayn K, Rose MT, Kecskés M, Pereg L, Nguyen HT, Kennedy IR (2012). Plant growth promoting characteristics of soil yeast (*Candida tropicalis* HY) and its effectiveness for promoting rice growth. Appl. Soil Ecol. 61:295-299.

Bahr AA, Gomaa AM (2002). The integrated system of bio-and organic fertilizers for improving growth and yield of triticale. Egypt. J. Appl. Sci. 17(10):512-523.

Boraste A, Vamsi KK, Jhadav A, Khairnar Y, Gupta N, Trivedi S, Patil P, Gupta G, Gupta M, Mujapara AK, Joshi B (2009). Bio-fertilizers: A novel tool for agriculture. Int. J. Microbiol. Res. 1(2):23-31.

Botha A (2006). Yeast in soil. In: Rosa, C.A., Péter, G. (Eds.), The Yeast Handbook; Biodiversity and Ecophysiology of Yeasts. Springer-Verlag, Berlin, pp. 221-240.

Botha A (2011). The importance and ecology of yeasts in soil. Soil Biol. Biochem. 43:1-8.

Bradford MM (1976). A rapid and sensitive method for quantitation of microgram quantities of protein. utilizing the principle of protein-dye-binding. Anal Biochem. 72:248-254.

Cakir A, Kordali S, Zengin H, Izumi S, Hirata T (2004). Composition and antifungal activity of essential oils isolated from *Hypericum hyssopifolium* and *Hypericum heterophyllum*. Flavour Fragr. J. 19:62-68.

Castelfranco PA, Beale SI (1983). Chlorophyll biosynthesis recent advances and areas of current interest. Ann. Rev. Plant Physiol. 34:241-278.

Cho DH, Chae HJ, Kim EY (2001). Synthesis and characterization of a novel extracellular polysaccharide by *Rhodotorula glutinis*. Appl. Biochem. Biotech. 95:183-193.

Cloete K, Valentine A, Stander M, Blomerus L, Botha A (2009). Evidence of symbiosis between the soil yeast *Cryptococcus laurentii* and a sclerophyllous medicinal shrub, *Agathosma betulina* (Berg.) Pillans. Microb. Ecol. 57:624632.

Duncan DB (1955). Multiple range and multiple F-test Biometrics, II. pp. 1-42

El-Tarabily KA, Sivasithamparam K (2006). Potential of yeasts as biocontrol agents of soil-borne fungal plant pathogens and as plant growth promoters. Mycoscience 47:25-35.

Eman AAA, Saleh MMS, Mostaza EAM (2008). Minimizing the quantity of mineral nitrogen fertilizers on grapevine by using humic acid, organic and bio-fertilizers. Res. J. Agric. Biol. Sci. 4:46-50.

Gaballah MS, Gomaa AM (2004). Performance of Faba Bean Varieties Grown under Salinity Stress and Biofertilized with Yeast. J. Appl. Sci. 4:93-99.

Gomaa AM, Mohamed MH (2007). Application of bio-organic agriculture and its effects on guar (Cyamopsis tetragonoloba L.) root nodules, forage, seed yield and yield quality. World J. Agric. Sci. 3:91-96.

Hayat AEH (2007). Physiological studies on Hibiscus sabdariffa L. production in new reclamated soils. M.Sc. thesis, Faculty of Agriculture, Zagazig University.

Hesham A-L, Mohamed H (2011). Molecular genetic identification of yeast strains isolated from Egyptian soils for solubilization of inorganic phosphates and growth promotion of corn plants. J. Microbiol. Biotechnol. 21:55–61.

Homme PM, Gonzalez B, Billard J (1992). Carbohydrate content, frutane and sucrose enzyme activities in roots, stubble and leaves of rye grass (Lolium perenne L.) as affected by sources/link modification after cutting. J. Plant Physiol. 140:282-291. http://www.seedbuzz.com/knowledge-center/article/organic-farming-standards-certification

Hussain T, Anjum AD, Tahir J (2002). Technology of beneficial microorganisms. Nat. Farm. Environ. 3:1-14.

Joubert L-M, Botha A, Wolfaardt GM (2003). Feeding relationships in yeast-ciliate biofilms. In: McBain A, Allison D., Brading M, Rickard A, Verran, J, Walker J. (Eds.), BBC 6, Biofilm Communities: Order from Chaos. Bioline, Cardiff. pp.409-415.

Khalil SE, Ismael EG (2010). Growth, Yield and Seed Quality of Lupinus termis as Affected by Different Soil Moisture Levels and Different Ways of Yeast Application. J. Americ. Sci. 6(8):141-153.

Kurtzman CP, Fell JW (1998). The Yeasts, a Taxonomic Study, fourth edn. Elsevier, Amsterdam. p.1055.

Kurtzman CP, Fell JW (2005). Biodiversity and Ecophysiology of Yeasts (In: The Yeast Handbook, Gabor P, de la Rosa CL, eds) Berlin, Springer. pp. 11-30.

Lichtenthaler HK (1987). Chlorophylls and carotenoids: pigments of photosynthetic biomemranes. Meth. Enzymol. 148:350–382.

Mahdi SS, Hassan GI, Samoon SA, Rather HA, Showkat AD, Zehra B (2010). Bio-fertilizers in organic agriculture. J. Phytol. 2(10):42-54.

Mekki BB, Ahmed, AG (2005). Growth, Yield and Seed Quality of Soybean (Glycine max L.) As Affected by Organic, Bio-fertilizer and Yeast Application. Res. J. Agric. Biol. Sci.1(4):320-324.

Mirabal Alonso L, Kleiner D, Ortega E (2008). Spores of the mycorrhizal fungus Glomus mosseae host yeasts that solubilize phosphate and accumulate polyphosphates. Mycorrhiza 18:197-204.

Mitchell GA, Vanderbist MJ, Meert FF (1982). Gas-liquid chromatographic determination of adipate content of acetylated di-starch adipate. J. Assoc. Anal. Chem. 65(2):238-240.

Nakayan P, Shen FT, Hung MH, Young CC (2009). Effectiveness of Pichia sp. CC1 in decreasing chemical fertilization requirements of garden lettuce in pot experiments. As. J. Food. Ag-Ind. Special. S66-S68.

Nassar A, El-Tarabily K, Sivasithamparam K (2005). Promotion of plant growth by an auxin-producing isolate of the yeast Williopsis saturnus endophytic in maize (Zea mays L.) roots. Biol. Fert. Soils. 42:97-108.

Omran YA (2000). Studies on histophysiological effect of hydrogen cyanamide (Dormex) and yeast application on bud fertility, vegetative growth and yield of "Roumi Red"' grape cultivar. Ph. D. Thesis, Fac of Agric Assiut Univ Egypt.

Pandya U, Saraf M 2010: Application of Fungi as a Biocontrol Agent and their Bio-fertilizer Potential in Agriculture. J. Adv. Dev. Res. 1(1):90-99.

Pandya U, Saraf M (2010). Application of fungi as a biocontrol agent and their biofertilizer potential in agriculture. J. Advan. Develop. and Res. 1(1): 90-99.

Saber MSM (1994). Bio-organic farming systems for sustainable agriculture. Inter-Islamic Network on Genetic Engineering and Biotechnology, INOGE Publ. 3, Cairo, Egypt.

Sampaio JP (1999). Utilization of low molecular weight aromatic compounds by heterobasidiomycetous yeasts: taxonomic implications. Can. J. Microbiol. 45:491-512.

Sansone G, Rezza I, Calvente V, Benuzzi D, Tosetti MISD (2005). Control of Botrytis cinerea strains resistant to iprodione in apple with rhodotorulic acid and yeasts. Postharvest Biol. Tech. 35:245-251.

Sass JA (1961). Botanical Microtechnique. Third ed. The Iowa State Univ. Press. Ames. Iowa, USA.

Simirgiotis MJ, Favier LS, Rossomando PC, Tonn CE, Juarez A, Giordano OS (2000). Phytochemical Study Conyza Sophiaefolia Antiinflammatory Activity. Molecules 5:605-607.

Steenbergen JN, Nosanchuk JD, Malliaris SD, Casadevall A (2003). Cryptococcus neoformans virulence is enhanced after growth in the genetically malleable host Dictyostelium discoideum. Infect. Immun. 71:4862-4872.

Stino RG, Mohsen AT, Maksouds MA, Abd El- Migeed MMM, Gomaa AM, Ibrahim AY (2009). Bioorganic fertilization and its Impact on Apricot young trees in newly reclaimed soil. American- Eurasion. J. Agric. Environ. Sci. 6(1):62-69.

Vishniac HS (1995). Simulated in situ competitive ability and survival of a representative soil yeasts, Cryptococcus albidus. Microbial. Ecol. 30:309-320.

Wali Asal MA (2010). The combined Effect of mineral organic and bio-fertilizers on the productivity and quality of some wheat cultivars. Ph.D. Thesis, Fac. Agric. Alex. Univ., Egypt.

Warring PE, Phillips IDG (1973). The control of growth and differentiation in plants. E L B S ed., Pub by Pergamon Press Ltd. VK.

Yemm EW, Willis AJ (1994). The respiration of barley plants. IX. The metabolism of roots during assimilation of nitrogen. New Phytol. 55:229-234.

Soil properties and tomato agronomic attributes in no-tillage in rotation with cover crops

Roberto Botelho Ferraz Branco[1]*, Denizart Bolonhezi[1], Fernando André Salles[1],
Geraldo Balieiro[1], Eduardo Suguino[1], Walter Seiti Minami[2] and Ely Nahas[3]

[1]Agency Paulista Agribusiness Technology, APTA, 14030-670 Ribeirao Preto, Sao Paulo (SP) - Brazil.
[2]Moura Lacerda University Center, 14085-420 Ribeirão Preto, SP - Brazil.
[3]Department of Microbiology, Universidade Estadual Paulista (UNESP), 14884-900 Jaboticabal, SP - Brazil.

Cover crops associated with no-tillage improves soil fertility by the production of mulch on the soil surface. This experiment was conducted in 2008 and 2009 to evaluate the potential of different cover crops grown in rotation with tomato in no-tillage in the soil and agronomic attributes of tomato. The treatments were velvetbeans (*Mucuna deeringiana* [Bort] Merr.), sunn hemp (*Crotalaria junceae* L.), pearl millet (*Pennisetum americanum* Leeke), fallow with free growth of weed, and maize crop in conventional tillage as control. Shoot dry biomass of the cover crop, weed establishment, concentration of the nutrient in tomato leaves, fertility and microbiology of soil and tomato yield were evaluated. Maize had the greatest shoot dry mass yield, but it was incorporated into the soil by tillage, being the control treatment. Regarding the crops for which the residue remained on the soil surface, millet and sunn hemp were the most productive. Millet and sunn hemp, as well as corn in tillage, were the most efficient in the suppression of the weed establishment in the tomato crop. Sunn hemp increased potassium content and nitrification activity of the soil nitrate. Tomato yield was higher when grown on straw of sunn hemp.

Key words: *Lycopersicon esculentum* Mill., weeds, soil microbiology, soil fertility, leaf nutrient.

INTRODUCTION

Modern agriculture aims at high yields along with the use of conservationist practices that reduce the environmental impact on natural resources, including soil, water, and organic matter content of soil. Vegetables are crops which require sophisticated technology to obtain economic profitability, and the intensive cultivation in a single area without proper soil management often decreases soil fertility (Bonanomi et al., 2011). Staked tomato cultivation in Brazil is predominantly carried out in areas of steep slopes subject to erosion by excessive storm water runoff on the soil surface. Cover crops are

excellent tools for soil protection and for the maintenance of the chemical, physical and biological balance of the soil, supporting the sustainability of environment production (Abdul-Baki et al., 1997a; Castro et al., 1993). Grass and legumes are frequently used as cover crops, because they satisfy the essential requirements of cover crops, such as ruggedness, vigorous vegetative growth, and high shoot dry matter yield (Wutke et al., 2009). Another advantage of cover crops in no-till is the improvement of soil microbial activity by the increase of organic matter (Duda et al., 2003; Castro et al., 1993), reduction of weeds (Carrera et al., 2004; Campiglia et al., 2010), improvement of the soil fertility (Perin et al., 2004; Wang et al., 2009), and consequently, an increase in crop yield (Kieling et al., 2009; Sainju et al., 2002). No-till also contributes to reduce erosion by avoiding the soil

---

*Corresponding author. E-mail: branco@apta.sp.gov.br.

exposure through harrowing and ploughing. For successful no-tillage, it is necessary to keep a certain amount of mulch on the soil surface which contributes to reduce erosion and improve soil fertility (Argenton et al., 2005; Colla et al., 2000). For this reason, the interaction between no-tillage and cover crops is very important for the quality of the technology. Therefore, this experiment aimed at evaluating the performance of no-tillage tomato and cover crops towards the suppression of weeds, improvement of soil fertility, soil microbial activity, nutrient content of tomato leaves and tomato yield.

MATERIALS AND METHODS

The experiment was conducted over two consecutive years, from 2007 to 2009 in Ribeirão Preto, São Paulo, located in the tropics at 21° 12' 26" S and 47° 51' 48" N, mean altitude of 646 m. The annual mean rainfall is 1427 mm, concentrated from November to March, and annual mean maximum and minimum temperatures are 25 and 19.3°C. The soil is classified as Oxisol udic eutrophic, which consists 10.2% sand, 32.1% silt, and 57.7% clay, presenting the following chemical fertility: pH = 5.5, organic matter (OM) = 23 g dm^{-3}, P = 43 mg dm^{-3}, K = 3.8 $mmol_c$ dm^{-3}, Ca = 26 $mmol_c$ dm^{-3}, Mg = 11 $mmol_c$ dm^{-3}, cation exchange capacity (CEC) = 70 $mmol_c$ dm^{-3}, V% = 58%, sum of the bases (SB) = 41 $mmol_c$ dm^{-3}, H + Al = 29 $mmol_c$ dm^{-3}, B = 0.27 mg dm^{-3}, Cu = 6.7 mg dm^{-3}, Mn = 34.8 mg dm^{-3} and Zn = 1.0 mg dm^{-3}. Limestone was applied to the soil at the beginning of the experiment to raise the base saturation to 80%, followed by harrow plowing. In the following two years, no more limestone or tillage were used during the experiment, characterizing it as a no-tillage soil.

The experimental design was complete randomized block with five treatments and five replications, with plots of 7 × 10 m. The treatments consisted of velvetbeans (Mucuna deeringiana [Bort] Merr.), sunn hemp (Crotalaria junceae L.), millet (Pennisetum americanum Leeke) and fallow with unrestricted growth of weed, followed by no-tillage tomato crop. Treatments were compared to a maize crop as a control, with conventional tillage for tomato. Cover crops were sowed in December 2007, with 0.50 m row spacing for velvetbeans and sunn hemp at seeding rates of eight and 27 seeds per linear meter, respectively. Millet was seeded at 0.30 m row spacing and 35 seeds per linear meter. The natural soil fertility was considered sufficient for cover crops growth; no weed control was applied. In the fallow experimental plots, weeds grew freely. Maize was sown at 0.90 m row spacing and seeding rate of five plants per linear meter in the conventional tillage plots. Cover crops were mowed at 80 days after sowing, remaining the whole biomass on the soil surface. To control the remnant weeds, 2.0 L ha^{-1} of glyphosate was applied before transplanting of tomato. In the plots of conventional tillage the soil was prepared with the rotary hoe to 0.20 m of depth. At mowing, the shoot dry biomass of cover crop was evaluated by sampling 1.0 m^{-2} sites of each experimental plot. The samples were oven-dried at 65°C until constant weight before evaluation. Seedlings of the tomato hybrid Débora Victori were transplanted in April at 1.20 × 0.40 m spacing. The crop was drip irrigated and fertilized with 300 kg ha^{-1} N, 500 kg ha^{-1} P_2O_5, and 300 kg ha^{-1} K_2O in irrigation water in the tomato crop cycle. Tomato was grown with a single stem up and with six bunch fruit.

The establishment of weeds was evaluated by counting the number of individuals per species at two randomized sites of 0.250 m^2 of each experimental plot 30 days after tomato transplanting. The nutritional conditions of tomato plants were determined by analysis of leaf nutrient concentration from mature leaves collected between the third and fourth fruit bunches. After collection, the leaves were washed with distilled water and neutral detergent, dried

in a forced air circulation oven at 65°C to constant weight. Later, the samples were milled and sent for laboratory analysis of nitrogen, phosphorus, potassium, calcium, magnesium, and sulphur (Malavolta et al., 1989). Tomato yield was assessed including fruits of the sixth bunch by harvesting ripe fruit and fruit changing color from green to red, indicating the beginning of ripening. After each harvest, the number and fresh fruit mass of market-quality and non-market-quality fruit and mean fruit mass of market-quality fruit were determined. To study the soil fertility, samples were taken at four times during the experiment, in the beginning and at the final stage of tomato growth in the two years, 2008 and 2009, in the soil profile from 0 to 0.20 m depth at three points of each experimental unit, which formed a representative sample. The pH, OM, CEC, phosphorus (P), potassium (K), calcium (Ca) and magnesium (Mg) of soil was analyzed according to Raij et al. (1997).

For the microbiological analysis, the samples were taken just once, in the final stage of tomato growth in 2009. They were then sent to the microbiology laboratory for analysis of soil microorganisms. For counting bacteria and fungi in the soil samples (10 g, dry weight), 95 ml of sodium pyrophosphate 0.1% were added and stirred for 30 min in an orbital shaker. After serial dilution, some volumes of this suspension were added to the culture medium and distributed in Petri dishes. A Bunt and Rovira (1955) medium containing per liter: 5.0 g glucose, 0.4 g K_2HPO_4, 0.5 g $(NH_4)_2HPO_4$, 0.05 g $MgSO_4$, 0.1 g $MgCl_2$, 0.01 g $FeCl_3$, 0.1 g $CaCl_2$, 1.0 g peptone, 1.0 g yeast extract, 250 ml soil extract (1 kg soil L^{-1} H_2O, sterilized for 15 min), 15 g agar and pH 7.4 was used throughout this study for bacteria counting and Martin (1950) medium containing per liter: 10.0 g glucose, 0.5 g K_2HPO_4, 0.5 g KH_2PO_4, 0.5 g $MgSO_4.7H_2O$, 5.0 g peptone, 0.5 g yeast extract, 0.03 g rose bengal, 15 g agar, pH 5.5 and streptomycin (0.03 g L^{-1}) was added in the melted and cooled medium to pour onto the plates for fungi counting. The incubation time was 72 h for bacteria and 96 h for fungi at a temperature of 30°C, and counts made according to Vieira and Nahas (2005). Nitrification activity was determined after incubation of soil for 30 days, with moisture content adjusted to 60% of water retention capacity (WRC) and with or without the addition of 160 mg of $(NH_4)_2SO_4$ g dry $soil^{-1}$. The nitrate produced was extracted and determined by the Keeney and Nelson (1982) method. The respiratory activity in the amount of 100 g of dry soil in accordance with Rezende et al. (2004) was determined with humidity corrected to 60% of WRC. The urease activity of soil was determined using 2.0 g of soil and, as substrate, 1.0 ml of 10% urea (McGarity and Myers, 1967).

The average maximum temperatures during the growth of the cover crops in 2008 and 2009 were 29.9 and 30.1°C, respectively, and the average minimum temperature in both years was 19.1°C. The accumulated rainfall during the cover crop grown in 2008 and 2009 were 813 and 810.5 mm, respectively. During the growth of tomato in the two years, the average maximum temperatures were 27.4 and 27.1°C, and the average minimum temperatures were 13.0 and 14.0°C, respectively. Rainfall during the period was 138.4 and 301.9 mm, respectively. The second tomato crop had greater and better distributed rainfall. The weather was favorable for the development, growth and production of tomato according to Nuez (1995).

The effects of the different cover crops, time, and their interactions with weed, fruit yield and tomato plant nutrition were tested using the software PROC MIXED of SAS (Littel et al., 2006) with data from a randomized block experimental design with repeated measures in time. Soil microbiology activity was measured only once during the experiment, so time and time × treatment interaction were not included in its model. The degrees of freedom were calculated using the Kenward-Roger correction. The most appropriate co-variance structure for each variable was chosen based on the Akaike and Schwarz criterion. Treatments, time and treatments × time interactions were considered significant when P ≤ 0.05. Differences among means were tested for statistical

Table 1. Shoot dry biomass of cover crops.

Cover rop	Shoot dry biomass (Mg ha^{-1})
Velvetbean	7.02^b
Sunn hemp	10.05^{bc}
Millet	14.40^{ab}
Fallow	6.40^c
Maize (Conventional)	18.26^a
Year	
2008	7.59^b
2009	14.87^a
ANOVA (P value)	
Treatment	0.0018
Year	0.0004
Treatment versus Year	0.5626

Values followed by the same letter in the column are not significantly different (Tukey test, P < 0.05).

Table 2. Weed suppression represented by the number of individuals per 0.25 m^2 in no-tillage cover crop treatment.

Cover crop (CC)	2008	2009
Velvetbean	24^{b†}	119^{a*}
Sunn hemp	27^{ab}	66^b
Millet	18^b	42^b
Fallow	62^a	156^{a*}
Maize (Conventional)	40^{ab}	49^b
ANOVA (P values)		
Treatment	<0.0001	
Year	<0.0001	
Treatment versus Year	0.0185	

†Values followed by the same letter in columns are not significantly different (Tukey test, P ≤ 0.05).*In lines are significantly different (Tukey test, P ≤ 0.05).

significance using Tukey's test.

RESULTS AND DISCUSSION

The treatments had different shoot dry biomass yields, ranging from 6.4 to 14.4 mg ha^{-1}. Maize presented the highest production compared to that of velvetbeans, sunn hemp and weed (fallow), but it was not different from millet (Table 1). Although the highest dry biomass was of maize, the benefits of mulching were not realized due to the incorporation of their straw into the soil by tillage. The analysis of shoot biomass of cover crops by years showed a greater shoot dry biomass yield in the second year, 2009, but no significant interaction between cover crop and year. The greater yield of 2009 must be related

to the nutrients remaining from the 2008 tomato crop, and the better rainfall distribution over the period of cover crops growth in 2009. Perin et al. (2004) also reported similar shoot dry biomass yields for millet and sunn hemp. Millet and maize were more efficient in the production of shoot dry biomass because of their C4 status which have a greater capacity to incorporate CO_2 and produce dry mass (Taiz and Zeiger, 2002). The greater yield of shoot dry biomass of millet, in relation to legumes such as sunn hemp and velvetbeans, has also been reported (Torres et al., 2008; Suzuki and Alves, 2006).

The establishment of weeds, 30 days after tomato transplanting, was impacted by an interaction between cover crop treatment and year (Table 2). In 2008, millet and velvetbeans had fewer weeds than fallow, but did not differ in relation to the other treatments. In 2009, millet, sunn hemp, and corn showed better results than velvetbeans and fallow in the suppression of the weed. The suppression of weeds varied between years for velvetbeans and fallow, as these treatments were less efficient in 2009. The weeds with the greatest occurrence in all treatments were *Alternanthera ficoidea* (L.) R. Br. and *Lepidium virginicum* L. This demonstrated the benefit of millet as a cover crop; its fast initial growth and greater biomass yield suppressed weeds in tomato crops. However, legumes such as sunn hemp, hairy vetch, and soybeans also suppress the weeds when they are used as cover crops (Campiglia et al., 2010; Carrera et al., 2004; Silva et al., 2009). No-till also contributes to the suppression of weeds by minimizing the dissemination of seeds (Sutton et al., 2006), which is not the case of the present work. Soil microbiological activity was more intense in sunn hemp and millet cultivation in comparison with conventional tillage for the nitrification activity of nitate. For all other microbiological characteristics analyzed, nitrification activity of ammonia, urease, total bacteria and fungi, and soil respiration rates did not differ among treatments (Table 3). The stimulus of nitrification activity of nitrate provided by sunn hemp and millet compared to conventional tillage indicates improvement in the process of transformation of nitrogen compounds and even the mineralization of organic nitrogen, provided for the maintenance of plant residue on the soil surface by no-tillage (Babujia et al., 2010). On the other hand, the other microbiological activities did not differ among treatments, contradicting the results of increased microbial activity in the no-till situation. We believe that the natural microbial activity of this soil was responsible for supressing the increase in these properties by cover crops. Hamido and Kpomblekou-A (2009) and Buyer et al. (2010) reported that the discrepancy in results of soil microbiology is related to several factors such as species of cover crop, climate and time and depth of soil sampling. According to Bonanomi et al. (2011), intense agricultural activity with excessive use of agricultural input deteriorates the microbiological quality of soil, which

Table 3. Nitrification activity of ammonia, nitrification activity of nitrate, urease activity, quantity of total bacteria, quantity of total fungi and soil activity respiration of experimental treatments of cover crops.

Cover crop	Nitrification activity NH4 (mg NH$_4$-N g^{-1} ds*)	Nitrification activity NO3 (mg NO$_3^-$-N g^{-1} ds)	Urease (mg NH$_4$-N 3 h^{-1} g^{-1} ds)	Bacteria (UFC g^{-1} ds)	Fungi (UFC g^{-1} ds)	Respiratory Activity (mg CO$_2$ 100 g ds)
Velvetbean	53.4[a]	50.7[ab]	26.2[a]	8.7E+06[a]	8.6E+04[a]	5.72[a]
Sunn hemp	49.5[a]	53.7[a]	22.6[a]	1.0E+07[a]	7.0E+04[a]	6.18[a]
Millet	51.2[a]	53.9[a]	30.5[a]	9.1E+06[a]	7.1E+04[a]	3.52[a]
Fallow	52.2[a]	39.8[ab]	26.5[a]	1.1E+07[a]	1.1E+05[a]	6.16[a]
Maize (Conventional)	58.2[a]	37.4[b]	28.8[a]	1.1E+07[a]	7.3E+04[a]	9.68[a]
ANOVA (P values)	0.3062	0.0491	0.4742	0.6267	0.2475	0.2111

†Values followed by the different small letter in columns are significantly different (Tukey test, P ≤ 0.05). *ds, Dry soil.

was not the case for this work.

No difference was detected among treatments for pH, organic matter and cation exchange capacity, probably due to high natural fertility of the soil regarding these characteristics. Potassium soilconcentration increased with the cultivation of sunn hemp in relation to velvetbean, fallow and maize (conven-tional tillage), but on the other hand, nutrients phosphorus, calcium and magnesium did not differ among the treatments. Millet had the same performance as sunn hemp to the potassium content in the soil (Table 4). Potassium is the nutrient extracted in greater quantity by tomato plants (Fontes et al., 2004), which may have favored the performance of sunn hemp grown after the sunn hemp. Silva et al. (2002) and Wang et al. (2009) reported significant amounts of potassium recycled for use of sunn hemp as cover crops. However, in this experiment, it was not detected that there was any increase in the levels of potassium in tomato plant leaves when grown after the sunn hemp (Table 5). Nutrient content of tomato leaves did not show any significant interaction between cover crops and years for nitrogen, phosphorus and potassium (Table 5). The leaf concentrations of nitrogen and potassium in tomato plants did not differ among cover crop treatments, but in 2009 the concentrations of these nutrients were greater than 2008, including phosphorus. The concentration of phosphorus in tomato leaves was higher when it was grown on the straw of fallow than velvetbean, sunn hemp and millet, but not significantly different from maize straw (conventional tillage). Calcium, magnesium and sulphur showed interactions (Table 6). The calcium concentration did not differ significantly between treatments during the experiment, but it was observed that in 2009, velvetbeans and fallow had greater concentrations than in 2008. Magnesium concentration in tomato leaves was higher in velvetbean, sunn hemp and millet treatments in the first year, and in the second year, fallow had a higher concentration than maize and sunn hemp. In 2009, magnesium concentrations were higher than in 2008 for all treatments. In relation to sulphur, velvetbean had a lower concentration in leaves of tomato in 2008, and millet, in 2009. The comparison of the two years shows that millet and fallow had greater sulphur concentrations in the leaves of tomato in 2008, and velvetbeans, in 2009.

Due to the crop rotation and green manure with velvetbean and sunn hemp, we had expected an increase in the leaf nitrogen concentration, which was not observed in this experiment. According to Thönnissen et al. (2000), the recovery of nitrogen from green manure by tomato crop is in the order of 9 to 15%, being more expressive in low fertility soils, which was not the case in this experiment and explains why the nitrogen concentrations in leaves did not increase in the legume crop treatment. In this experiment, the tomato crop had great nutrient leaf content in comparison with other reports of tomato status leaf nutrient (Fontes et al., 2004; Abdul-Baki et al., 1997b). Total fruit yield and marketable fruit yield of tomato was higher when grown on the straw of sunn hemp compared to other treatments; but it was not different from that of millet. The fallow treatment had the lowest total fruit yield and marketable fruit yield of tomato. The number of total marketable fruit was similar among treatments; only the fallow had the lower numbers of fruit than the other treatments. The highest fresh mass of marketable fruit of tomato was with sunn hemp, but not differing from millet (Table 7). All measured parameters were higher in 2009 than in 2008, possibly due to the greater leaf nutrient concentration in the tomato. Tomato yield results demonstrated the efficiency of no-tillage, mainly in

Table 4. Phosphorus (P), potassium (K), calcium (Ca) and magnesium (Mg) content in the soil in the tretaments of cover crops, Ribeirão Preto.

Cover crop	P (g dm^{-3})	K (mmol$_c$ dm^{-3})	Ca (mmol$_c$ dm^{-3})	Mg (mmol$_c$ dm^{-3})
Velvetbean	71.7^a	3.4^b	26.8^a	14.4^a
Sunn hemp	64.2^a	4.2^a	27.4^a	15.0^a
Millet	60.3^a	3.8^{ab}	27.9^a	14.7^a
Fallow	60.5^a	3.2^b	27.8^a	15.0^a
Maize (Conventional)	59.2^a	3.3^b	28.3^a	14.3^a
ANOVA (P value)	0.167	0.054	0.863	0.931

Values followed by the different letter in columns are significantly different (Tukey test, P ≤ 0.05).

Table 5. Concentration of macronutrients (N, P and K) in the leaves of tomato in the treatments of cover crops.

Cover crop (CC)	N	P	K
	(g kg^{-1} dry mass)		
Velvetbean	31.7^a	2.5^b	45.7^a
Sunn hemp	33.2^a	2.4^b	44.8^a
Millet	33.9^a	2.5^b	45.5^a
Fallow	34.8^a	2.9^a	45.3^a
Maize (Conventional)	34.3^a	2.7^{ab}	47.7^a
Year			
2008	31.9^b	1.7^b	40.2^b
2009	35.3^a	3.4^a	51.5^a
ANOVA (P ≤ 0.05)			
CC	0.1074	0.0282	0.6927
Year	<0.0010	<0.0010	<0.0010
PC versus Year	0.7084	0.1443	0.4846

Values followed by the same letter in columns are not significantly different (Tukey test, P ≤ 0.05). Samples of leaves were collected between 3rd and 4th fruit bunches.

Table 6. Concentration of macronutrients (Ca, Mg and S) in the leaves of tomato in the treatments of cover crops.

Cover crop (CC)	Ca (g kg^{-1} dry mass)		Mg (g kg^{-1} dry mass)		S (g kg^{-1} dry mass)	
	2008	2009	2008	2009	2008	2009
Velvetbeans	37.3^{a†}	42.3^{a*}	9.5^a	11.3^{ab*}	5.1^b	5.8^{a*}
Sunn hemp	40.2^a	40.9^a	8.6^a	10.6^{bc*}	5.9^a	5.8^a
Millet	37.2^a	39.6^a	8.6^a	11.1^{abc*}	6.0^{a*}	5.3^b
Fallow	36.0^a	44.2^{a*}	7.1^b	11.9^{a*}	6.3^{a*}	5.9^a
Maize (Conventional)	38.9^a	38.3^a	8.0^b	10.1^{c*}	6.2^a	5.9^a
ANOVA (P ≤ 0.05)						
PC	0.4076		0.0043		0.0022	
Year	0.0047		<0.0010		0.0798	
PC vs. Year	0.0153		0.0028		0.0009	

Samples of leaves were collected between 3rd and 4th fruit bunches. Values followed by the same letter in columns are not significantly different (Tukey test, P≤0.05). *In lines are significantly differ rent (Tukey test, P ≤ 0.05).

Table 7. Total fruit yield (TFY), marketable fruit yield (MFY), total fruit number (TFN), marketable fruit number (MFN) and average fresh mass of marketable fruit (AFMMF) of tomato in the treatments of cover crop.

Cover crop (CC)	TFY (Mg ha^{-1})	MFY (Mg ha^{-1})	TFN (1.000 ha^{-1})	MFN (1.000 ha^{-1})	AFMMF (g)
Velvetbean	73.5^{bt}	67.4^b	631^a	575^a	117.1^b
Sunn hemp	79.2^a	73.4^a	645^a	598^a	122.0^a
Millet	74.5^{ab}	67.5^b	611^a	558^{ab}	120.3^{ab}
Fallow	62.9^c	58.7^c	569^b	520^b	111.7^c
Maize (Conventional)	72.8^b	66.3^b	626^a	569^a	115.7^{bc}
Year					
2008	65.4^b	58.6^b	507^b	507^b	114.7^b
2009	79.8^a	74.6^a	726^a	620^a	120.0^a
ANOVA (P value)					
CC	0.0004	0.0017	0.0211	0.0316	0.0034
Year	<0.0001	<0.0001	<0.0001	<0.0001	0.0076
PC versus Year	0.4542	0.5390	0.6383	0.6976	0.5657

[t]Values followed by the same letter in columns are not significantly different (Tukey test, $P \leq 0.05$).

rotation with cover crops like sunn hemp and millet, thus eliminating the need of tillage for growth. Wang et al. (2009) also reported better yields for tomato when grown after sunn hemp, with an average yield of marketable fruits similar to that produced in this experiment. The results of the tomato yield in this experiment in no-till were similar to those of Campiglia et al. (2010), Abdul-Baki et al. (1996) and Lenzi et al. (2009). However, Sainju et al. (2002) reported that the no-tillage tomato yield was lower than the minimum and conventional tillage, which was attributed to soil com-paction, but in the following year, the tomato yield was similar to those of tillage treatments due to the better physical conditions of the soil provided by the no-tillage treatment.

Conclusion

This field study demonstrated that tomato no-tillage in rotation with cover crops is technologically viable over two years. Millet was the best cover crop to produce dry biomass together with sunn hemp; it was also the best to suppress the establishment of weeds. Sunn hemp increased tomato yield and ensured a high quality of fruits. In the soil, potassium content and nitrification activity were enhanced by sunn hemp. The cycling of nutrients by cover crops did not impact on the nutritional satus of tomato except in fallow soil, which contributed to increase phosphorus leaf content in tomato. However, the growth of tomato in no-till in rotation with cover crops provides good practices for the growers who want to contribute to the development of sustainable agriculture around the world.

ACKNOWLEDGEMENT

The Foundation for Research Support of São Paulo (FAPESP) for the research grant, project 2007/03328-6 is acknowledge.

REFERENCES

Abdul-Baki AA, Morse RD, Devine TE, Teasdale JR (1997a). Broccoli Production in Forage Soybean and Foxtail Millet Cover Crop. HortScience 32:836-839.

Abdul-Baki AA, Teasdale JR, Korcak R (1997b). Nitrogen Requirements of Fresh-market Tomatoes on Hairy Vetch and Black Polyethylene Mulch. HortScience 32:217-221.

Abdul-Baki AA, Teasdale JR, Korcak R, Chitwood Dj, Huettel RN (1996). Fresh-Market Tomato Production in a Low-input Alternative System Using Cover-crop Mulch. HortScience 31:65-69.

Argenton J, Albuquerque Ja, Bayer C, Wildner LP (2005). Comportamento de atributos relacionados com a forma da estrutura de Latossolo Vermelho sob sistemas de preparo e plantas de cobertura. R. Bras. Ci. Solo 29:425-435.

Babuji LC, Hungria M, Franchini JC, Brookes PC (2010). Microbial biomass and activity at various soil depths in a Brazilian oxisol after two decades of no-tillage and conventional tillage. Soil Biol. Biochem. 42:2174:2181.

Bonanomi G, D'ascoli R, Antignani V, Capodilupo M, Cozollino L, Marzaioli R, Puopolo G, Rutigliano FA, Scelza R, Scotti R, Rao Ma, Zoina A (2011). Assessing soil quality under intensive cultivation and tree orchards in Sourthen Italy. Appl. Soil Ecol. 47:184-194.

Bunt JS, Rovira AD (1955). Microbiological studies of some subantarctic soils. J. Soil Sci. 6:119-128.

Buyer JS, Teasdale JR, Roberts DP, Zasada IA, Maul JE (2010). Factor affecting soil microbial community structure in tomato cropping systems. Soil Biol. Biochem. 42:831-841.

Campiglia E, Caporali F, Radicetti E, Mancinelli R (2010). Hairy Vetch (Vicia villosa Roth.) cover crop residue management for improving weed control and yield in no-tillage tomato (Lycopersicon esculentum Mill.) production. Eur. J. Agron. 33:94-102.

Carrera LM, Abdul-Baki AA, Teasdale JR (2004). Cover Crop Management and Weed Supression in No-tillage Sweet Corn Production. Hort. Sci. 39:1262-1266.

Castro OM, Prado H, Severo ACR, Cardoso EJBN (1993). Avaliação da atividade de microrganismos do solo em diferentes sistemas de manejo de soja. Sci. Agric. 50:212-219.

Colla G, Mitchel JP, Joyce BA, Huych LM, Wallender Ww, Temple SR, Hsiao TC, Poudel DD (2000). Soil Physical Properties and Tomato Yield and Quality in Alternative Cropping Systems. Agron. J. 92:924-

932.

Duda GP, Guerra JGM, Monteiro MT, De-Polli H, Teixeira MG (2003). Perennial herbaceous legumes as live soil mulches and their effects on C, N and P of the microbial biomass. Sci. Agric. 60:139-147.

Fontes PCR, Loures JI, Galvão JC, Cardoso AA, Mantovani EC (2004). Produção e qualidade do tomate produzido em substrato, no campo e em ambiente protegido. Hort. Bras. 22:614-619.

Hamido AS, Kpomblekou-A K (2009). Cover crops and tillage effects on soil enzyme activities following tomato. Soil Till. Res. 105:269-274.

Keeney DR, Nelson DW (1982). Nitrogen inorganic forms. In: Page, AL, Miller RH, Keeney DR. Method of soil analysis: chemical and microbiological proporties. 2 ed. Madison: American Society Agronomy. pp. 646-698.

Kieling AS, Comim JJ, Fayad JA, Lana MA, Lovato PE (2009). Winter Cover Crops in Ni-Tillage System without Herbicides: Effects on Weed Biomass and Tomato Yield. Ci. Rural 39:2207-2209.

Lenzi A, Antichi D, Bigongiali F, Mazzoncini M, Migliorini P, Tesi R (2009). Effect of different cover crops on organic tomato production. Renew. Agric. Food Syst. 24:92-101.

Littel RC, Milliken GA, Stroup WW, Wolfinger RD, Schabenberger O (2006). SAS for Mixed Models, Second Edition, Cary, NC: SAS Institute Inc.

Malavolta E, Vitti GC, Oliveira AS (1989). Avaliação do estado nutricional de plantas: princípios e aplicações. Piracicaba: Potafós. p. 210.

Martin JP (1950). Use of acid, rose bengal, and streptomycin in the plate method for stimate soil fungi. Soil Sci. 69:215-232.

Mcgarity JW, Myers MG (1967). A survey of urease activity in soil of Northen new south wales. Plant Soil 27:217:238.

Nuez F (1995). El cultivo del tomate. Madrid : Mundi-Prensa. p. 793.

Perin A, Santos RHS, Urquiaga S, Guerra JGM, Cecon PR (2004). Produção de fitomassa, acúmulo de nutrientes e fixação biológica de nitrogênio por adubos verdes em cultivo isolado e consorciado. Pesq. Agropec. Bras. 39:35-40.

Raij B Van Cantarella H, Quaggio JA, Furlani AMC (1997). Recomendações de adubação e calagem para o estado de São Paulo. Campinas: Instituto Agronômico/Fundação IAC. (Boletim Técnico, 100).

Rezende LA, Assis LC, Nahas E (2004). Carbon, nitrogen and phosphorus mineralization in two soils amended with distillery yeast. Bioresour. Technol.94:159-167.

Sainju MM, Singh BP, Yaffa S (2002). Soil Organic Matter and Tomato Yeld Following Tillage, Cover Crop and Nitrogen Fertilization. Agron. J. 94:594-602.

Silva JAA, Vitti GC, Stuchi ES, Sempionato OR (2002). Reciclagem e incorporação de nutrientes ao solo pelo cultivo intercalar de adubos verdes em pomar de laranjeira – 'Pera'. Ver. Bras. de Frut. 24:225-230.

Silva AC, Hirata EK, Monquero PA (2009). Produção de palha e supressão de plantas daninhas por plantas de cobertura, no plantio direto do tomateiro. Pesq. Agropec. Bras. 44:22-28.

Sutton KF, Lanini WT, Mitchell JP, Miyao EM, Shrestha A (2006). Weed Control, Yield, and Quality of Processing Tomato under Different Irrigation, Tillage, and Herbicide Systems. Weed Tech. 20:831-838.

Suzuki LES, Alves MC (2006). Fitomassa de plantas de cobertura em diferentes sucessões de culturas e sistemas de cultivo. Bragantia 65:121-127.

Taiz L, Zeiger E (2002). Plant Physiology. Sunderland: Sinauer Associates. p. 690.

Thönnissen C, Midmore DJ, Ladha JK, Holmer RJ, Schimidhalter U (2000). Tomato Crop Response to Short-Duration Legume Green Manures in Tropical Vegetable Systems. Agron. J. 92:245-253.

Torres JLR, Pereira MG, Fabian AJ (2008). Produção de ftomassa por plantas de cobertura e mineralização de seus resíduos em plantio direto. Pesq. Agropec. Bras. 43:421-428,

Vieira FCS, Nahas E (2005). Comparison of microbial numbers in soil by using various culture media and temperature. Microbiol. Res. 160:196-202.

Wang Q, Klassen W, Li Y, Codallo M (2009). Cover Crops and Organic Mulch to Improve Tomato Yields and Soil Fertility. Agron. J. 101:345-351.

3

Spatial variability in the physico-chemical properties of soils affected by animal wastes in Uyo, Akwa Ibom State of Nigeria

B. Ndukwu[1]*, S. U. Onwudike[1], M. C. Idigbor[1], C. E. Ihejirika[2] and K. S. Ewe[1]

[1]Department of Soil Science and Technology, Federal University of Technology Owerri, Imo State, Nigeria.
[2]Department of Environmental Technology, Federal University of Technology Owerri, Imo State, Nigeria.

Variability in the physico-chemical properties of soils affected by animal wastes in Uyo, Akwa- Ibom State, Nigeria was investigated in this study. A free survey technique guided field sampling. Five profile pits were dug on five studied sites namely: Site severely affected with poultry manure, site moderately affected with poultry manure, site severely affected with swine manure, site moderately affected with swine manure, and control site (50 m away from the affected sites). Soil samples were collected at different depths and subjected to routine laboratory analysis. Data collected were analyzed statistically. Results of the investigation showed that variations existed among some physico-chemical properties of soils that are affected with animal wastes when compared to control. There were no variations in the silt and clay content of the affected soil. Among the affected soils, the pH and effective cation exchange capacity decreased down the profile. Results also showed that soils affected with poultry manure and swine manure had the highest exchangeable cations when compared to control. Appropriate measure should therefore be taken in the disposal of these wastes to avoid environmental hazards.

Key words: Variability, Epipedon, profile pits, environment, Akwa-Ibom State.

INTRODUCTION

With an increasing rate of human population worldwide, the rate of animal husbandry has been on the increase in order to meet up with food demand. Consequently, large quantities of animal wastes are produced which are deposited on soil as wastes. Research has shown that deposition of these animal wastes on soil increase soil organic matter and carbon fractions and enhances soil quality and productivity (Kingery et al., 1994). Application of animal wastes increase nutrient supplying capacity of the soil (Webster and Gouiding, 1989; Rochette and Gregorich, 1998), vegetative and reproductive growth of plants (Azam Shah et al., 2009; Suthar, 2009, Maftoun and Moshiri, 2008 and Sawyer et al., 2006), enhance moisture retention capacity and infiltration rate (Erikson et al., 1999), improve physical conditions of soil such as bulk density, aggregate stability and aeration (Yuksel and Orhan, 2004) as well as reduce the pH of an acid Ultisol (Bauer and Black, 1994) and crusting and runoff (Rochette and Gregorich, 1998).

Animal wastes are deposited on soil due to their nutrient value (Jackson and Bertsch, 2001; Garbarino et al., 2003), their effect on the environment has become an issue of interest (Mohammad et al., 2010). Studies have shown that long term deposition of animal waste on soil has resulted into ground and surface water pollution as a result of leaching and runoff of nutrients as well as accumulation of excessive soluble salts and the buildup of micronutrients (Mohammad et al., 2010; Zachary et al., 2008).

In Uyo, Akwa Ibom State Nigeria, large quantities of poultry manure and swine wastes abound in these areas and the effect of long term deposition on these materials on soil need to be investigated for proper environmental management and sustainability. Based on this premise

*Corresponding author. E-mail: onwudikestanley@yahoo.com.

therefore, this study was aimed at investigating the variability in the physico-chemical properties of soils affected by poultry manure and swine waste deposits.

MATERIALS AND METHODS

Study area

Uyo, Akwa Ibom State lies between latitude 4' 02'N and longitude 8' 21'E. The area has a relative humidity of about 70 to 80%. Soils were derived from coastal plain sand (Benin formation). It has a humid tropical climate with a mean annual rainfall of about 2000 to 2500 mm and an annual temperature of about 26 to 30°C (Uwah et al., 2011). The vegetation is that of the rainforest characterized by a variety of plants forms arranged in tiers. Soils have low nutrient content as a result of heavy rainfall that encourages leaching of basic cations. The socio-economic activity of people in this study site is growing of tropical crops like maize, cassava, plantain, okra, oil palm etc.

Field study

A free soil survey technique was used in this study. Five soil units were examined, namely soils severely affected with poultry manure, soils severely affected with swine waste, soils moderately affected with poultry manure (50 m away from severely affected poultry manure), soils moderately affected with swine waste (50 m away from severely affected swine waste), and control which was 1.5 km away from soils affected with the animal manure. Five soil profile pits were sited in the entire studied area, representing a soil unit. The profile pits were dug and described according to the procedures of FAO (1990) with depths of 1 to 20, 20 to 40, 40 to 60 and 60 to 80 cm, thereby making a total of 20 soil samples. These samples were air dried and sieved using 2 mm mesh sieve and subjected to routine laboratory analysis.

Laboratory analysis

Particle size distribution was determined by hydrometer method (Gee and Or, 2002) using Sodium hexametaphosphate (Calcon) as dispersant. Soil pH was determined electronically in a 1: 2.5 soil: solution ratio (Hendershot et al., 1993). Bulk density was estimated using core sampler calculated by mass of oven dried soil divided by volume of core sampler (Foth, 1984). Soil moisture content was determined gravimetrically. Total Nitrogen was determined by Kjeldal digestion method using concentrated H_2SO_4 and Sodium Copper Sulphate as catalyst mixture (Bremner and Yeomans, 1988). Organic carbon was determined by wet oxidation method (Nelson and Sommers, 1982). Exchangeable bases were determined using 1N ammonium acetate solution according to Jackson (1964). Exchangeable Calcium (Ca) and Magnesium (Mg) were determined using Ethylene Diamine tetracetic acid (EDTA) while exchangeable Sodium (Na) and Potassium (K) were determined flame photometrically. Exchangeable acidity (Al^{3+} and H^+) was determined according to Mclean (1982). Available phosphorus was determined by extraction with Bray II solution and determined calorimetrically on spectrophotometer according to Nelson and Sommers (1982). Effective Cation Exchange Capacity (ECEC) was computed by the summation of all exchangeable bases and exchangeable acidity while percentage base saturation was calculated by dividing total exchangeable bases with effective cation exchangeable capacity and the quotient multiplied by 100. Soil data were subjected to mean descriptive statistics and correlation and coefficient of variation analysis were used to

ascertain the variability according to Aweto (1982).

RESULTS AND DISCUSSION

Some selected physical properties of the soil are shown in Table 1. The soils are texturally more of sandy loam. There was more clay fraction in soils moderately affected with poultry manure than soils severely affected with either poultry or swine waste. Dominance of sand fractions in the studied site could be attributed to the parent material (coastal plain sand) from which the soils are formed as well as humid rainfall characteristic that promote leaching of silt and clay fractions down the macro-porous soils. Since most of the secondary minerals are domicile in clay fractions of the soil, moderate application of swine waste on soil with resultant increase in clays could boost soil fertility because of an increase in ion exchange reactions.

There were variations in the bulk density of the soil among the treatments. There was a lower bulk density on soils affected with animal wastes when compared to the control. This could be attributed to the role of organic wastes in reducing the bulk density of the soil and this was in concord with Brady and Weil (1999) who noted that application of animal wastes either for plant nutrient supply or for disposal purposes reduces soil bulk density because of their ability in forming soil aggregates.

Soil moisture retention also varied among the soils but the highest moisture content was recorded on soils severely affected with either poultry manure or swine waste and the moisture content increases down the soil. This could be attributed to the mulching effect of these materials as well as improved structure and macro-porosity. This observation was in agreement with Aluko and Oyedele (2005) who noted that application of organic wastes increases the water retention capacity of soil. However, the mobility and retentivity of water may depend on the activation energy of the clays which could be influenced by the type of basic cations (Logsdon and Laird, 2004). This observation could be vital in the management of surface and ground water pollution since severe application of animal waste could cause ground water pollution (Mohammad et al., 2010).

The chemical properties of the soils are shown in Table 2. In each of the studied area, soil pH decreases down the profile with control location having the lowest results due to low accumulation of organic matter and soils severally affected with poultry manure recorded the highest values. The acidity of the soils could be attributed to the acid parent material from where the soils are formed and the mineralization of organic wastes which releases organic acids (fulvic and humic acid) which are leached down the profile hence, decreasing the pH value. Soils affected with poultry manure and swine waste recorded higher exchangeable Calcium and Magnesium than the control. This could be attributed to the improvement in soil pH, since pH range of 5.5 to 7.0

Table 1. Some physical properties of the studied soils.

Depth	Sand (g/kg)	Silt (g/kg)	Clay (g/kg)	BD (g/cm3)	MC (%)	Total porosity (%)	Textural class
		Profile Pit 1 (Severally affected with poultry manure)					
0 - 20	785.2	152	62.8	1.20	1.80	55	Sandy loam
20 - 40	775.2	132	92.8	1.26	2.00	52	Sandy loam
40 - 60	735.2	192	82.8	1.43	2.07	46	Loamy sand
60 - 80	755.7	172	72.7	1.38	2.42	48	Sandy sand
Mean	762.7	162	77.8	1.32	2.07	50	
		Profile Pit 2 (moderately affected with poultry manure)					
0 - 20	825.2	132	42.7	1.24	1.58	53	Sandy loam
20 - 40	765.2	192	42.8	1.30	1.55	51	Sandy loam
40 - 60	715.2	252	32.7	1.19	1.89	55	Sandy clay loam
60 - 80	672.2	292	32.8	1.13	1.49	57	Sandy clay
Mean	694.5	217	37.8	1.22	1.63	54	
		Profile Pit 3 (Severely affected with Swine manure)					
0 - 20	735.2	142	102.8	1.08	2.57	59	Sandy loam
20 - 40	805.2	162	32.7	1.07	2.27	60	Sandy loam
40 - 60	725.2	252	22.8	1.26	1.44	52	Sandy clay loam
60 - 80	705.2	262	32.8	1.13	2.17	57	Sandy clay
Mean	742.7	204.5	47.4	1.14	2.11	57	
		Profile Pit 4 (Moderately affected with Swine manure)					
0 - 20	775.2	152	72.8	1.14	1.31	57	Sandy loam
20 - 40	755.2	182	62.6	1.27	1.15	52	Sandy loam
40 - 60	712.2	242	42.9	1.19	1.17	55	Sandy clay loam
60 - 80	735.2	202	62.8	1.25	1.39	53	Sandy clay loam
Mean	744.5	194.5	60.3	1.21	1.26	54	
		Profile Pit 5 (Control)					
0 - 20	685.2	252	62.8	1.92	1.26	28	Loamy sand
20 - 40	655.2	272	72.8	1.81	1.27	32	Sandy clay loam
40 - 60	635.2	292	72.7	1.76	1.22	34	Sandy clay loam
60 - 80	645.2	302	52.9	1.87	1.15	29	Sandy clay loam
Mean	655.2	279.5	65.3	1.84	1.23	31	

BD, Bulk density; MC, moisture content.

favors the availability of exchangeable cations. The observation was in concord with Khaleel et al. (1981) who reported that high soil acidity lowers the availability of Calcium, Sodium, Phosphorus and Potassium due to the production of nitrate.

There was no effect on exchangeable Sodium in all the pedons indicating unavailability of soluble salts in the area or in the animal wastes used in this study. High precipitation common in the area could lead to dissolution and leaching of any soluble salt that might have been introduced into the soil through the animal foods. In all the pedons, Calcium dominated other cations. This was in agreement with Braver et al. (1978). Calcium saturated soils have lower activation energy, hence its water retention is high and therefore its hydration energy is low. Increase in the concentration of Calcium therefore increases aggregate stability of soil particles since Calcium act as a binding agent holding soil particles and organic polymers together (Baver et al., 1978). Application

of animal waste therefore increases soil aggregation.

The effective cation exchange capacity (ECEC) decreases down the profile in all the soil. This could be due to decrease in organic matter content down the profile since increase in organic matter increases the cation exchange capacity (Onwudike, 2010). The same trend was observed in percentage base saturation. There is a positive correlation between organic matter and base saturation as also reported by Bell and Moody (1998).

There was moderate organic carbon in the epipedon (mainly A- horizon) in all the soil. This was due to organic matter availability but higher organic carbon was found in soils affected with animal wastes than in control. Low total Nitrogen and organic carbon recorded in this study could be as a result of high mineralization of organic manure associated with tropical environment which leads to leaching of available organic matter due to high precipitation. Application of poultry manure and swine wastes increased the availability of Phosphorus which was

Table 2. Soil chemical properties of the study site.

Depth (cm)	pH (H2O)	pH (KCl)	Al2+/H+ (Cmol)	Ca2+ (Cmol)	Mg2+ (Cmol)	Na+ (Cmol)	K+ (Cmol)	TEB (Cmol)	ECEC (Cmol)	BS (%)	OC (%)	TN (%)	Av.P (ppm)
										Profile Pit 1			
0 - 20	6.43	5.60	0.95	1.56	0.80	0.01	0.03	2.40	3.35	72	1.7	.04	17.5
20 - 40	5.55	4.88	1.10	1.50	0.86	0.02	0.03	2.41	3.51	69	1.6	.03	14.7
40 - 60	5.25	4.35	1.30	0.90	0.56	0.01	0.01	1.50	2.80	54	1.2	.02	13.6
60 - 80	4.80	4.01	0.90	0.68	0.21	0.02	0.01	0.92	1.82	51	1.2	.03	12.9
Mean	5.51	4.71	1.60	1.16	0.61	0.02	0.02	1.81	2.87	62	1.4	.03	14.7
							Profile Pit 2						
0 - 20	5.21	4.14	0.65	1.34	1.10	Tr	Tr	1.10	1.75	63	1.6	.03	15.4
20 - 40	4.52	3.64	0.55	0.90	0.76	0.01	Tr	0.77	1.27	61	1.2	.03	13.3
40 - 60	4.35	3.55	0.45	0.70	0.56	0.01	Tr	0.57	1.02	56	1.2	.02	9.8
60 - 80	4.48	3.58	0.50	0.62	0.20	0.01	Tr	0.29	1.71	30	1.1	.01	8.4
Mean	4.64	3.73	0.54	0.89	0.66	0.01		0.66	1.44	53	1.3	.02	11.7
						Profile Pit 3							
0 - 20	5.42	3.77	0.75	1.42	1.36	0.02	0.03	2.83	3.58	79	1.8	.04	22.4
20 - 40	4.36	3.29	0.50	1.38	0.90	0.02	0.04	2.34	2.84	82	1.3	.04	16.1
40 - 60	4.24	3.20	1.30	1.26	0.44	0.01	0.03	1.74	3.04	57	1.2	.02	14.0
60 - 80	4.30	3.22	0.70	0.96	0.34	0.01	0.02	1.33	2.03	66	1.2	.02	13.3
Mean	4.58	3.37	0.81	1.26	0.76	0.02	0.03	2.06	2.87	72	1.4	.03	16.5
						Profile Pit 4							
0 - 20	5.86	4.82	0.55	1.24	0.70	Tr	0.02	0.72	1.27	57	1.7	.03	16.8
20 - 40	4.79	3.94	0.96	0.86	0.36	0.01	0.01	0.74	1.70	44	1.4	.03	13.8
40 - 60	4.47	3.63	1.00	0.38	0.38	0.01	0.01	0.48	1.78	44	1.3	.02	9.1
60 - 80	4.12	3.23	0.72	0.18	0.18	0.01	0.01	0.38	1.10	35	1.1	.01	6.9
Mean	4.81	3.91	0.59	0.54	0.41	0.01	0.01	0.66	1.46	45	1.4	.02	11.7
						Profile Pit5							
0 - 20	4.79	3.75	1.20	0.56	0.18	0.01	Tr	0.19	1.39	14	1.5	.03	14.0
20 - 40	4.34	3.57	1.35	0.50	0.32	0.01	Tr	0.33	1.68	20	1.4	.02	11.2
40 - 60	4.10	3.23	0.80	0.66	0.12	Tr	Tr	0.12	0.92	13	1.2	.01	9.1
60 - 80	3.80	3.95	1.15	0.60	0.20	0.01	Tr	0.22	1.67	13	1.2	.01	6.7
Mean	4.26	3.39	1.20	0.58	0.21			0.22	1.42	15	1.3	.02	10.3

TN, Total Nitrogen; OC, organic carbon; Av.P, available Phosphorus; TEB, Total exchangeable bases; ECEC, effective cation exchange capacity; BS, base saturation; Tr, trace.

found higher than in control. This could be attributed to favourable pH range and total N which complements each other. Some of the relationships existing among soil physico-chemical properties are shown in Tables 3 and 4. Results showed that positive correlation existed (P = 0.01) among the selected physico-chemical properties in the studied locations (Table 3). No significant difference existed between moisture,

Table 3. Correlation coefficient among physico-chemical properties of soil in the study site.

Parameter	Sand (g/kg)	Clay (g/kg)	Silt (g/kg)	BD (g/cm³)	MC (%)	TN (%)	pH (H₂O)	pH (KCl)	OC (%)	Av. P (ppm)	TEA (ppm)	TEB (Cmol/kg)	ECEC (Cmol/kg)
Clay (g/kg)	0.95**												
Silt (g/kg)	0.04 NS	0.43 NS											
BD(g/cm³)	0.68*	0.53*	0.04 NS										
MC (%)	0.27 NS	0.19 NS	0.21 NS	0.58*									
TN (%)	0.70*	0.87**	0.45*	0.33 NS	0.14 NS								
pH (H₂0)	0.58**	0.76**	0.52*	0.37 NS	0.19 NS	0.86**							
pH (KCl)	0.59**	0.77**	0.46**	0.39 NS	0.21 NS	0.88**	0.98**						
OC (%)	0.43 NS	0.67**	0.54*	0.04 NS	0.19 NS	0.81**	0.82**	0.83**					
Av. P (ppm)	0.62**	0.76**	0.36 NS	0.22 NS	0.09 NS	0.93**	0.76**	0.74**	0.76**				
TEA (Cmol/kg)	0.41 NS	0.22 NS	0.29 NS	0.24 NS	0.12 NS	0.14 NS	0.49**	0.01*	0.01*	0.12 NS			
TEB (Cmol/kg)	0.72**	0.73**	0.23 NS	0.29 NS	0.07 NS	0.79**	0.80**	0.61**	0.66**	0.81**	0.26 NS		
ECEC (Cmol/kg	0.29 NS	0.48*	052*	0.0 NS	0.06 NS	0.91**	0.52**	0.53**	0.64**	0.58**	0.53**	0.68**	
BS (%)	0.77**	0.65**	0.05 NS	0.39 NS	0.02 NS	0.63**	0.47**	0.47**	0.48**	0.64**	0.62**	0.86**	0.26 NS

* , Significant at 1%; *, significant at 5%; NS, not significant at 5%. BD, bulk density; MC, moisture content; TN, total Nitrogen; OC, organic carbon; Av. P, available Phosphorus; TEA, total exchangeable acidity; TEB, total exchangeable bases; ECEC, effective cation exchange capacity; BS, base saturation.

Table 4. Variability in the physico-chemical properties of the studied soils.

Depth (cm)	Sand (g/kg)	Silt (g/kg)	Clay (g/kg)	BD (%)	MC (%)	pH (H₂0)	Al³⁺H⁺ (Cmol/kg)	TEB (Cmol/kg)	ECE ECEC (Cmol/kg)	pH (KCl)	BS (%)	OC (%)	TN (%)	Av. P (ppm)
Profile Pit 1 (Severely affected with poultry manure)														
Mean	762.7	162	77.8	1.3	1.7	5.5	5.7	1.1	1.8	2.9	61.1	1.4	0.03	14.7
%CV	2.62	16.6	15.9	8.04	24.7	12.5	14.7	16.9	46.7	22.8	33.8	19.9	22.8	13.9
Rank	L	M	M	L	M	L	L	M	H	M	H	M	M	L
Profile Pit 2 (Moderately affected with poultry manure)														
Mean	694.5	217	37.8	1.2	1.6	4.6	3.7	0.5	0.7	1.4	52.2	1.3	0.02	11.7
%CV	8.7	15.3	32.3	6	11	8.3	7.5	15.8	56.2	37	29.5	10.7	36.6	27.3
Rank	L	M	H	L	L	L	L	M	H	H	H	L	H	H
Profile Pit 3 (Severely affected with Swine manure)														
Mean	742.7	207.5	47.4	1.1	2.1	4.6	3.9	0.8	2.1	2.9	71.1	1.4	0.03	16.5
%CV	5.8	17.3	30	8	22.7	12.3	10.8	42.2	59.4	31.7	42.2	19.7	36.2	25.2
Rank	L	M	H	L	M	L	L	H	H	H	H	M	H	M
Profile Pit 4 (Moderately affected with Swine manure)														
Mean	744.5	194.5	60.3	1.2	2.3	4.8	3.9	0.6	0.7	1.5	44.7	1.4	0.02	11.7
%CV	3.4	20.9	19.4	4.9	7.7	15.6	18.9	24.4	50.9	21.4	36.7	17.3	33.7	38.6

Table 4. Contd.

Rank	L	M	L	M	L	Profile pit 5 (Control)								
	L	M	L	M	L	M	M	M	H	M	H	M	M	H
Mean	655.2	279.5	1.84	65.3	1.23	4.3	3.4	1.2	0.2	1.4	14.9	1.3	0.02	10.3
%CV	3.3	14.7	8.3	16.9	27.4	9.3	11.9	23.8	40.6	25.2	21.4	10.8	48.9	32.1
Rank	L	L	L	M	H	L	L	M	H	H	M	L	H	H

L, Low; M, medium; H, high; BD, bulk density; MC, moisture content; TN, total nitrogen; OC, organic carbon; Av. P, available Phosphorus; TEB, total exchangeable bases; ECEC, effective cation exchange capacity; BS, base saturation; CV, coefficient of variation.

bulk density content and selected soil properties except between moisture content and bulk density. In all the studied locations, variability existed among the physico-chemical properties except sand, bulk density and total exchangeable bases (Table 4). Variability existing among these soil properties may be attributed to land use system, (Esu et al., 1991) and differences in the mineral composition of these organic wastes.

CONCLUSION AND RECOMMENDATION

Hitherto, dumping of animal wastes on soils has been one of the effective measures of improving soil fertility and productivity. However, long term dumping of these wastes has started posing an environmental and management problem such as surface and ground water pollution, eutrophication, as well as leaching of mineral salts. Results of this investigation have shown that some degree of variability exists among soil properties that are affected with poultry manure and swine waste when compared to unaffected soils. Some of these variations are prominent down the profile due to leaching of mineral elements. From the findings of this study, it is recommended that long term deposition of organic wastes should be discouraged. A sustainable approve such as utilizing them as organic fertilizers should be adopted than dumping them as means of disposal.

REFERENCES

Aluko OB, Oyedele DJ (2005). Influence of organic waste incorporation on changes in selected soil physical properties. J. Appl. Sci. 5(2);357-362.

Aweto AO (1982). Variability of upper slope soils developed under sand stone in Southeastern Nigeria. J. Geol. 25:27-37.

Azam Shah SS. Mohmood S, Mohammad W, Shafi M, Nawaz H (2009). N. uptake and yield of wheat as influenced by integrated use of organic and mineral nitrogen. Int. J. Plant Prod. 3 (3):45-56.

Bauer A, Black AL (1994). Quantification of the effect of soil organic matter content on soil productivity. Am. J. Soil. Sci. 5:185-193

Braver I.D, Gardner WH, Gardner WR (1978). Soil physics. In: John and Sons, fourth Ed., New York. p. 498.

Bremner JM, Yeomans JC (1988). Laboratory Techniques in J.R. Wilson (ed.). Advances in Nitrogen cycling in Agricultural ecosystem. C.A.B Int. England.

Bell MJ, Moody PW (1998). The role of active fraction of soil organic matter in physical and chemical fertility of ferrosols. Aust. J. soil Res. 36:309-819.

Brady NC, Weil RR (1999). The nature and properties of soils. 12th Ed., Prentice Hall Inc., New Jersey. p. 860.

Erikson G, Coale F, Bellero G (1999). soil nutrient dynamics and maize production in organic waste amended soil. Soil Sci. Soc. Am. J. 15:85–92.

Esu IE, Odunze AC, Morberg, JP (1991). Physico-chemical and mineralogical properties of the soils in Talata-Mafara area of Sokoto. Samaru J. Agric. Res. 8:41–56. FAO (Food and Agricultural Organisation) 1990. Guidelines for soil profile description. 3th Ed, FAO, Rome. p. 70.

Foth HD (1984). Fundamentals of soil Science. 7th Ed., John Wiley and Sons, New York. p 435.

Garbarino JR, Bednar AJ. Rutherford DW, Beyer RS, Wershaw R.L (2003). Environmental fate of roxarsone in poultry litter. I. Degradation of roxarsone during composting. Environ. Sci. Technol. 37:1509-1514.

Gee GW,.Or D (2002). Particle Size analysis. In: Dane, J. H and G.C.Topps (ed.). Methods of soil analysis, Part 4, Physical Methods. Soil Science Society of America, Book Series. No. 5, asa and SSSA Madison WI. pp.255– 293.

Hendershot WH, Lalande H, Duquette M (1993). Soil reaction and exchangeable acidity. In : Carter, M.R (Ed.). Soil Sampling and Methods of Soil Analysis. Canadian Society of Soil Science. Lewis Publishers, London. pp.141- 145.

Jackson ML (1964). Soil Chemical Analysis: Prentice Hall Englewood Cliffs, N.J: pp.86-92.

Jackson BP, Bertsch PM (2001). Determination of arsenic speciation in poultry wastes by IC-ICP-MS. Environ. Sci. Technol.35: 4868-4873.

Khaleel R, Reddy KR, Overcash MR (1981). Changes in soil physical and chemical properties due to application of animal manure. J. Environ. Quality. 326:126-178.

Kingery WL, Wood CW, Delaney DP, Williams JC, Mullins GI (1994). Impact of long-term land application of broiler litter on environmentally related soil properties. J. Environ. Quality. 23:139– 147.

Logsdon S, Laird D (2004). Cation and water content on dopole rotation activation energy of smectites. Soil Sci. Soc. Am. J. 64:54-61.

Maftoun M, Moshiri F (2008). Growth, mineral nutrition and selected soil properties of lowland rice, as affected by soil application of organic wastes and phosphorus. J. Agric. Sci. Technol. 10:481-492.

Mclean EV (1982). Aluminum. In: Page. A.I., R. H. Miller and D.R. Keeney (ed.). Method of soil analysis, Part 2. 2nd ed., Agron,Monpgr p 9, ASA and SSSA Madison, WI. pp. 978-998.

Spatial variability in the physico-chemical properties of soils affected by animal wastes in Uyo...

25

Mohammad RM, Serda A, Kobra K (2010). Using animal manure for improving soil chemical properties under different leaching conditions. Res. J. Soil Water Manage. (1):34-37. Nelson DW, Sommers EL (1982). Total carbon, organic carbon and organic matter. In: Page, A.L. (ED.). Methods of soil analysis. Part 2. 2nd Ed. Madison, WI, ASA and SSSA. pp 149-157.

Onwudike SU (2010). Effectiveness of cow dung and mineral fertilizer on soil properties, nutrient uptake and yield of sweet potato (*Ipomoea batatas*) in Southeastern Nigeria. Asian. J. Agric. Res. 4 (3):148-154.

Rochette P, Gregorian EG (1998). Dynamics of soil microbial biomass carbon, soluble organic carbon and CO$_2$ evolution after three years of manure application. Can. J. Soil Sci. 78:283-290.

Sawyer J, Helmers M, . Mallarino A, Lamkey K, Baker J (2006). Environmental Protection Commission-Alternative Considerations Regarding Liquid Swine Manure Application Rates to Soybean. Iowa State University, Ames, IA.

Suthar S (2009). Impact of vermicompost and composted farmyard manure on growth and yield of garlic (Allium sativum L.) field crop. Int. J. Plant Prod. 3:27-38.

Uwah DF, Udoh AU, Iwo GA (2011). Effect of Organic and Mineral Fertilizers on Growth and Yield of Cocoyam (*Colocasia Esculenta* (L.) SCHOTT). Int. J. Agric. Sci. 3(1):33-38.

Webster CP, Goulding KWT (1989). Influence of soil Carbon content on denitrification from fallow land during autumn. J. Sci. Food Agric. 49:131-142.

Yukse, BX, Orhan Y (2004). Effect of swine manure on some chemical characteristics of clay soil. J. Agric. 3(1):43– 45.

Zachary NS, Ermson ZN,, A.T. Ivenus AT, Chandra KR (2008). Tillage cropping systems and nitrogen fertilizer source; effects on soil carbon sequestration and fraction. In: J. Environ. Quality.37:880-885

4

Effects of *Faidherbia albida* on the fertility of soil in smallholder conservation agriculture systems in eastern and southern Zambia

Bridget B. Umar[1,2]*, Jens. B. Aune[1] and Obed. I. Lungu[3]

[1]Department of International Environment and Development Studies, Norwegian University of Life Sciences, Box 5003, 1432. Aas, Norway.
[2]Geography and Environmental Studies Department, University of Zambia, Box 32379. Lusaka, Zambia.
[3]Soil Science Department, University of Zambia, Box 32397, Lusaka, Zambia.

This study explored the benefits of *Faidherbia albida* on soil fertility in farmers' fields in areas which were suitable for Conservation Agriculture (CA) and where mature stands of *F. albida* already existed. It investigated the effects of *F. albida* on soils by testing for differences in the soil reaction (pH), total nitrogen, potassium, phosphorus and organic carbon at increasing radial distance from the tree trunk. Soil samples were collected from under and outside the canopies of 102 *F. albida* trees in four districts situated in the Southern and Eastern provinces of Zambia. The results showed evidence of a negative linear relationship between distance from *F. albida* and total nitrogen (p = 0.003), organic carbon (p = 0.0001), and potassium levels (p = 0.0001) but not for available phosphorus (p = 0.708) and soil reaction pH (p = 0.88). The nutrient levels were 42, 25 and 31% higher under the tree canopies than away for total nitrogen, potassium, and organic carbon respectively. *F. albida* added significant amounts of the agriculturally important nutrients which resource constrained households had difficulties replenishing to the soils through mineral fertilizer amendments because of their limited ability to purchase mineral fertilizers. It was concluded that *F. albida* improved soil fertility in farmers' fields and could be promoted in smallholder CA systems in Zambia.

Key words: *Faidherbia albida*, soil fertility, conservation agriculture, biological nitrogen fixation, Zambia.

INTRODUCTION

Soil nutrient mining is considered to be a major threat to food security and natural resource conservation in Sub-Saharan Africa (SSA). According to Bationo et al. (2006), Africa loses US$4 billion per year due to soil nutrient mining. The problem is pervasive among mixed crop and livestock farming systems of the region where competing uses for crop residues such as livestock fodder, or household fuel mean that nutrients are not sufficiently replenished into the soil. Nutrient replacement using mineral fertilizers is a limited option for many smallholder farming households of the region. At only eight kilograms per hectare, the region has the lowest mineral fertilizer application rates in the world and concomitantly, much lower crop yields than achieved in other developing regions (Morris et al., 2007). In Southern Africa, the consequent downward spiral of soil fertility has contributed to a corresponding decline in crop yields, an increase in food insecurity, food aid and environmental degradation (Mafongoya et al., 2006). In order to mitigate

---

*Corresponding author. E-mail: brigt2001@yahoo.co.uk.

nutrient mining, agroforestry (or fertilizer tree) systems have been proposed as an innovation especially suited to resource poor farming households. Fertilizer tree systems add biologically fixed nitrogen and other agriculturally important nutrients to the soils. This is done in a way that complements the crops grown in association with the trees (Akinnifesi et al., 2010). The use of the fertilizer tree *F. albida* (Del) A. Chev has been documented in semi-arid Africa, north and south of the equator, from southern Algeria to Transvaal and from the Atlantic to the Indian Ocean (Kirmse and Norton, 1984). It has been promoted in agroforestry as its characteristic reverse phenology allows satisfactory production of crops under a full stand of the species (Roupsard et al., 1999). The leaves are shed at the onset of the rainy season which significantly reduces the shade cast beneath the trees and reduces competition for water (Kirmse and Norton, 1984), light and nutrients (Kho et al., 2001) with associated crops grown during the rainy season.

Conservation Agriculture (CA) promoters contend that integrating *F. albida* trees into CA systems based on the three principles of minimum tillage, diversified crop rotations and permanent soil surface cover enhances the soil improving benefits of CA as not only does *F. albida* fix nitrogen, it also returns other nutrients to the soil and increases Soil Organic Matter (SOM) content through the shedding of its nutrient-rich leaves and the subsequent decomposition of its leaf litter at the onset of rains (Saka et al., 1994). The increased SOM improves soil structure, enhances soil microfauna populations and minimizes excessive evapo-transpiration and soil temperatures (Mokgolodi et al., 2011).

Incorporation of *F. albida* into smallholder CA systems has been promoted in Zambia for more than 10 years. The Conservation Farming Unit (CFU) has been encouraging farmers to plant 100 of the tree per hectare as a long term means of boosting soil fertility. It has been claimed that with trees planted at this density, the nutrient-rich leaves can supply the equivalent of 300 kg ha^{-1} nitrogen and 30 kg ha^{-1} of phosphorus per year (Dancette and Poulin, 1969), a valuable asset for the many farmers who cannot afford to buy these nutrients in form of mineral fertilizers. The tree has an extensive tap root system which develops rapidly and taps groundwater at large depths in the soil (Kirmse and Norton, 1984; Roupsard et al., 1999) and brings up nutrients from deeper soil layers to surface layers which are within the crop root systems. Supplements of soil nitrogen are thus possible under *F. albida* canopy through underground release of nitrogen in decomposing roots and nodules (Mokgolodi et al., 2011). In this way, it is argued, *F. albida* reduces the requirements of externally procured mineral fertilizers. Combining *F. albida* with the three CA principles supplemented by locally adapted agronomic practices such as dry season land preparation and precision in input application is regarded as being able to

improve crop yields and farmers' incomes, and to redress the soil fertility decline associated with farming as conventionally practiced by smallholders in Zambia.

Analysis of soils from under *F. albida* in Senegal indicated a remarkable fertility gradient from bare soils to soils under the canopy and yields of millet near trees increased two to four folds (Charreau and Vidal, 1965). Studies by Dougain (1960) in Niger indicated that on a 10 cm depth basis, which represented about 1500 tons of soil ha^{-1}, the nutrient increases due to the presence of *F. albida* were equivalent to 300 kg nitrogen, 31 kg phosphorus as P_2O_5 and 24 kg Magnesium. Soil water retention also increased under the canopy of *F. albida* (Radwanski and Wickens, 1967; Kamara and Haque, 1992) with increases of as much as 43% reported (Charreau and Vidal, 1965). On-station research in Malawi showed that mature trees could sustain unfertilized maize yields of 2.5 to 4 tons ha^{-1}, 200 to 400% more than the national averages (New Agriculturalist, 2010). Research station trials conducted in Zambia on nine year old *F. albida* trees concluded that the tree supplied 150 kg ha^{-1} of nutrients and 100 kg ha^{-1} of lime to the soil (GART, 2007).

F. albida is widely distributed in the southern half of Zambia particularly the semi-arid valley areas. It is locally known as *Musangu* and is increasingly being appreciated by smallholder farmers for its fertilizing effects on crops grown under its canopy and for its nutritious pods that serve as fodder during the dry season. The CFU has been promoting the tree through demonstrations of the yield effects of the tree on crops by conducting on-farm trials under and outside the canopies and holding field days where the results are publicized. CFU also distributes free *F. albida* seeds to CA farmers on its Conservation Agriculture Programme (CAP) and conducts training sessions with them on how to plant and look after *F. albida*. CFU had a goal of providing 120 000 farmers with starter packs of seed and sleeves and training them on how to raise and transplant seedlings into crop fields at a density of 100 plants ha^{-1} throughout its project areas in Southern, Central, Lusaka and Eastern Provinces of Zambia (CFU, 2006).

Although a lot of research has been conducted on *F. albida* in the Sahel, there is a paucity of on-station and on-farm studies on the effects of *F. albida* on soil fertility in the Zambian and regional context. A deeper understanding of the supply of nutrients by *F. albida* is important in the development of crop - tree systems that suit the needs of smallholder farmers. The objective of this study was, therefore, to explore the benefits of *F. albida* on soil fertility in farmers' fields in areas which are suitable for CA and where mature stands of *F. albida* already existed. This study investigated the effects of *F. albida* by assessing the effects of the trees on soil reaction (pH), N, K, P and soil organic carbon at increasing radial distance from the tree. We also compared the results based on district and province, whether the soils under the trees had been cultivated recently and whether livestock had deposited dung under the trees which could also contribute to the improved nutrient status of the soils. We briefly discuss some issues associated with the management of *F. albida* in this study area.

MATERIALS AND METHODS

This study was conducted between June and August 2008, and May and August 2009. Soil samples were collected from 102 *F. albida* trees in districts situated in the Southern and Eastern Provinces of Zambia. These districts were selected as they were the ones known to have *F. albida* trees among the 12 districts were CA was being promoted. The location of the sampled trees were recorded using a GPS and have been plotted on the maps presented as Figure 1a, b and c.

Description of study sites

Monze, Chipata and Petauke districts are located in Agro-Ecological Region (AER) IIa while Sinazongwe is found in AER I. AERs are categorized based on mean annual rainfall, growing season and elevation. AER IIa is characterized by mean annual rainfall of between 800 mm and 1000 mm and an average growing season of 100 to 140 days. AER I is the driest and most drought prone part of the country. It normally receives less than 700 mm of rainfall annually and has a growing season of between 80 and 120 days. Sinazongwe is located in a valley area with a mean elevation of 536 m above sea level (a.s.l). Monze is at an elevation of 1080 m a.s.l, Chipata is at 1011 m a.s.l. while Petauke is at a lower elavation of 850 m a.s.l. The climate of this study area is typified by a uni-modal rain season which usually lasts from November to April, followed by a cool and dry season lasting from May to August and a hot and dry season between September and November (GRZ, 2007). Monze, Chipata and Petauke received average annual rainfall of 732, 1033 and 932 mm respectively during the 2009/2010 farming seasons (GRZ, 2010).

Soils types in AER I range from slightly acidic Nitosols to alkaline Luvisols with pockets of Vertisols, Arenosols, Leptosols and Solonetz (MACO/JICA, 2007). The use of these soils for agricultural production is limited by lack of adequate water and high soil erosion potential. AER IIa soils are largely classified as Lixisols, Luvisols, Alisols, Acrisols and Leptosols. These are some of the best agricultural soils in Zambia and they host much of the country's commercial farming sector (MACO/JICA, 2007).

Selection of trees and measurement of dimensions

Trees were selected based on canopy size and ability to sample at three different radial distances. Very young trees (less than ten years old) were not included in the sampling frame and neither were the trees whose canopies overlapped with those of other trees. This was because the canopies were too small for the former and as we were unable to collect samples from 5 m outside the canopies for the latter. Random sampling was then used to collect the predetermined number of trees from a population of *F.albida* trees from Sinazongwe and Monze while all the trees known to exist in Chipata and Petauke Districts were included in the sample. The total sample size was 102 trees. This included 52 trees from

Monze District, 35 trees from Sinazongwe District, 8 trees from Chipata District and 7 trees from Petauke District. Tree heights were estimated using clinometers while the tree circumferences were measured at breast height (137 cm) with a 10 m measuring tape. Estimations of the ages of the trees were provided by owners of the fields in which the trees were found. The canopies were determined using 30 m measuring tapes.

Soil sampling

Three composite samples were taken at three different radii (inner, middle, outer) per tree. The radial distance of the inner and middle radii depended on the canopy size of the tree while the outer radius was 5 m from the edge of the tree canopy. While the inner and middle radii fell under the tree canopies, the outer radius was outside the canopies, at an average distance of 14.5 m from the tree trunk. Ten sub-samples were taken at each radius and at soil depths of 0 to 20 cm using an auger and mixed thoroughly to obtain a composite sample. Assessments were made on the presence of dung and cultivation of crops under the sampled *F. albida* trees. Laboratory analyses were conducted on the following parameters: soil reaction (pH_{CaCl2}), total nitrogen (Kjeldahl Method, Bremner, 1965), available phosphorus (Bray 1, Bray and Kurtz, 1945), organic carbon (Walkley and Black, 1934; Walkley and Black, 1934) and potassium (Ammonium acetate buffered at pH 7 for Potassium, Chapman, 1965).

Statistical analyses

Differences in the mean nutrient levels found in the soils at the three different radii and in the four districts were tested using ANOVA and linear mixed models using the statistical software Minitab 15 (Minitab, 2009) and R Studio (R Development Core Team 2011) respectively. The linear mixed models were used as they capture and control for both fixed and random effects. The fixed effects are specific variables that are directly recorded in the course of an experiment while random effects result from repeated measurements made on the same unit. In this instance, the variations due to the trees were the random effects that were controlled for. This enhanced the probability of detecting variations that were due to the predictors, and not due to the variations between one tree and the next. Several different models were tested for each of the four nutrients. Model selection started with the inclusion of four predictors namely distance from tree, district, dung under canopy, and presence of crops under canopy. Predictors that were insignificant were dropped from the model and the final model only contained variables that were significant at the five percent level. The models were checked using the normal q-q plot and were found to be acceptable. Simple linear regressions were used to determine fertility gradients for N, P, K and OC. All the statistical analyses were conducted at probability level of $p \leq 0.05$.

RESULTS AND DISCUSSION

The sampled trees had an average height of 21.0 m and diameter at breast height of 0.99 m. The average crown diameter was 18.6 m. The age range was 25 to 150 years. The average inner, middle and outer radii for the 102 trees were 1.7, 8.7 and 14.5 m respectively. The results showed that after controlling for the variations between the trees, the radial distance from the *F. albida*

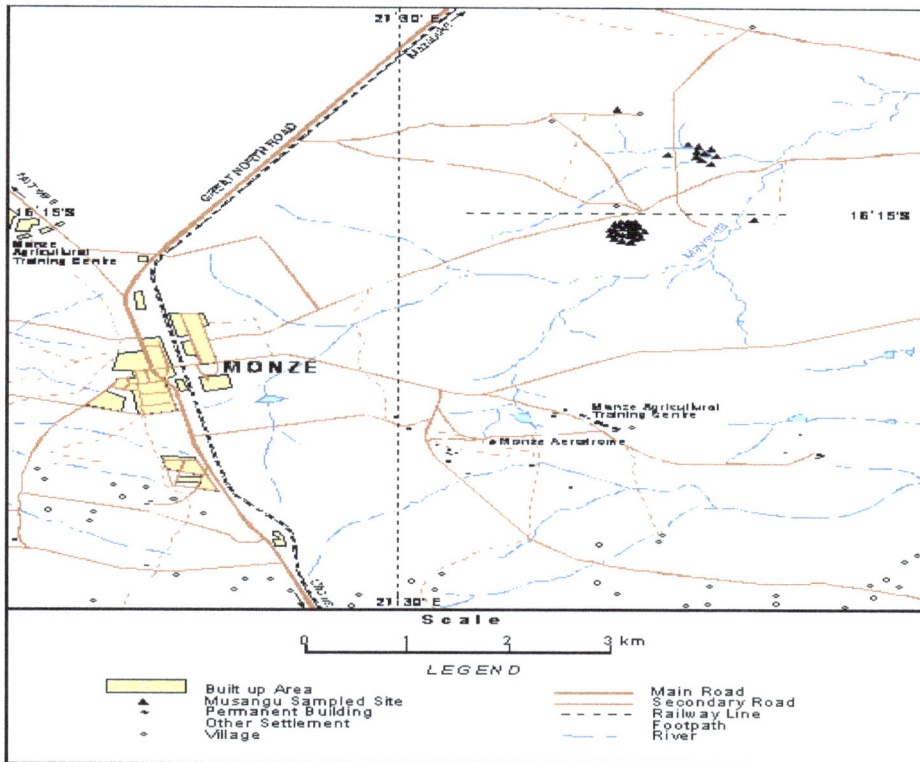

Figure 1a. Location of sampled *Faidherbia albida* trees in Monze District, Zambia.

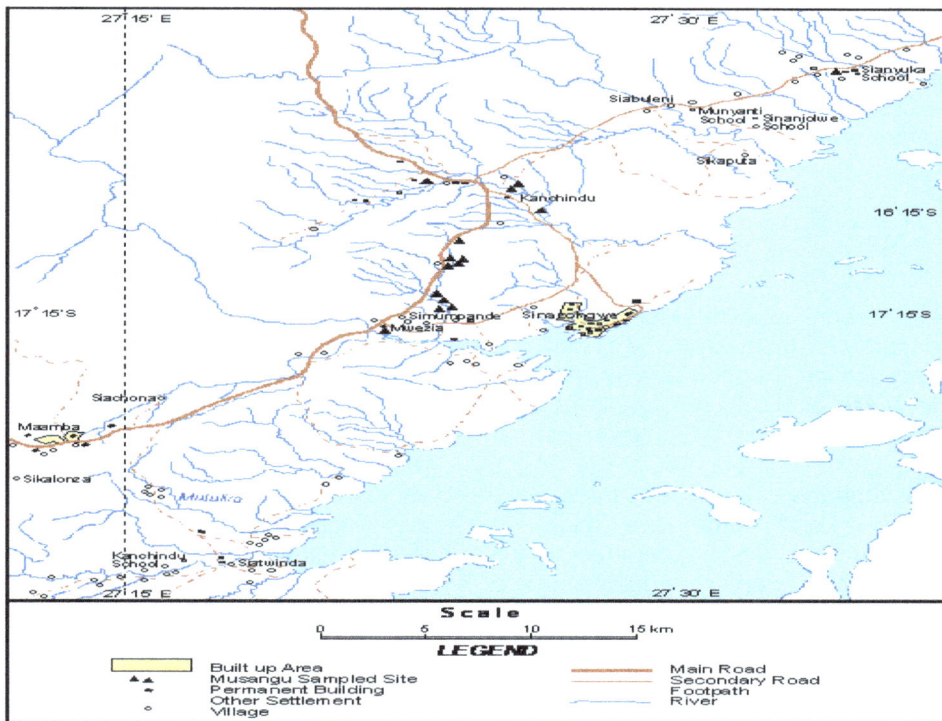

Figure 1b. Location of sampled *F. albida* trees in Sinazongwe District, Zambia.

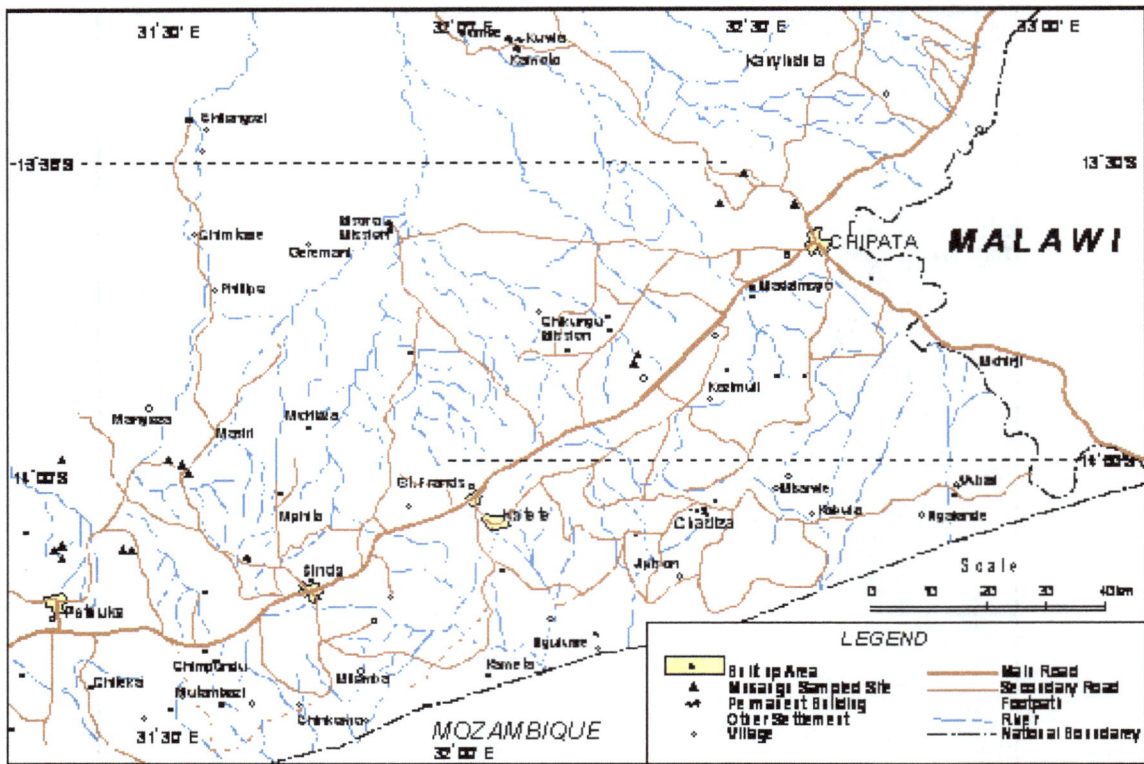

Figure 1c. Location of sampled *F. albida* trees in Eastern Zambia.

tree and the district in which the tree was located were important in determining the levels of organic carbon, total nitrogen, and potassium (p ≤ 0.05). The model used was: Nutrient level ~Distance from tree + District + 1(1| Tree). For phosphorus, only the effect of district was significant (p ≤ 0.05).

Effect of radius

The levels of organic carbon, total nitrogen, potassium and phosphorus reduced with increasing radial distance from the *F .albida* tree. Radial distance of 1 m from the tree resulted in a decrease of 0.3% for organic carbon, 0.003% for total nitrogen, and 0.015 cmol kg^{-1} of potassium (Table 1).

The random effects variance showed that there were large variations between and within the trees as evidenced by the large tree variance compared to the residual variance for organic carbon, potassium and phosphorus (Table 1).

The results showed evidence of a negative linear relationship and existence of a fertility gradient between radial distance from *F. albida* and N, OC, and K, but not for P and soil reaction (pH) (Figure 2 and Table 2). The N, OC and K levels were significantly higher under the

F. albida canopies than 5 m from the edge of the canopies (Table 3). The N levels were 0.17% in the inner radius and 0.12% in the outer radius. Soil N levels were 42% higher under the canopies than outside. Assuming a density of 100 mature *F. albida* trees per hectare, this would be equivalent to 4420 kg N ha^{-1} compared to 3120 kg N ha^{-1} for a field of the same area without the trees. N levels would be 1300 kg ha^{-1} higher in the field with *F. albida* compared to that without. Using a decomposition constant of 0.03 as for soils in Savanna areas (Young, 1989), this would give an annual mineralization due to the presence of *F. Albida* of 39 kg N ha^{-1}. This is equivalent to the N content in 390 kg of D- Compound fertilizer (N: P$_2$O$_5$: K$_2$O, 10:20:10). The C: N ratio was 11 under the canopies and 14 outside the canopies' edges, and reflected the higher N levels under the canopies. This shows that the nitrogen is more readily available under the canopy than outside. The OC level was 1.58% in the inner radius and 1.21% in the outer radius. Organic carbon was therefore 30.5% higher under the canopy as compared to outside the canopy. This would be equivalent to 41080 kg OC in a 1 hectare field with a 100 mature *F. albida* trees compared to 31460 kg OC for a similar field without the trees. These results were similar to those obtained by Charreau and Vidal (1965) who reported a remarkable fertility gradient from bare soil to

Table 1. Estimates for effect of radial distance from tree for four types of nutrients using linear mixed model.

Organic C/ nutrient	Fixed effects			Random effects variance	
	Estimate	Standard error	P-value	Tree (intercept)	Residual
Organic carbon	-0.2985	0.0032	<0.0001	0.2305	0.0938
Nitrogen	-0.003033	0.001028	0.0032	0.005	0.01
Potassium	-0.015573	0.001389	<0.0001	0.0375	0.0178
Phosphorus	-0.04577	0.1221	0.7077	429.29	136.43

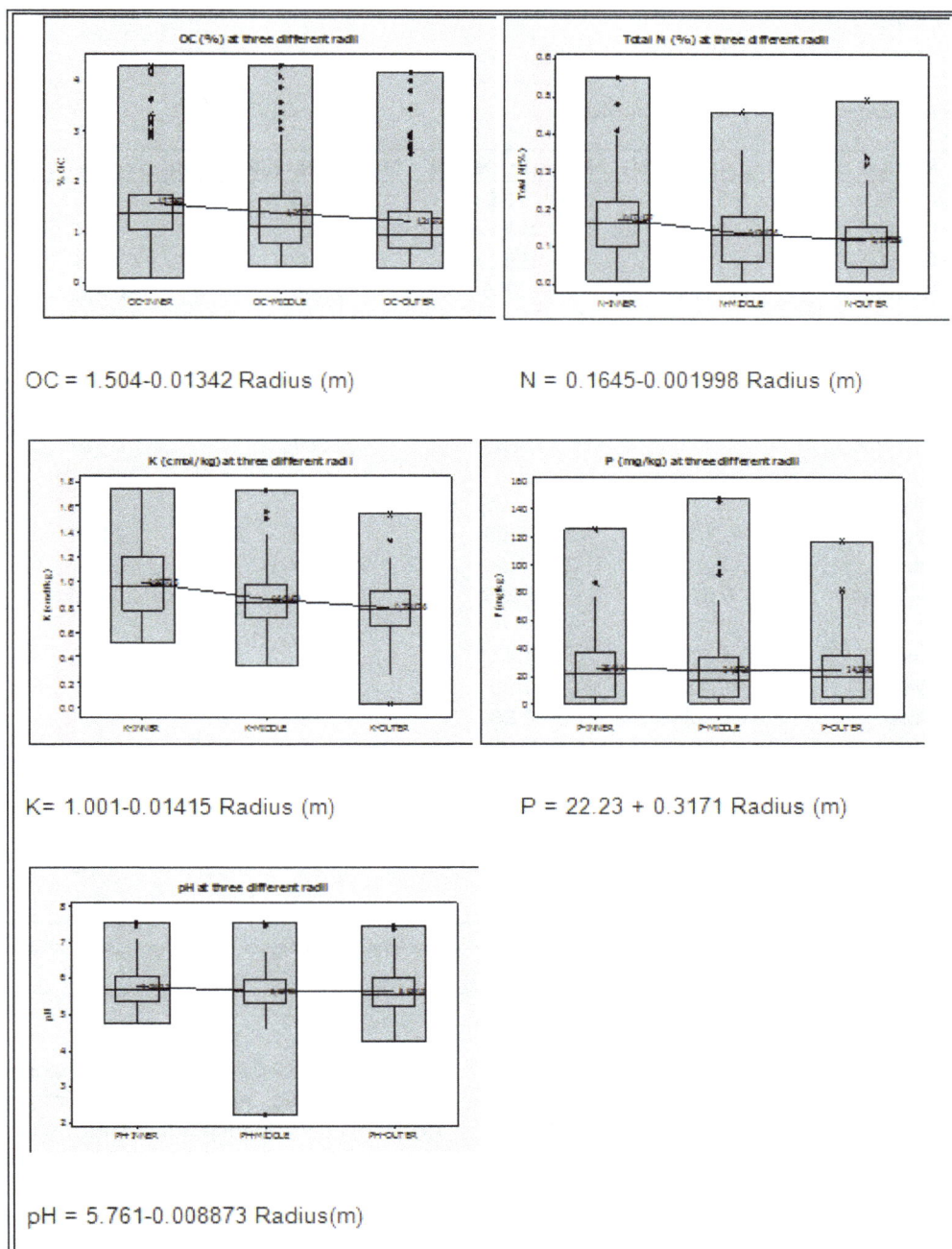

OC = 1.504-0.01342 Radius (m)

N = 0.1645-0.001998 Radius (m)

K= 1.001-0.01415 Radius (m)

P = 22.23 + 0.3171 Radius (m)

pH = 5.761-0.008873 Radius(m)

Figure 2. Box plots for P, K, OC, N and pH showing mean levels and distribution.

Table 2. Nutrient levels at different radii from *F. albida*.

Parameter	Inner radius (N=102)	Middle radius (N=102)	Outer radius (N=102)	p - value
pH (Cacl$_2$)	5.77^a (0.095)	5.66 ^{a}(0.126)	5.63^a (0.128)	0.283
Nitrogen (%)	0.17^a (0.014)	0.14^b (0.009)	0.12^b (0.009)	0.0001
Organic carbon (%)	1.58^a (0.074)	1.36^{ab} (0.061)	1.21^b (0.057)	0.009
C:N ratio	10.95^a (1.38)	12.92^{ab} (1.20)	13.84^b (1.76)	0.031
Phosphorus (mg/kg)	25.41^a (3.80)	24.67^a (5.24)	24.31^a (3.82)	0.944
Potassium (cmol/kg)	0.99^a (0.041)	0.86^b (0.030)	0.79^c (0.028)	0.0001

The values in parentheses are standard errors. ^{a,b} means in the same column followed by the same letter were not significantly different at p≤ 0.05.

Table 3. Nutrient levels under *F. albida* based on district.

Parameter	Radius	Chipata (N= 24)	Petauke (N=21)	Monze (N=156)	Sinazongwe (N= 105)	p- value
pH (Cacl$_2$)	Inner	5.76	5.85	5.70	5.75	
	Middle	5.79	5.74	5.58	5.64	0.884
	Outer	5.65	5.82	5.57	5.63	
	P- Value	*0.174*	*0.816*	*0.596*	*0.429*	
Organic Carbon (%)	Inner	2.78	3.62	1.28	1.35	
	Middle	2.61	3.37	1.14	1.0	0.0001
	Outer	2.15	3.39	0.94	0.94	
	P- Value	*0.111*	*0.993*	*0.001*	*0.007*	
Phosphorus (mg/kg)	Inner	22.19	31.06	18.9	30.12	
	Middle	21.1	35.06	22.23	24.55	0.260
	Outer	21.1	33.56	19.14	27.67	
	P- Value	*0.615*	*0.929*	*0.879*	*0.255*	
Potassium (cmol/kg)	Inner	0.81	1.11	0.10	0.95	
	Middle	0.73	0.90	0.85	0.85	0.002
	Outer	0.62	0.88	0.79	0.79	
	P- Value	*0.056*	*0.136*	*0.0001*	*0.003*	
Nitrogen (%)	Inner	0.22	0.29	0.15	0.17	
	Middle	0.17	0.25	0.12	0.12	0.0001
	Outer	0.15	0.26	0.10	0.10	
	P- Value	*0.008*	*0.819*	*0.555*	*0.012*	

The vertical p values are from comparisons among the four districts for the listed parameters while the horizontal ones are from the mean separation of the nutrient levels for each district.

soils under the canopies of *F. albida* in Senegal. OC improves the water holding capacity of the soil (Katyal et al., 2001; GART, 2008). A study conducted in the low rainfall areas of Monze and Lisutu of Southern Zambia reported organic matter levels that were on average 1% higher under the canopies than outside (GART, 2008). The N levels were 43% higher under the canopies than outside. The authors argued that this contributed to improved productivity under the canopies.

Soil organic matter (SOM) improves water holding capacity, increases plant nutrient and moisture availability and reduces soil erosion (Woomer and Swift, 1994; Lal 2006). Continuous cropping, crop residue removal and tillage reduce SOM. In light of the challenges associated with crop residue retention reported in smallholder farming systems in SSA (Giller et al., 2009; Lal, 2006;

Chivenge et al., 2007) and Zambia in particular (Umar et al., 2011), incorporation of F. albida into CA could go a long way in increasing SOM through leaf and pod litter. SOM content of 3.2 and 2.4% under and outside the canopies respectively reported during this current study were higher than what has generally been reported in Zambian soils (Stromgaard, 1984; Lungu and Chinene, 1999). GART (2008) reported SOM levels of 2.2% under F. albida canopies and 1.1% at 5 m outside the canopies. Such high SOM levels coupled with the low C: N ratio under the canopies would significantly improve crop yields under the canopies by making N more available and improving soil moisture retention. This would help resource constrained smallholder farmers to increase their crop production while reducing dependency on externally procured mineral fertilizers.

In a study of F. albida and its effects on Ethiopian Highland Vertisols, Kamara and Haque (1992) found a significant inverse relationship between SOM, N, P, K concentration and distance from the tree. The F. albida did not seem to influence soil reaction (pH) and the exchangeable cations Na, Ca and Mg. The build-up of SOM, N, P and K under the tree canopies was attributed to the F. albida litter fall and accumulation. They found N and P contents in the fresh leaves and twigs to be 3.85% N and 0.3% P for the leaves and 1.27% N and 0.2% P for twigs. Hadgu et al. (2009) found clear differences in total nitrogen, SOM and available phosphorus under and away from the canopy in their study of traditional F. albida based land use systems of northern Ethiopia. GART (2008) reported similar N content (3.6%) in F. albida leaves. GART (2007) estimated that a six year old tree was capable of shedding off 500 kg of leaf dry matter and equated this to the application of 18 kg nitrogen per hectare. The trees in this study were older with much larger canopies and shed off more leaves than the trees reported on by GART (2008). This coupled with the biological nitrogen fixation in the roots resulted in the higher value of 39 kg N ha^{-1} attributed to the presence of F. albida in this current study.

At several sites in Malawi, carbon and total nitrogen were from 3 to 30% and from 5 to 29% respectively higher beneath F. albida canopies while exchangeable K, Ca and Mg were also higher beneath tree canopies (Saka et al., 1994). In an on-farm field experiment in Niger, Kho et al. (2001) estimated the nitrogen and phosphorus availability under F. albida trees to be more than 200 and 30%, respectively higher under the canopies than outside.

Potassium levels in the soils under the canopies were higher than those outside the canopies by 25%. This represented levels of 1007 kg K ha^{-1} and 802 kg K ha^{-1} under and outside the canopies respectively. Under a full stand of the tree, K levels would on average be higher by 202.8 kg K ha^{-1} compared to a field without the trees. This is equivalent to the K contained in 2443 kg of D-

Compound fertilizer which is the nationally recommended basal dressing fertilizer in Zambia. In a study on the ecology of F. albida in Sudan, Radwanski and Wickens (1967) found that the uppermost horizons of the soil profiles under F. albida contained more OC, N and P but less K that those outside the canopies.

In this study, the P levels were high regardless of distance from the tree (p = 0.944) and there was no clear effect of distance from the canopy and the P level in the soil. Dabin (1980) and von Uexkull (1986) contended that the nature of the parent rock, soil reaction (pH), the presence of P fixing compounds such as sesquioxides, the nature and amount of organic matter in the soil were important factors in the P available dynamics of soils. Review of the results of the on-farm trials which had been conducted by GART (2008) in Magoye and Lusitu areas of Southern Zambia during this current study found no statistically significant differences in the P levels from soils under and outside the F. albida canopies (p = 0.44). The P levels under and outside the canopies were as high as 27.65 and 29.54 mg kg^{-1} respectively. The authors attributed this to the increased solubility of P at high pH which at between 6 and 7 was reported to be optimal. The contrasted effects of the P in this study and those of others (for example, Radwanski and Wickens, 1967) may point to the importance of other factors in the environment which may mask the effects of the F. albida. For instance, no significant differences were found in the soil nutrient pools beneath and outside the F. albida canopies on the lakeshore plains of Malawi. This was attributed to the high natural fertility of the alluvial soils which along with the soil mixing during tillage activities may have masked the nutrient enrichment associated with the trees (Rhoades, 1995). P levels could also vary based on individual soil types although we did not analyze this aspect during this current study.

Many of the reported benefits of the nutrient accumulation and yield benefits under the canopy over F. albida assume that there is a complete canopy cover of the tree. This is, however, rarely the case and it is, therefore, easy to overestimate the real benefits of F. Albida when calculating nutrient accumulation and production benefits per unit area. For this study, the average number of F. albida trees per household was 18 during the 2006/2007 farming season. This figure doubled by the 2008/2009 farming season but the added trees were still too immature to contribute any significant amounts of nutrients. Thus, the estimated amounts of N, OC and K added to the soil based on the actual number of mature trees is 234 kg ha^{-1} N, 1732 kg ha^{-1} OC and 37 kg ha^{-1} K. It can thus be seen that the actual accumulated nutrients levels were less than would be expected under full canopy. The households found it challenging to achieve the recommended 100 trees per hectare due to low access to planting material (either seeds or seedlings); low survival rate and susceptibility to browsing by livestock of the young trees; and termite attacks.

Crop production in the tropics is often limited by the availability of soil P. Highly weathered soils in the tropics, such as Ferralsols, Acrisols, and Luvisols, as well as Gleysols and Vertisols are generally deficient in P. Cambisols show average P deficiency while Fluvisols and Nitosols often show low to no deficiency (Dabin, 1980). Potassium levels greater than 10 mg kg^{-1} are considered high for tropical soils (Aune and Lal, 1997). The P levels reported in this study were, therefore, relatively high. The high available P in this study may be attributed to the presence of carbonaceous minerals in the parent rock as interviews with the owners of the fields in which the trees were found revealed almost a complete lack of mineral fertilizer application under and outside the tree canopies and the high P could, therefore, not be attributed to mineral fertilizer amendments. A search of the literature on available P levels in Zambia did not reveal any results showing low available P levels. Yet numerous projects aimed at *inter alia* increasing the low levels of available P in Zambian soils have been and continue to be implemented. This alludes to the need for knowledge on local soil conditions before promotion of technologies premised on regional or national soil nutrient estimates.

The soil reaction (pH) showed insignificant variations with increasing radius from tree with an average of 5.8 under the canopy and 5.6 for outside the canopy. Since the values were higher than 5 at all radii, the soils are suitable for production of most tropical crops (Aune and Lal, 1997). These results are similar to those obtained by GART (2008) which found that soil reaction (pH) under the canopies was generally similar to that away from the canopies and ranged between 5.3 and 7.0. Kho et al. (2001) estimated that N available from the *F. albida* caused a production increase of 26% in millet while the phosphorus availability was estimated to cause a production increase of up to 13%. GART (2008) reported yield responses due to *F. albida* in groundnuts, sorghum, cowpeas and maize. Maize yields were 3 tons ha^{-1} under *F. albida* canopies and only 2 tons ha^{-1} outside the canopies. From Senegal it was reported that the yield of pearl millet was 2.5 times higher under the canopy of *F. albida* than outside the canopy (Charreau and Vidal, 1965). The N and OC additions from *F. albida* reported in this current study could result in similarly higher yields under the canopies. These yields would be higher than the average yields of only one ton ha^{-1} currently achieved by smallholder farmers without the benefits of *F. albida* and low levels of mineral fertilizer application.

Effect of district

There were statistically significant differences in the N, OC, and K based on district (and province) but not for soil reaction (pH) and P. Chipata and Petauke districts (Eastern Province) had significantly higher N in soil compared to Monze and Sinazongwe (Southern Province) (p = 0.0001). The mean N level for Chipata and Petauke was 0.18 and 0.25% respectively while that for Monze and Sinazongwe was 0.14 and 0.13% respectively (Table 3).

The OC levels in the samples from Chipata and Petauke were significantly higher than those from Monze and Sinazongwe (p = 0.0001). With mean values of 2.51 and 3.31%, these soils are considered to be rich and productive. For K, Chipata had significantly lower levels than the other three districts (p = 0.002) while no significant differences were observed among the other districts. The higher levels of plant nutrients observed in Eastern Province could be attributed to the higher rainfall received there compared to Southern Province which promoted higher crop and biomass production and resulted in higher soil water holding capacity (GART, 2008). Moisture stress probably reduced biological N yield in Southern Province as the province is a drought prone region. It could also be possible that there were genetic differences among the trees found in the two provinces and more research on this could help better explain this observation. Values of over 20 mg/kg available P were observed in all the districts. The variations within districts were also quite large.

F. albida trees can be particularly important in the Southern Province as the soil organic carbon and N levels were the lowest here. In this Province, there is also more livestock and more competition for use of crop residues as fodder. The SOM additions from the tree are therefore more critical in enhancing soil moisture retention in this Province. Another opportunity for promotion of *F. albida* in the Southern Province is that most local communities there are already aware of its soil fertility improving benefits and as a source of dry season fodder for their livestock. This is different from the situation in the Eastern Province where it was reported that most CA farmers were unaware of the tree's benefits and most of them had cut it down from their fields (personal communication with senior CFU staff).

Effect of cultivation and presence of dung

No significant differences in nutrient levels were observed due to cultivation under the trees (p = 0.711). Out of the 102 trees sampled, 97 had soils which had been cultivated under them while no evidence of cultivation was found for the soils under the remaining 5 trees. Mean values for OC, N and K for soils under *F. albida* trees where dung was observed under the tree canopies were not statistically different from the values from under the tree canopies with no visual evidence of dung at 5% level of significance (p = 0.059). Only 11 out of the 102 trees were observed to have dung under their canopies. The low incidence of dung found under the *F. albida* trees could have been because, in some cases, the fields in which the trees were found had been fenced off to keep livestock out, the dung may have been incorporated into the soil during tillage, or the trees were very far from homesteads and out of reach of free range livestock. Sturmheit (1990) contended that most farmers of Southern Zambia perceived leaf drop from *F. albida* less important for soil improvement compared to the effects of livestock dung which accumulated under the tree when livestock sought shade there and relished its pods during the hot dry season. However, during this study, all the farmers talked to and anecdotal evidence attributed the better performance of crops under *F. albida* to the tree itself.

In this study the results seem to show that the higher N, OC and K levels observed under the *F. albida* trees' canopies were due to the presence of the trees which improved the nutrient economy of the soils through biological nitrogen fixation and litter fall and not due to presence of dung. Similarly, Charreau and Vidal (1965) reported crop yield increase in an area from which livestock had been excluded for a number of years and attributed this to the leaf litter provided at the start of the rainy season when it could readily be decomposed or mineralized.

Conclusion

The results revealed that mature *F. albida* trees supplied significant amounts of N, OC and K to the soils under their canopies resulted in a clear fertility gradient for these nutrients. The N, OC and K levels were 42, 31 and 25% respectively higher under the canopies than outside. There were no significant differences in the P levels and soil reaction (pH) based on radius and district. The SOM addition through litter fall is an important alternative to the SOM from crop residues which are routinely removed from the fields after harvest and put to alternative uses. For these reasons, use of the tree could be promoted among smallholder conservation agriculture farmers in the areas where mature stands exist and planting of new trees in other areas could also be encouraged as is already being done by the CFU in Zambia. This would supplement the nutrient additions that are made through mineral fertilizer application and rotations with annual leguminous crops. More efforts could be expended on the promotion of *F. albida* in areas where mature stands already exist than in the promotion of new stands of *F. albida* since establishment of trees is time consuming in terms of planting and protection of new stands. In addition, it takes at least 15 to 30 years before the full benefit of new planting can be reaped. However, new planting can still be an alternative where population density of *F. albida* is low. Insecure land tenure may also

discourage farmers from planting and protecting new stands of *F. albida*. Planting of *F. albida* trees is thus an option for the households that have secure tenure to their land. The existence of already mature stands would save labour and time commitments inherent in the establishment of young *F. albida* seedlings and the long time it takes for the tree to mature and provide substantial benefits to farmers that are eager for immediate gains and may not have the land tenure security to invest in soil improvements that take years to show benefits. There is also a need for more local soil analyses to have a better idea of what nutrients are in very limited supply and in need of immediate replenishment instead of relying on estimates calculated from elsewhere.

ACKNOWLEDGMENTS

The study was conducted with funding provided under the Conservation Agriculture Programme. This programme is funded by the Norwegian Ministry of Foreign Affair through the Royal Norwegian Embassy in Lusaka. All the views expressed in the paper are solely those of the authors and not the sponsors.

REFERENCES

Akinnifesi FK, Ajayi OC, Sileshi G, Chirwa PW, Chianu J (2010). Fertiliser trees for sustainable food security in the maize-based production systems of East and Southern Africa. A review. Agron. Sustain. Dev. 30(3):615-629.

Aune JB, Lal R (1997). Agricultural Productivity in the Tropics and Critical Limits of Properties of Oxisols, Ultisols, and Alfisols. Trop. Agric. (Trinidad) 74(2):96-103.

Bationo A, Waswa B, Kihara J, Kimetu J (2006). Advances in integrated soil fertility management in sub Saharan Africa: challenges and opportunities. Nutrient Cycling Agroecosyst. 76(2-3):1-2.

Bray RH, Kurtz LT (1945). Determination of total and available phosphorus in soils. Soil Sci. 59:39-45.

Bremner JM (1965). Inorganic forms of nitrogen. Methods of Soil Analysis. A. Black Agron. 9:1179-1237.

CFU (2006). Reversing Food Insecurity and Environmental Degradation in Zambia through Conservation Agriculture. Lusaka, Conservation Farming Unit.

Chapman HD (1965). Cation Exchange Capacity. Methods of Soil Analysis. A. Black. Madison, American Society of Agronomy Inc. 9:891-901.

Charreau C, Vidal P (1965). Influence de l'*Acacia albida* Del. sur le sol, nutrition minerale et rendements des mils Pennisetum au Senegal. L' Agronomie Trop. 20:600-626.

Chivenge PP, Murwira HK, Giller KE, Mapfumo P. Six J (2007). Long-term impact of reduced tillage and residue management on soil carbon stabilization: Implications for conservation agriculture on contrasting soils. Soil. Tillage Res. 94(2):328-337.

Dabin B (1980). Phosphorus defiency in tropical soils as a constraint on agricultural output. Soil related constraints to good production in the tropics. Colloquim. Los Banos, ORSTOM.

Dancette C, Poulain, JF (1969). Influence of Acacia albida on pedoclimatic factors and crop yields. Afri. Soils. 14:143-184.

Dougain E (1960). Report de mission au Niger. Dakar, ORSTOM.

GART (2007). Golden Valley Agriculture Research Trust: 2006 Yearbook. . Lusaka, GART.

GART (2008). Golden Valley Agricultural Research Trust: 2007 Yearbook. Lusaka, GART.

Giller KE, Witter E, Corbeels M, Tittonell P (2009). Conservation agriculture and smallholder farming in Africa: The heretics' view." Field Crops Res. 114(1):23-34.

GRZ (2007). The National Adaptation Programme of Action (NAPA). Ministry of Tourism Environment and Natural Resources. Lusaka, Government of Republic of Zambia/Global Environment Facility/United Nationsa Development Fund.

GRZ (2010). Daily Rainfall Records. Meteorological Department. Lusaka., Meteorological Department.

Hadgu K, Kooistra L, Rossing W, van Bruggen A (2009). Assessing the effect of <i>Faidherbia albida</i> based land use systems on barley yield at field and regional scale in the highlands of Tigray, Northern Ethiopia. Food Secur. 1(3):337-350.

Kamara CS, Haque I (1992). The potential of agroforestry to increase primary production in the Sahalian and Sudanian zones of West Africa. Agroforestry. Syst. 16:17-29.

Katyal J, Rao N, Reddy M (2001). Critical aspects of organic matter management in the Tropics: the example of India. Nutrient Cycling Agroecosyst. 61(1):77-88.

Kho RM, Yacouba B, Yaye M, Katkore B, Moussa A, Iktam A, Mayaki A (2001). Separating the effects of trees on crops: the cases of *Faidherbia albida* and millet in Niger. Agroforestry Syst. 52(3):219-238.

Kirmse RD, Norton BE (1984). The potential of Acacia albida for desertification control and increased productivity in Chad. Biol. Conserv. 29(2):121-141.

Lal R (2006). Enhancing crop yields in the developing countries through restoration of the soil organic carbon pool in agricultural lands. Land Degradation. Dev. 17(2):197-209.

Lungu OIM, Chinene VRN (1999). Carbon Sequestration in agricultural soils of Zambia. Zambia Status Paper. Presented at Workshop on Carbon Sequestration in Soils and Carbon credits: Review and Development of options for Semi-Arid and Sub-Humid Africa. Sioux Falls, SD., Earth Resources Observation Systems Data Center.

MACO/JICA (2007). Sustainable Agriculture Practices. PaViDIA Field Manual Lusaka. p. 3.

Mafongoya P, Bationo A, Kihara J, Waswa B (2006). Appropriate technologies to replenish soil fertility in southern Africa. Nutrient Cycling Agroecosyst. 76(2):137-151.

Minitab (2009). Minitab 15 Statistical Software. Minitab. State College, USA. www.minitab.com, Minitab Inc.

Mokgolodi N, Setshogo M, Shi L, Liu Y, Ma C (2011). Achieving food and nutritional security through agroforestry: a case of *Faidherbia albida* in sub-Saharan Africa. Forestry Studies China 13(2):123-131.

Morris M, Kelly VA, Kopicki RJ, Byerlee D (2007). Fertilizer use in African agriculture. Lessons learned and good practice guidelines. Washington, DC., The World Bank.

New Agriculturalist (2010). Faidherbia - Africa's Fertilizer Factory. January 2010 Issue. [Retrieved 27th November 2010].

R Development Core Team (2011). R: A language and environment for statistical computing. Vienna, R Foundation for Statistical Computing.

Radwanski SA . Wickens GE (1967). The Ecology of Acacia albida on Mantle Soils in Zalingei, Jebel Marra, Sudan. J. Appl. Ecol. 4(2):569-579.

Rhoades C (1995). Seasonal pattern of nitrogen mineralization and soil moisture beneath *Faidherbia albida* (syn *Acacia albida*) in central Malawi. Agroforestry. Syst. 29:133-145.

Roupsard O, Ferhi A, Granier A, Pallo F, Depommier D, Mallet B, Joly HI, Dreyer E (1999). Reverse Phenology and Dry-Season Water Uptake by Faidherbia albida (Del.) A. Chev. in an Agroforestry Parkland of Sudanese West Africa. Funct. Ecol. 13(4):460-472.

Saka AR, Bunderson WT Itimu OA, Phombeya HSK, Mbekeani Y (1994). The effects of Acacia albida on soils and maize grain yields under smallholder farm conditions in Malawi. Forest Ecol. Manage. 64(2-3):217-230.

Stromgaard P (1984). The immediate effect of burning and ash-fertilization. Plant. Soil 80(3):307-320.

Sturmheit P (1990). Agroforestry and soil conservation needs of smallholders in Southern Zambia. Agroforestry. Syst. 10 (3): 265-289.

Umar BB, Aune JB, Johnsen FH, Lungu OI (2011). Options for Improving Smallholder Conservation Agriculture in Zambia . J. Agric. Sci. 3(3):52-62.

von Uexkull HR (1986). Efficient fertilizer use in acid upland soils of the humid tropics. Rome, Food and Agricultural Organization (FAO) Land and Water Development Division. Bulletin. p.10.

Walkley A, Black IA (1934). An examination of the Degtjareff method for determining soil organic matter and a proposed modification of the chromic acid titration method. Soil Sci. 37:29-38.

Woomer PL, Swift MJ Ed (1994). The importance and management of soil organic matter in the tropics. The Biological Management of Tropical Soil Fertility. Chichester, U.K., Wiley-Sayce.

Young A (1989). Agroforestry for Soil Conservation. Oxon, UK, C.A.B. International.

5

Soil depth inferred from electromagnetic induction measurements

Johan VAN TOL[1]*, Johan BARNARD[2], Leon VAN RENSBURG[2] and Pieter LE ROUX[2]

[1]Department of Agronomy, University of Fort Hare, Alice, 5700, South Africa.
[2]Department of Soil, Crop and Climate Sciences, University of the Free State, Bloemfontein, 9300, South Africa.

There is a need for rapid cost effective methods to obtain spatially distributed data of soil depth. Soil depth determines the subsurface topography, a major control on the distribution of flowpaths in landscapes. An EM38 survey was conducted on a 12 ha site at Bloemfontein, South Africa. A significant linear relationship between soil depth and EC_a were obtained with multiple linear regression (Soil depth = 149 - 29 $CV_{0.5}$ + 34 CV_1). It was found that the equation can reasonably accurately (RMdAE = 20%, REF = 0.49) estimate soil depth from EC_a readings. This made it possible to estimate 15,000 soil depths across the study area, which contributed to the successful characterization of subsurface topography. Consequently the following conclusions could be made. There was a close correlation between surface and subsurface topography. Overland flow seems to be high causing erosion on higher elevations and deposition of sediments and accumulation of water in lower lying areas. From flow accumulation maps, sites possibly controlling the hydrology of the study area were identified. The methodology developed should contribute towards characterising hydrological research sites.

Key words: EM38, soil depth, soil water, hydrological response, thresholds, water flow.

INTRODUCTION

Hillslopes are considered fundamental landscape elements and the smallest entity for a holistic study of hydrological processes. The hydrological response of watersheds depends on responses of individual hillslopes in the watershed (Sivapalan, 2003a; Weiler and McDonnell, 2004). The complexity of hydrologicaln processes, driven by heterogeneities in landscape characteristics, diminishes the applicability of hillslopes as basic elements for watershed models. It was therefore argued that instead of focussing on unconventional behaviour of different hillslopes one should search for common threads, concepts and patterns in the hydrological response of hillslopes, to be able to intercompare hillslopes from various regions, geologies, with different soils and vegetation (Sivapalan, 2003a, b; McDonnell et al., 2007).

The threshold response of hillslopes to precipitation is proposed as a unifying concept of how hillslopes function and a suitable tool for intercomparisons of subsurface processes between hillslopes (Tromp-van Meerveld and McDonnell, 2006a). These thresholds might be the formation of a saturated wedge at the discharge face, expanding upslope (Weyman, 1973); threshold pre-event water contents, favouring the generation of macropore flow (Uchida et al., 2005); groundwater ridging (Sklash and Farvolden, 1979) and/or saturated excess overland flow, that is, variable source areas (Dunne and Black, 1970); and more recently, the connectivity of the hillslope in terms of transient saturation. Isolated patches of saturation should first be connected before significant subsurface flows are generated (Tromp-van Meerveld and McDonnell, 2006b; Lexartza-Artza and Wainwright, 2009; Hopp and McDonnell, 2009). The *'fill and spill'* hypothesis was suggested by Tromp-van Meerveld et al. (2006b), where depressions in the bedrock topography ought to be 'filled' first before 'spilling' over

---

*Corresponding author. E-mail: jvantol@ufh.ac.za.

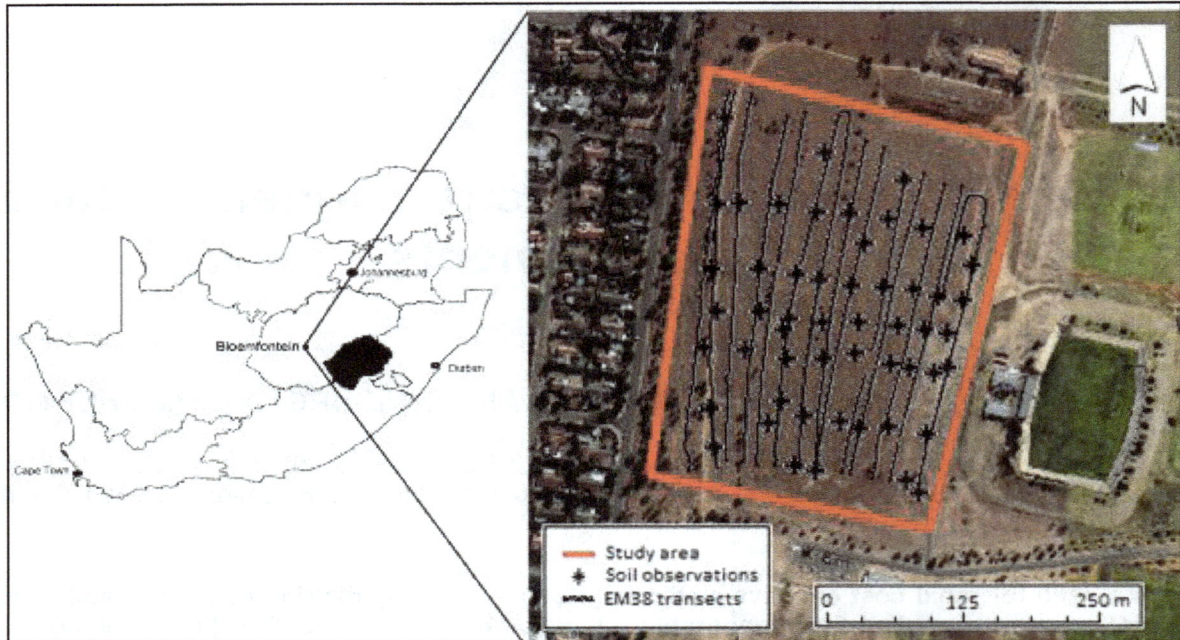

Figure 1. Location of study area, EM38 transects and soil depth observations.

microtopographic relief at the soil/bedrock interface, connecting subsurface saturated areas and increasing the generation of subsurface stormflow.

The spatial variability of flowpaths makes generalisation from transect data to hillslope scale a difficult task. Tromp-van Meerveld et al. (2006b) agreed that they probably would not have noticed 'connectedness' in their Panola hillslope if not for a detailed network of spatially distributed wells and detailed soil depth measurements. We agree that more measurements of the surface and subsurface lateral flow paths, water table fluctuations, connectivity of the various water bodies and the residence flow time of water through the landscape would be ideal for an enhanced understanding and modelling of the hydrological behaviour of hillslopes. Detailed spatial measurements are however only feasible on relatively small hillslopes (e.g. Panola is approximately 0.1 ha), but it becomes more expensive and impractical as the size of the study area increases. Electromagnetic induction (EMI) is a non-invasive, cost- and time efficient technique, able to produce large quantities of data about subsurface conditions. EMI have been used *inter alia* to estimate depth to clay layers (Doolittle et al., 1994), soil salinity (Hendrickx and Kachanoski, 2002), water table depths and soil water contents (Sherlock and McDonnell, 2003) and soil texture (Abdu et al., 2008). Indeed, geospatial measurement of apparent soil electrical conductivity (EC_a), generally applied in site specific crop management or precision agriculture, has become one of the most reliable and frequently used measurements to characterise soil variability. From the available sensors, the mobile non-invasive electromagnetic induction (EMI)

Geonics EM38 and EM31 sensor is the most popular.

Since EC_a is influenced by everything in the soil that conducts an electrical induced current, EC_a survey data is focused toward ensuring that acquired EC_a data correlate with the specific soil variable. Differences in the conducting capacity of consolidated material and un-weathered bedrock will influence EC_a measurements to such an extent that depths to soil/bedrock transitions can be determined.

The aim of this study was firstly to assess whether EMI data can be used to predict soil depth and secondly to determine the subsurface topography and to advocate the applicability and importance of EMI interpretations in hydrological studies.

STUDY AREA AND METHODOLOGY

An EM38 survey was conducted on an open area on the western part of the University of the Free State (UFS) campus, Bloemfontein, South Africa (Figure 1). The selected area is approximately 12 ha in size and receives an average rainfall of around 550 mm year[-1], predominantly in the form of high intensity thunder storms during the summer months that is November to March. The elevation ranges between 1440 and 1420 masl with very gentle slopes. Beaufort shales, mudstones and dolerite are the dominant geological formations in the area. The soils exhibit different degrees of weathering due to the variation in parent material and different water regimes, resulting in a variety of depths. Surface crusting resulting in overland flow is expected to govern the hydrological behaviour of the study site. Although the area is not located in a hydrological research site it was selected for this study due to its accessibility and because the soils are relatively shallow (<1500 mm), falling within the maximum effective reading depth of the EM38.

Table 1. Statistical analysis of non-parametric quantile regression between measured soil depth and EC_a ($CV_{0.5}$ and CV_1).

Soil depth	Coefficient	Standard error	t	P>t
$CV_{0.5}$	-29.144	5.171	-5.640	0.000*
CV_1	34.495	3.923	8.790	0.000*
Constant	148.569	47.594	3.120	0.003*

*, Significance level of 0.01; **, significance level of 0.05.

The survey was done on the 13th of September 2011 following the dry winter months. A calibrated GeoNics EM38 was pulled behind a quad bike in north-south transects over the study area. EC_a measurements were taken on 1 s intervals totalling more than 15 000 readings. Two EC_a readings were taken simultaneously namely $CV_{0.5}$ (0.5 m between coils) and CV_1 (1 m between coils). The difference between the coils determines the effective reading depth, the smaller the distance, the shallower the reading depth. $CV_{0.5}$ and CV_1 are integrated conductivity values in mS m^{-1} over a depth of 0 to 750 and 0 to 1500 mm, respectively.

Soil depths were measured up to the bedrock with a 1 m cone-penetrometer. Refusal that is soil/bedrock interface was defined as the depth where 5 blows with a 4.6 kg hammer did not result in a 1 cm downward movement of the penetrometer. For the soils deeper than 1 m, the soils were hand-augured to the bedrock, and the depths recorded. A total of 65 soil depth observations, spread over the study area, were made (Figure 1).

To test for a linear relationship between measured soil depth (dependent variable) and EC_a ($CV_{0.5}$ and CV_1, two independent variables) a multiple linear regression was done. EC_a and soil depth measurements were not normally distributed. Transformation of the data did result in a normal distribution of EC_a measurements, but was unsuccessful for soil depth measurements. For this reason it was decided to use non-parametric quantile regression. This method estimates the median (not mean) of the soil depth measurements (dependent), conditional on the values of the EC_a measurements (independent). The method finds a line through the data that minimizes the sum of absolute residuals. The EC_a measurement closest to the measured observation was used in the regression.

From the function that was developed, soil depths were calculated for surveyed transects. These depths were then interpolated using the inverse distance weighted (IDW) technique and a soil depth map was created for the study site. Surface elevations were also obtained from the EM survey and a Digital Elevation Model (DEM) was created. Soil depth and surface elevation rasters were converted to point layers, spatially joined based and the difference between the surface elevation and the soil depth is equal to the subsurface elevation; for which another DEM was created. To infer surface and subsurface flowpaths, flow accumulation rasters were created where a stream channel has more than 350 cells draining into it. These flowpaths are considered to be important localities for measuring surface and subsurface hydrological processes. ArcMap™ 10.1 (ESRI, 2012) software was used for all the GIS related operations.

RESULTS AND DISCUSSION

Relationship between soil depth and EC_a

The results of the regression between measured soil depth and EC_a ($CV_{0.5}$ and CV_1) are presented in Table 1. Satisfactorily was the fact that there was a significant relationship (y=149-29x+34x) between soil depth and EC_a.

Additionally, the proportion of variation in soil depth, that could be estimated by knowing the EC_a and the coefficients for the equation of the line, was high (R^2 = 0.53).

To evaluate the accuracy of the equation, estimations of soil depth from EC_a measurements were compared against measured soil depths (Figure 2). The relative median absolute error (RMdAE) shows that there was a 20% over- and/or under-estimation of soil depth by the function. The estimated soil depths compared well to the median of the measurements, with a relative modelling efficiency (EF) of 0.49. A value below 0 would have meant that the median measured soil depth would have been a better estimator than the function, which was not the case here. The quantile regression of measured soil depth versus estimated soil depth, showed that the slope (0.94) and intercept (23) did not differ significantly from 1 and 0, respectively, which are good indicators of accuracy (Bellocchi et al., 2010). Unfortunately the reliability of the function could not be assessed, due to the fact that an independent data set (a data set other than the one used to develop the function) was required. This was however beyond the scope of this paper. Thus, it can be concluded that there is a significant linear relationship between soil depth and EC_a, which can be used to estimate soil depth in the study area, with reasonably accuracy, from the 15 000 EC_a readings. Because no reliability asessment could be completed, the authors are under no illusion that the equation is a universal soil depth estimation equation for EM38 measurements. Variations in soil conditions (e.g. salt and water content) will modify the conducting capacity of the soil, and consequently alter the calibration equation, possibly decrease or the accuracy.

Soil depth and surface topography

The interpolated soil depths are presented in Figure 3a. The majority of the soils are shallower than 800 mm; isolated pockets of deeper soils can clearly be identified. In general, the soil depth follows the surface evaluation inversely, that is, deeper soils are found at lower elevations (Figure 3b). The inverse relationship between soil depth and surface elevation is presented in Figure 4.

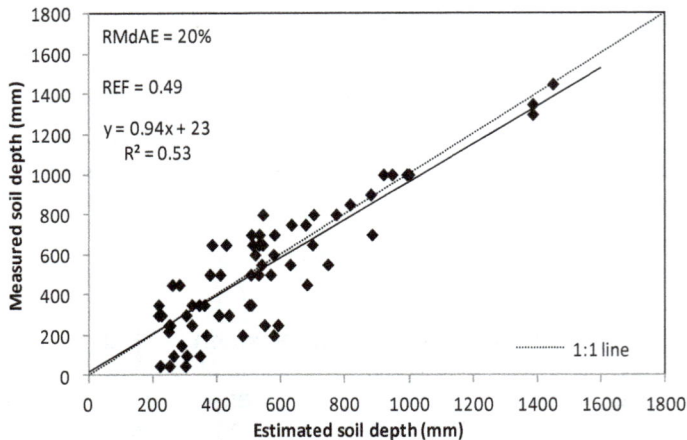

Figure 2. Measured soil depths vs. soil depths estimated from EC_a measurements.

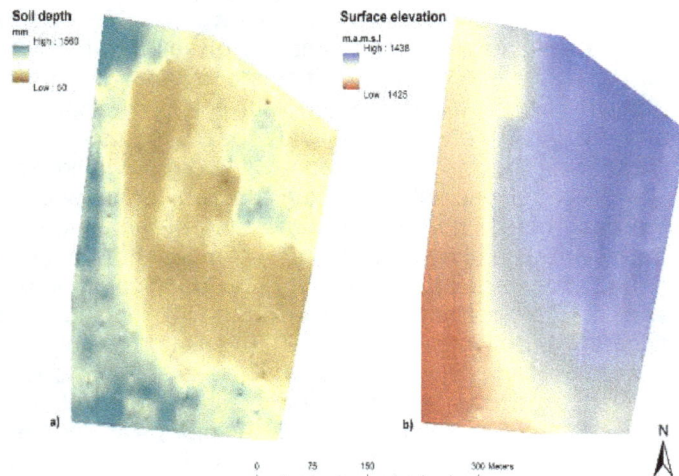

Figure 3. (a) Soil depths as calculated from regression (Table 1) and interpolated from measured transects and (b) surface elevation, also interpolated from surveyed transects.

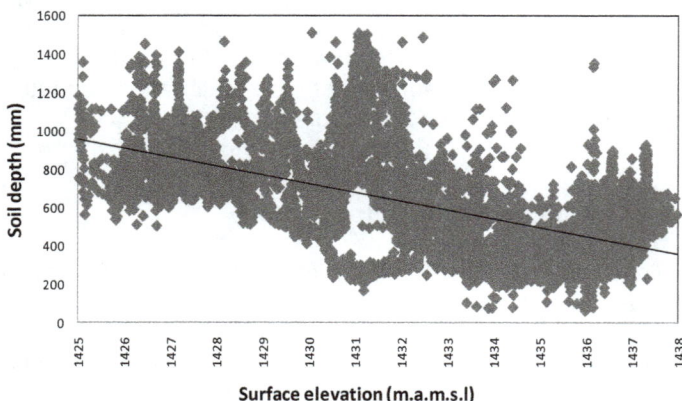

Figure 4. Relationship between soil depth and surface elevation.

The isolated pockets of deeper soils which are visible in Figure 3a can be observed as peaks in Figure 4 that is deep soils at relatively high elevations. The correlation between surface and subsurface topography and therefore soil depth, can be attributed to hydrological controlled soil genesis. Overland flow is an important flowpath controlled by surface topography where higher elevations are eroded. Limited infiltration results in less weathering and the combination of these processes results in shallower soils on high elevations compared to soils of lower lying land where water and sediments accumulate. Accumulation of colloidal material and more chemical weathering due to increased water contents result in deeper soils present in the lower lying areas (Figure 3a and b).

Subsurface topography and flow accumulation

The inverse correlation between surface topography and soil depth results in similar trends between surface topography and subsurface topography. The subsurface topography (surface elevation – soil depth), follow the surface topography very closely in this study site (Figure 5). In semi-arid arid areas where overland flow and associated process (erosion and accumulation of water and sediments) are dominant, the subsurface topography will be amplified by the surface topography that is in lower lying surface elevation areas, soils will be deeper.

This interaction between surface and subsurface topography is explained in Figure 5, where the flow accumulation (cells) for the surface and subsurface topography respectively, were calculated. Figure 5 represents the number of cells contributing to a specific cell based on the location and the relative elevation of the cell in the study area. Note that Figure 4 does not represent all the points in the study area (>70,000). The maximum and minimum flow accumulation values were selected and then approximately 20,000 randomly selected locations. Lower lying cells would receive more water and result in a higher flow accumulation value and *vice versa*. One cell is approximately 1.6 m^2 in size. There is a very good correlation between the surface flow accumulation and the subsurface flow accumulation on a specific location in the study area (R^2 = 0.98 in Figure 5). This supports the visual interpretation of Figures 3a and b, stating that the surface topography controls the soil depth and thus the subsurface topography. The deviation of the trendline from the 1:1 line in Figure 5, indicates that there is more water accumulating in the lower lying areas of the subsurface topography than compared to that of the surface topography. The increase in soil depth with a decrease in the relative elevation, for the reasons given earlier, might be the reason for more cells contributing to specific areas in the subsurface topography.

The highest number of cells accumulating on a certain point is 21,160 and 18,832 for the subsurface and surface elevation layers, respectively. This represent 3.38

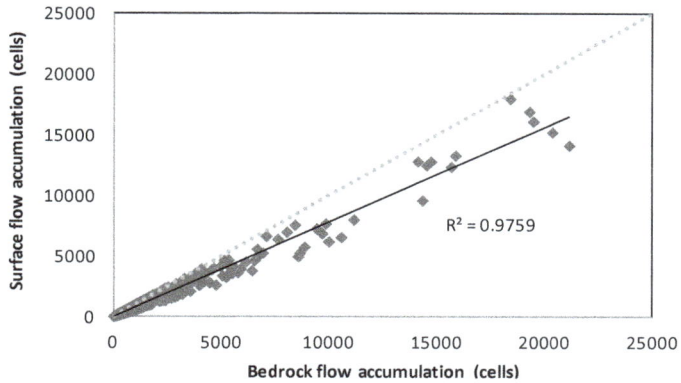

Figure 5. Flow accumulation calculated from surface and subsurface topography.

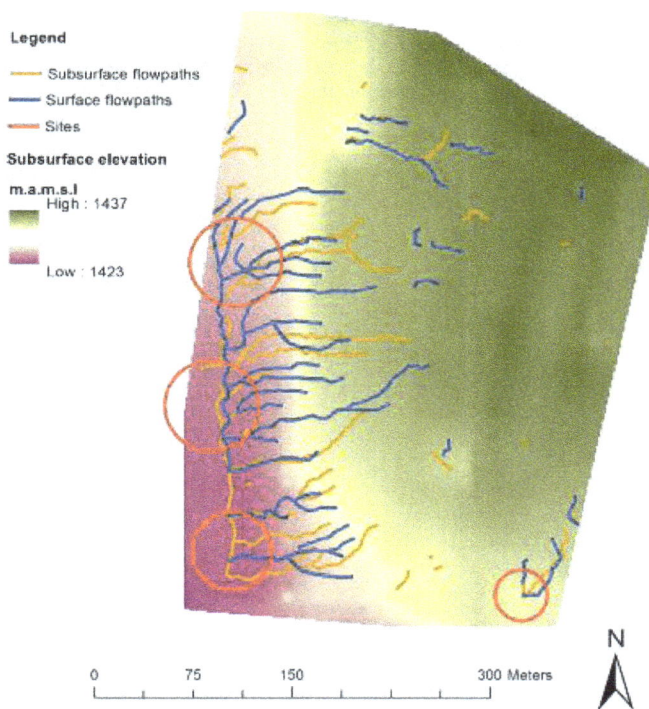

Figure 6. Subsurface elevation, subsurface and surface flowpaths and suggested research sites of the study area.

Suggested sites for future research and instrumentation

In Figure 6, four areas encircled in red, are suggested for further investigation, based on the origin and confluence of hydrological pathways. Although future research is unlikely on the selected site (it was only selected for the development of the methodology), we believe that these areas represent some of the key hydrological mechanisms occurring in the study area. In catchment and hillslope hydrological studies, the major aim is to quantify outflow in the form of streamflow exiting the catchment or the contribution of a hillslope to streamflow. It would be expensive and ultimately futile to study and instrument the entire area when flows are only generated on small portions. We believe that areas where flow pathways originate and converge are the important areas prompting the hillslope or catchment to respond hydrologically.

The traditional way of studying and instrumenting hillslopes in the form of transects perpendicular to the contours, might therefore not reflect the dominant control mechanisms of that hillslope. Also, inflection points in the surface topography are often identified as the ideal location for detailed investigation of hydrological processes. However, in areas with soils with high infiltration capacities (the majority of hydrological study sites), the subsurface topography controls the response of the hillslopes, and surface and subsurface topographies often do not correlate well. This might lead to incorrect selections of "representative" sites and therefore erroneous interpretations of the hydrological response of that hillslope.

Conclusions

A significant linear relationship (equation) between soil depth and EC_a were obtained with multiple linear regressions. Measured soil depths compared well (RMdAE = 20%, REF 0.49) to estimations made with the equation from EC_a measurements. Thus, the equation proved to be accurate, from where 15,000 soil depths could be estimated across the study area. This contributed to the successful characterization of subsurface topography, which made the following conclusions possible.

The soil depth of the study area shows a close inverse association with the surface topography, as is evident from visual interpretations and flow accumulation correlations. The reason for the close correlation between soil depth and surface topography can be attributed to overland flow following the surface topography removing soils from higher lying areas and deposition in lower lying areas. Flow accumulation maps indicate that accumulation of water in lower lying areas might result in a higher degree of weathering in the lower areas.

The flow accumulation maps, based on the subsurface

ha or 28% of the study area for the subsurface topography and 3 ha or 25% of the study area for the surface topography. The location of this high accumulation of cells is found in the south western (SW) corner of the study site (Figure 6). This area is marked by low elevation and deep soils. Figure 6 also higlightes the difference between surface and subsurface flowpaths (flow accumulation of more than 350 cells). Flowpaths of the surface topography are more connected and occur more frequently compared to that of the subsurface topography.

topography, suggest that only small portions of the study area are involved in the generation of flow. These areas are important to investigate as they will ultimately determine the hydrological response of the study area.

We suggest that a 3-D survey of any research site should prelude any effort to instrument new research sites. These surveys will also improve interpretations on existing research sites. In this study inference of soil depths from EMI measurements proved to be a valuable, time and cost efficient contribution to the understanding of the hydrology of the research site.

REFERENCES

Abdu H, Robinson DA, Seyfried M, Jones SB (2008). *Geophysical imaging of* watershed subsurface patterns and prediction of soil texture and water holding capacity. Water Resour. Res. p. 44, W00D18, doi:10.1029/2008WR007043.

Bellocchi G, Rivington M, Donatelli M, Matthews K (2010). *Validation of biophysical models: issues and methodologies. A review.* Agron. Sustain. Dev. 30:109-130.

Doolittle JA, Sudduth KA, Kitchen NR, Indorante SJ (1994). *Estimating depths to claypans using electromagnetic induction methods.* J. Soil. Water Conserv. 49:572-575.

Dunne T, Black RD (1970). *Partial area contributions to storm runoff in a small New England watershed.* Water Resour. Res. 28:1926-1311 (1970).

ESRI (2012). ESRI® ArcMap™ 10.1. ESRI, Redlands, CA.

Hendrickx JMH, Kachanoski RG (2002). *Nonintrusive electromagnetic induction*, in Dane JH, Topp CC (eds) Methods of Soil analysis. Part 4 Physical Methods. Soil Sci. Soc. Am. Book Ser. 5:1297- 1306.

Hopp L, McDonnell JJ (2009). *Connectivity at the hillslope scale: Identifying interactions between storm size, bedrock permeability, slope angle and soil depth.* J. Hydrol. 376:378-391.

McDonnell JJ, Sivapalan M, Vaché K, Dunn S, Grant G, Haggerty R, Hinz C, Hooper R, Kirchner J, Roderick ML, Selker J, Weiler M (2007). *Moving beyond heterogeneity and process complexity: A new vision for watershed hydrology.* Water Resour. Res. p. 43, W07301, doi:10.1029/2006WR005467.

Sklash MG, Farvolden RN (1979). The role of groundwater in storm runoff. J. Hydrol. 43:45 – 65.

Sherlock M, McDonnell JJ (2003). *A new tool for hillslope hydrologists: Spatially distributed measurements of groundwater and soil water using electromagnetic induction.* Hydrol. Processes 17:1965-1978.

Sivapalan M (2003a). *Process complexity at hillslope scale, process simplicity at watershed scale: is there a connection.* Hydrol. Process. 17:1037-1041.

Sivapalan M (2003b). *Prediction in ungauged basins: a grand challenge for theoretical hydrology.* Hydrol. Process. 17:3163-3170.

Tromp-van Meerveld I, McDonnell JJ (2006a) *Threshold relations in subsurface stormflow: 1. A 147-storm analysis of the Panola hillslope.* Water Resour. Res. p.42, W02410, doi.10.1029/2004WR003778.

Tromp-van-Meerveld I, McDonnell JJ (2006b). *Threshold relations in subsurface stormflow: 2. The fill and spill hypothesis.* Water Resour. Res. p.42, W02411, doi.10.1029/2004WR003800.

Uchida T, Tromp-van-Meerveld I, McDonnell JJ (2005). *The role of lateral pipeflow in hillslope runoff response: an intercomparison of non-linear response.* J. Hydrol. 311:177-133.

Weiler M, McDonnell JJ (2004). *Virtual experiments: a new approach for improving process conceptualization in hillslope hydrology.* J. Hydrol. 285:3-18.

Weyman DR (1973). *Measurements of downslope flow of water in a soil.* J. Hydrol. 20:267-288.

6

The use of scientific and indigenous knowledge in agricultural land evaluation and soil fertility studies of two villages in KwaZulu-Natal, South Africa

N. N. Buthelezi[1]*, J. C. Hughes[2] and A. T. Modi[2]

[1]Soil Science, School of Agricultural and Environmental Sciences, University of Limpopo, Sovenga, South Africa.
[2]Crop Science, School of Agricultural Sciences and Agribusiness, University of KwaZulu-Natal, Pietermaritzburg, South Africa.

Local people and small-scale farmers have knowledge of their lands based on soil and land characteristics that remain largely unknown to the scientific community. It is therefore important for researchers to understand farmers' knowledge of soil classification and management. To address this, indigenous knowledge was elicited by questionnaires from 59 households in two villages (Ezigeni and Ogagwini), near Durban in KwaZulu-Natal. Farmer vernacular soil and land suitability evaluations were compared to scientifically obtained soil and land suitability maps. Yield was used as a quantifiable indicator to test the effect of fertility management practices. It was found that farmers' soil classification was based mainly on topsoil colour and texture. Slope position was the main factor determining land suitability. Crop yield, crop appearance, natural vegetation, soil colour, soil texture, and mesofauna were used to estimate soil fertility. Despite this, there was a correlation between farmers' indigenous evaluation and scientific evaluation, implying that there are similarities between the two approaches.

Key words: Local knowledge, scientific knowledge, soil properties, crop indicators, soil fertility.

INTRODUCTION

Land degradation is a threat to the sustainability of agricultural soils. Hence there is an urgent need for more sustainable management of this fundamental resource (Ingram, 2008). Barrera-Bassols et al. (2009) emphasize the importance and relevance of soil information in land use planning and land management. It is, however, evident that rural communities' local knowledge about land, in terms of soil and land characteristics, still remains largely unknown to the scientific community (Ingram, 2008). Local farmers derive this knowledge mainly from informal experiential learning as they interact with their and adapt to constantly changing natural environments. For example, farmers in West Java used traditional

agroforestry to improve biophysical properties of the soil and to sustain the economy of their villages (Christanty et al., 1986).environment (McGregor, 2004). This knowledge has assisted rural communities to manage natural resources Due to inadequate arable land available for crop production; farmers in the High Andes of Peru manipulated the agricultural potential of a highly elevated area using sectoral fallowing (Pestalozzi, 2000). He et al. (2007) also reported a case of farmers in the Sichuan region in China where farmers had developed slope-land management practices to avoid loss of soil through erosion. These case studies reveal how local farmers have developed innovative ways of survival using indigenous knowledge.Land evaluation processes have been done mainly through soil surveys which farmers may not fully understand and which exclude social and cultural aspects. Brinkman and Smyth (1972) reported that these factors are essential when conducting and

---

*Corresponding author. E-mail: musa.buthelezi@ul.ac.za.

interpreting land evaluation, because that approach allows people to contribute directly to land use planning. The exclusion of local knowledge from land evaluation systems often resulted in the failure of scientific interventions to improve land use, especially in rural areas where scientific logic is lacking (Sillitoe, 1998; Barrera-Bassols and Zinck, 2003a, b; Ingram, 2008; Ingram et al., 2010). According to Gadgil et al. (1993), indigenous knowledge can complement more general scientific knowledge with site-specific, highly contextualized knowledge. It would provide sustainable solutions by bringing a 'locally informed perspective' to development strategies which ensures the inclusion of socioeconomic factors as well as cultural diversity (Agrawal, 1995; Sillitoe, 1998; Crevello, 2004; Briggs, 2005; Sillitoe and Marzano, 2009).

Land use planning at farm level is mainly based on local perceptions of the quality of the natural resource base (Krogh and Paarup-Laursen, 1997). For example, farmers in many countries have comprehensive local soil quality indicators (Neef et al., 2006; Mairura et al., 2007; Kissing et al., 2009; Tesfahunegn et al., 2011). Having limited access to soil analytical services and synthetic fertilizers, farmers have used this soil knowledge to develop proper management plans that for many decades have ensured both agricultural and environmental sustainability (Materechera, 2008; Handayani and Prawito, 2010; Tesfahunegn et al., 2011). Therefore, more information is needed for better understanding of this knowledge and the consequent contribution it can make towards agricultural sustainability (Phillips-Howard and Oche, 2006). This will further secure agricultural sustainability in both subsistence and commercial farming. It is noteworthy that there is a wide spectrum of studies ranging from those proposing and testing possible integration methodologies (Norton et al., 1998; Braimoh, 2002; Gowing et al., 2004; Barrios et al., 2006) to others highlighting the benefits and challenges associated with the integration of these two knowledge systems (Nadasdy, 1999; Newton et al., 2005; Ocholla, 2007; Gagnon and Berteaux, 2009; Sileshi et al., 2009; Lynch et al., 2010; Raji et al., 2011).

Despite such information on the indigenous knowledge of soil, there have been few studies on farmers' knowledge of soils in South Africa. Materechera (2008) investigated and documented practices used by farmers in the North West Province of South Africa in soil fertility management. He found that farmers had developed contextualized and cost effective strategies such as using ash from veld fires, animal manure, agroforestry, fallowing and earthworms. Moreover, he reported that farmers also had the ability to assess the quality of manure using characteristics such as colour, wetness, presence of moulds and sand content. Using a participatory survey, Nethononda and Odhiambo (2011) studied the local classification of soils by farmers in Vhembe district of Limpopo province in South Africa. In the same location, Mutshinyalo and Siebert (2010) conducted a study that investigated the use of myth as a biodiversity conservation strategy. Hart and Vorste (2006) reported a number of innovations used by rural communities to achieve agricultural and livelihood sustainability. Phillips-Howard and Oche (2006) gave a comprehensive account of indigenous knowledge and fertilizer use in the Transkei (currently part of the Eastern Cape Province), South Africa.

The main objectives of this study were to (a) explore soil indigenous knowledge of small-scale subsistence farmers in relation to land evaluation, (b) compare indigenous and scientific land evaluation and (c) assess soil fertility management by the farmers using scientific methods.

MATERIALS AND METHODS

Study site

The study was conducted in two villages (Ezigeni and Ogagwini) of the uMbumbulu area (KwaZulu-Natal). The area is located inland from Durban at 29° 59' South, 30° 42' East between 394 and 779 m a.s.l. Members of the Ezemvelo Farmers Organization form part of the population of both Ezigeni and Ogagwini villages. This group of farmers was the first subsistence farmer organization certified to supply organic vegetables to supermarkets. Farmers rely on crop rotation, crop residues and animal manure for soil fertility management. Primary cash crops grown are amadumbe (taro), sweet potatoes and potatoes.

Indigenous land evaluation

A total of 59 farmers from both villages were interviewed to gain a general background of indigenous agricultural land evaluation and management. A questionnaire focused on local soil classification and its importance in land evaluation. Another questionnaire was produced to gather more detailed information on indigenous soil management from each farmer. To obtain this information, six (three from each village) of the 59 households were chosen. These were chosen based on their willingness and farming experience. The questionnaire required information on the cropping history, knowledge specific to the cultivated lands, and detailed soil description and fertility assessment. The information gathered from both sets of interviews was recorded and analyzed using SPSS version 15.

Scientific land evaluation

A general purpose free survey was conducted at a scale of 1:10 000. Soil forms and families were classified according to the Soil Classification Working Group (1991). Soils were classified for land suitability (for maize, taro and dry bean) and capability based mainly on soil form, depth and drainage (Davidson, 1992).

A total of 24 representative samples were taken from the six randomly chosen homesteads. These were taken from the two most common agricultural soils from both villages (that is, Hutton and Oakleaf). Pairs of samples were taken from each homestead. Each pair consisted of sub-samples taken from 0 to 30 cm and 30 to 60 cm depth. The samples were collected from different management practices (that is, fallow, veld, taro and vegetable production lands). However, not all six homesteads had all four land uses hence the total number of samples.

Table 1. Characteristics of the households interviewed.

Gender	No.	Education level	No.
Males	5	Not educated	6
Females	54	Grade 1 - 8	26
		Grade 9 - 12	21
		Higher education	6
Age		*Family size*	
< 30	15	1 - 5 members	11
31 - 45	15	6 - 10 members	31
46 - 55	13	> 10	11
>56	13		
Missing data	3		6

Table 2. Multivariate analysis showing the significance of age, gender and education level and their interactions on farmers' indigenous knowledge (n=59).

Factors	F	DF	P
Gender	0.180	1	0.894
Age	0.457	1	0.237
Level of education	1.163	4	0.345
Gender* age	0.589	2	0.560
Gender* education	0.456	1	0.504
Age* education	0.602	8	0.770
Gender* age*education	6.054	1	0.019

Soil samples were air-dried and passed through a 2 mm sieve before analysis. Soil pH was measured using a 1:2.5 ratio of soil:distilled water as well as a 1:2.5 ratio of soil:1 M KCl. Particle size distribution was determined using the pipette method (Gee and Bauder, 1986). The potassium dichromate oxidation procedure was used to determine organic carbon (Walkley, 1947). For soil fertility, the samples were analyzed by the Soil Fertility Analytical Service at Cedara (Riekert and Bainbridge, 1998). For microbial analysis, the soil samples were rewetted to 50% water holding capacity before carrying out microbial activity analysis. The 50% water holding capacity was calculated using texture and organic carbon (Smith, 1995; Smith et al., 2001). The samples were then incubated for four weeks to allow for the regeneration of microorganisms. They were then put in the refrigerator a day before the analysis. The analysis with two replications was done using the FDA (fluorescein diacetate) method (Schnürer and Rosswall, 1982).

Comparison methodology

Scientific and indigenous evaluation systems were compared based on the land suitability classification. The information provided by the scientific suitability maps was compared to the vernacular suitability evaluation provided by farmers. Farmers' fertility assessment was also compared with the scientific perception. Yield was used as a quantifiable indicator to test the effect of fertility management practices implemented by Ezigeni and Ogagwini farmers. This was measured in terms of total biomass of the dominant available crops including maize, amadumbe (taro) and dry beans. Data were analyzed using Genstat® Statistical Package 11. Multivariate analysis was used to determine correlations. Significant differences were determined at p ≤ 0.05 and least significant differences (LSD) are presented to indicate differences between treatments.

RESULTS AND DISCUSSION

Household characteristics

Household characteristics acquired are summarized in Table 1. There were a comparable number of farmers across the age groups. Only 5 of 59 respondents were men. The reason for this, according to the farmers, was that men work away and only comes home on weekends. Hence the responsibility for farming is mainly taken by the women.

Table 2 shows that there was no significant effect of gender in the knowledge gathered. This may be attributed to the fact that there were very few male compared to female farmers. Most farmers had either grade 8 or grade 12 as the highest level of education. The respondents with matriculation could not afford to go to institutions of higher learning and so these young people stay at home and are available to help in the fields. Even those that are still at school are taken to the fields during weekends and school vacations. Because of this exposure of young people to indigenous farming there was no significant effect of age and education on

Table 3. Local soil taxonomy used by farmers of Ezigeni and Ogagwini villages.

Local name	Texture	Colour	Location	Uses
Ugadenzima	Clayey (*ubumba*)	Reddish black	Midslope	Agriculture
Idudusi	Loam (*uthambile*)	Black	Lower slope	Agriculture
Isibomvu	Clayey (*ubumba*)	Dark Red	Upslope	Agriculture
Udongwe	Clayey (*ubumba*)	Grey	Footslope	Agriculture
Umgogodi	Clayey (*ubumba*)	Grey	Footslope	Plastering
Isdaka	Clayey (*ubumba*)	Black	Footslope	Agriculture
Umgubane	gravelly (*ungamatshe*)	black or red	Upslope	Construction
Ugwadule	Clayey (*ubumba*)	black or red	Upslope	NS*
Isduli	Clayey (*ubumba*)	Black	Footslope	Agriculture
Ugedle	Sandy (*isihlabathi*)	Red	Upslope	Agriculture

*NS = Not specified.

the knowledge elicited (Table 2). This is contrary to the results reported by Birmingham (2003) from Ivory Coast who found that older farmers had more detailed knowledge than younger farmers. However, the combination of gender, age and education had a significant effect on status of knowledge (Table 2). Younger people, because of their education, are able to easily grasp the knowledge being passed on to them, and may even be able to develop and make it better.

Indigenous soil management

The area cultivated by each farmer in both villages ranged from 0.6 to 4 ha. Farming is the livelihood for most of the households in uMbumbulu and food production is for both marketing and subsistence. All farmers from randomly selected households were involved in organic farming of mainly amadumbe (taro) and other crops such as dry beans, maize, potatoes, pumpkins and sweet potatoes. They only practiced traditional farming, although tractors were used for tilling the soil. Farmers in both villages, with a few exceptions, owned livestock and practiced similar management systems. These included mixed cropping and rotation systems (below-ground followed by above-ground type of crop) for fertility management. Respondents recommended frequent rotation in taro plots especially when planted in dark soils to avoid reduction in yield. A study by Asao et al. (2003) showed that this decrease in yield is attributed to a detrimental effect of taro roots exudates. In the current study, farmers rotated taro with either maize or beans, depending on the soil type and drainage in order to avoid this effect. They used kraal manure, stubble mulch and fallowing to replenish depleted nutrients. Farmers that have infertile soils in their fields had observed the positive response in yield when these soils are treated with large amounts of manure one or two months before planting. However, the scarcity of organic amendments has encouraged some of the farmers to try anaerobic composting suggested to them by an extension officer. Unfortunately, this alternative was not successful because of high interference from pests (birds, wild hogs and soil organisms). Overall, despite similar management practices, farmers have observed that crops yield more when planted in Ogagwini soils and hence consider these soils more fertile than the soils of Ezigeni village.

Famers' soil classification

Farmers recognized 10 soil types (Table 3). Farmers were only concerned with the topsoil, because they use this part of the profile for their agricultural activities. This follows a trend observed for other local classification systems (Sillitoe, 1998). Culture, which is an integral part of farmers' lives in this region, does not allow for digging as it is believed to anger the ancestral spirits. Subsoil is only seen when digging for a grave and hence is not important for a farmer's agricultural knowledge. This explained why farmers' land evaluation focuses only on suitability of the land for production systems (Ettema, 1994). Farmers were asked to critique the scientific approach and they were concerned with the time and labour involved in this approach. Hence their classification was based on descriptive soil characteristics rather than characteristics of the whole profile as in the scientific classification. Although their soil classification was based on different soil morphological attributes, soil colour and texture were key properties recognized by over 80% of the farmers. This is consistent with the findings of Sandor and Furbee (1996) and Talawar and Rhoades (1998). The use of texture may not be surprising due to the large influence it has on other soil properties such as structure, water holding capacity, permeability and drainage.

Farmers' land suitability assessment

In common with scientific evaluation, farmers recognized

Table 4. Crop suitability for the local soil names according to Ezigeni and Ogagwini farmers.

Local name	Fertility status*	Principal crops
Ugadenzima	Low to moderate	Potatoes, maize, beans
Idudusi	High	Maize, taro, beans
Isibomvu	Moderate to high	Sweet potatoes, maize, beans
Udongwe	Moderate	Beans, taro
Isdaka	Moderate to high	Spinach, taro
Isduli	Low to moderate	Taro, maize, beans
Ugedle	Low	Potatoes, sweet potatoes

*Fertility status estimated from farmers responses.

drainage status and soil depth (referred to as the amount of topsoil) as limiting factors for land use. However, slope was also considered an essential factor affecting land suitability (indicated by 60% of farmers). Farmers preferred footslope soils for agriculture as these are regarded as more fertile compared to upslope and midslope soils. They attributed this difference in fertility to the removal of soil from upslope and deposition downslope resulting in higher nutrient levels in footslope soils. Twenty percent of farmers used natural vegetation focusing mainly on vegetative growth and species diversity. Consistent with a healthy soil ecosystem, farmers in these villages associated agriculturally suitable land with high species diversity (Mäder et al., 2002). Some farmers' land suitability evaluation was based on the differences they observed between the soils in both villages, and hence they used 'villages' as a classification criterion. Farmers had an understanding of the effect of soil type on land suitability for different crops (Table 4). Farmers have observed the effect of soil type on yield differences between the two villages with higher yields from Ogagwini. Farmers thus regard Ogagwini soils as more fertile because they do not demand high supplementary fertilizer inputs.

Scientific land evaluation

Soil types mapped ranged from highly suitable, deep, well drained soils to the least suitable, shallow soils (Figures 1 and 2). Similar to the indigenous evaluation, scientific evaluation showed that the limited suitability of Ezigeni soils was mainly due to constraints which were rarely observed for Ogagwini. These included shallow soil depth, poor drainage and stoniness. Despite deep soils (> 120 cm), many at Ezigeni had a duplex character. Despite these differences between the villages, Table 3 shows that the soils in both villages are generally suitable for crop production. Moreover, land suitability maps showed higher agricultural potential for the Ogagwini than Ezigeni soils. This correlation between indigenous and scientific approaches shows that there are similarities between the local and scientific land use evaluation

(Habarurema and Steiner, 1997; Barrera-Bassols et al., 2006).

Soil fertility indicators

Farmers used a combination of indicators to rate the land as either 'good' or 'bad' (Figure 3). In scientific terms these lands will be either fertile or infertile, respectively. Soil colour and texture were used by 48% of farmers with dark soils indicating higher fertility than lighter soils. The abundance of mesofauna was used by 51% of farmers. Natural vegetation (18%), especially weed growth and diversity observed before planting, also gave a statement about soil fertility. However, the presence of weeds did not always reflect fertile soil conditions and led to errors by some farmers in their fertility assessment. Crop production factors are considered most reliable as they are said to clearly reflect soil fertility differences. These include crop colour and firmness (32%) during the establishment stages and crop yield (70%). This shows that crop yield forms a benchmark for soil quality in the indigenous approach (Gruver and Weil, 2006). It is clear that farmer fertility assessment is mainly concerned with food security which is highly dependent on land productivity.

Soil fertility analysis

Only two soil types (that is, Hutton 2200 and Oakleaf 1210) were sampled from the six homesteads chosen for the detailed questionnaire. The following discussion is based on the assumption that these two soils would behave similarly under similar management. Both Hutton and Oakleaf are red or brown indicating good drainage, they are both very deep soils (> 120 cm) and both are formed from dolerite, the Hutton on *in-situ* rock; the Oakleaf on doleritic colluvium.

Plant nutrients and soil pH

Tables 5 to 8 show that the average soil pH (H_2O) was

Legend

Soil depth
- D1: > 100 cm
- D2: 60 - 100 cm
- D3: 40 - 60 cm
- D4: 30 - 40 cm
- D5: < 30 cm

Soil form	Soil family
Clovelly (Cv)	2100 (Buckland)
	2200 (Leiden)
Glenrosa (Gs)	1111 (Dumisa)
	1211 (Tsende)
	2121 (Maringo)
Hutton (Hu)	1200 (Kelvin)
	2100 (Hayfields)
	2200 (Suurbekom)
Inanda (Ia)	2200 (Highlands)
Katspruit (Ka)	1000 (Lammermoor)
Mispah (Ms)	1100 (Myhill)
Oakleaf (Oa)	1110 (Ritchie)
	1210 (Caledon)
Sepane (Se)	1210 (Katdoorn)
Swartland (Sw)	1111 (Spreyton)
Valsrivier (Va)	1111 (Slykspruit)
	1211 (Wepener)
	1221 (Helvetia)
Willow brook (Wo)	1000 (Ottawa)

0   100   200   400 Meters

Source: Cartographic Unit, Discipline of Geography, University of KwaZulu-Natal, PMB Campus, 2009.

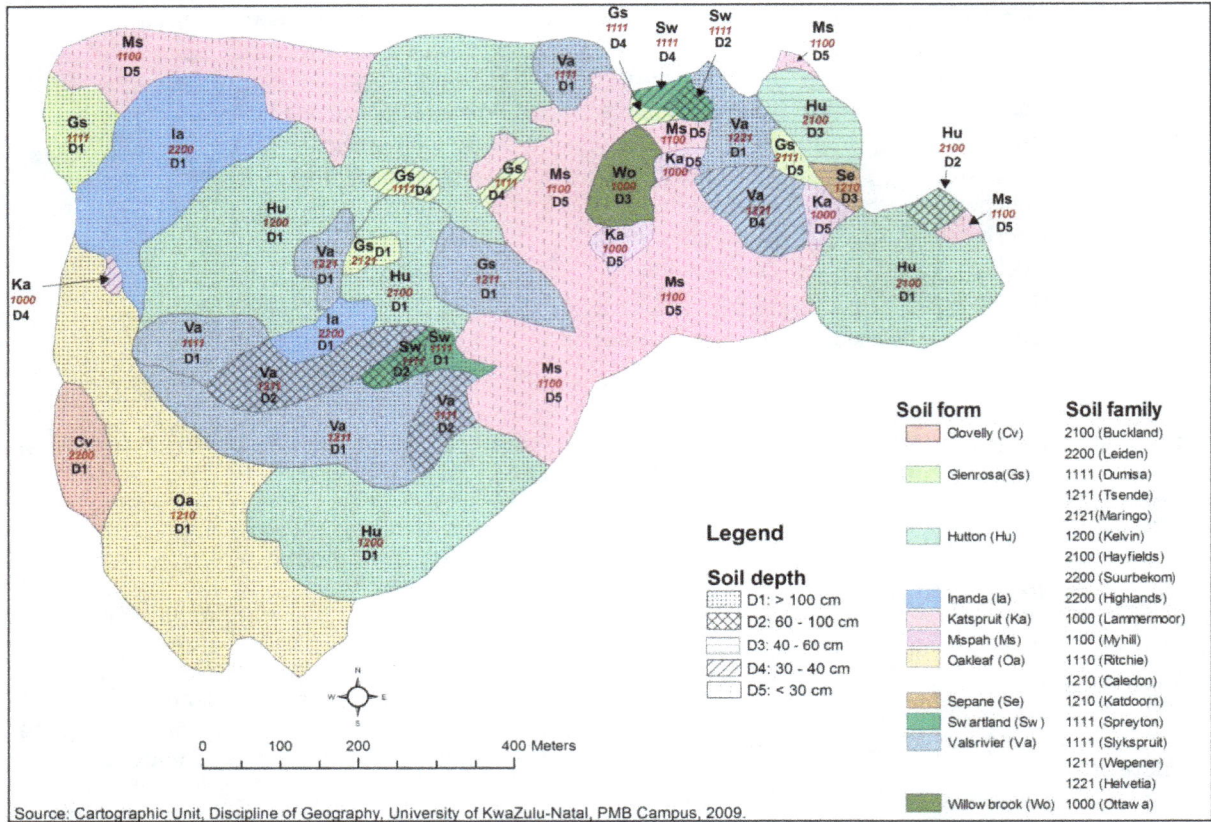

Figure 1. Soil map of Ezigeni.

comparable across the two villages for both A (5.79 and 5.93) and B (6.07 and 6.00) horizons. There was high acid saturation in both Ezigeni and Ogagwini soils. However, Ezigeni topsoils had a higher median acid saturation value of 44% compared to 30% in Ogagwini topsoils. This inevitably results in a decrease in exchangeable basic cations (Foth and Ellis, 1997). Ogagwini soils therefore had higher plant nutrient levels than Ezigeni soils. For example, Ogagwini topsoils had an average of 3.24 mg kg^{-1} available P (Table 7) compared to 1.88 mg kg^{-1} P in Ezigeni topsoils (Table 5) and the average effective cation exchange capacity (ECEC) of the soils from Ogagwini village was higher (5.09 and 5.03 cmol$_c$ kg^{-1} in the A and B horizon, respectively) than the soils from Ezigeni village (4.16 and 3.43 cmol$_c$ kg^{-1}, respectively). Calcium and P values were significantly different between homesteads (p < 0.05). This may be due to past soil management rather than intrinsic soil differences. In addition to the effect of soil pH and acid saturation, N and Mg were significantly affected by land use (p < 0.05). Although the overall nutrient levels were relatively low in the soils from both villages, there was moderately high N under cultivated land (that is, taro and vegetable fields). This can probably be attributed to the N retained in cereals and vegetable residues, especially those from legumes, which are recycled during

decomposition (Hartemink et al., 2000).

Although Ezigeni soils had higher amounts of organic C and were less sandy, the ECEC of the soils is on average lower than soils of Ogagwini. The high average acid saturation value for Ogagwini is inflated by the values of the Ngcamu homestead (86 and 71% in the A and B horizon, respectively). If this topsoil value is omitted, the mean acid saturation drops to 26% in the topsoils, compared to 37% in the Ezigeni topsoils. The same homestead has the highest available P value (6.25 mg kg^{-1}) but even this, and the generally higher P in Ogagwini soils, would be deficient for most crops. The pH (KCl) values are also similarly low in all soils from both villages. In general, it appears from these data that the soils in the two villages are not very different in terms of their fertility parameters and that the most marked differences are most likely due to differences in management practices. This is noticeable in the microbial activity results if the disturbed and undisturbed land uses are compared.

Yield

Both scientific and farmer suitability evaluation found Ogagwini village to be more highly suitable than Ezigeni.

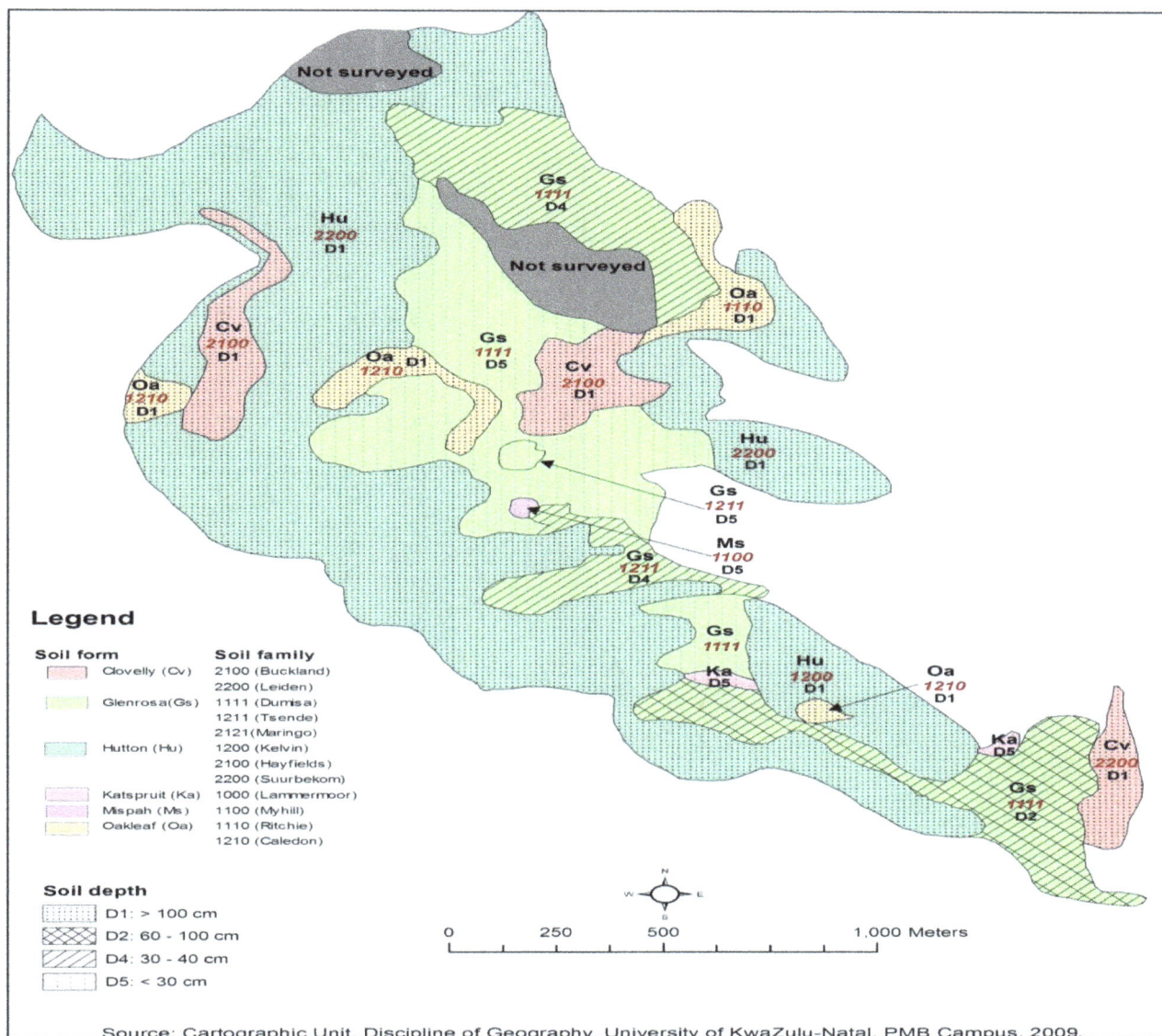

Figure 2. Soil map of Ogagwini.

This was further confirmed by yield measurements taken for beans, maize and taro (Figure 4). There was a significant difference ($p < 0.05$) in yield between homesteads. However, maize and taro yields in both villages were generally lower than those obtained on large commercial farms. Dry beans are known for their survival strategy of releasing citric acid which chelates Al in the rhizosphere thus preventing the detrimental effects of Al (Ma et al., 2001). Dry beans thus yielded satisfactorily. However, these yields were not very different between villages, a result again predicted by the similar fertility status of all soils analysed. This supports the scientific data (Tables 5 to 8) which showed that none of the sampled soils from either village is very fertile and that all have considerable constraints. Thus, while yields

from Ogagwini were higher than from Ezigeni, all were low as predicted by the scientific fertility data. The differences observed in yield may therefore reflect management factors (e.g. time of planting, weeding, availability of organic amendments, etc.) since most of the differences in soils in both villages are inherent. For example, although kraal manure was widely used in both villages not all homesteads own a herd of cattle. There was also only one tractor to assist farmers to till their soils at the beginning of the season. This sometimes led to delays in planting as farmers have to wait their turn and for the tractor driver to be available. Thus it is noteworthy that the soil properties (measured scientifically) follow and support the trend of the vernacular land suitability evaluation. The effect of management clearly plays the

Table 5. Soil chemical properties and particle size distribution of the A horizon of soils from Ezigeni village.

Homestead	Soil form and family	Land use	pH H₂O	pH KCl	N %	P mg kg⁻¹	K	Ca	Mg	H	ECEC*	Acid saturation	Organic carbon %	Clay	Silt	Sand	Microbial activity µg g⁻¹ h⁻¹
							cmol$_c$ kg⁻¹							%			
F. Mkhize	Oa 1210	Veld	5.65	4.40	0.60	1.92	0.04	2.58	0.88	1.13	4.63	24	7.6	11	54	35	17.21
Mbili	Oa 1210	Fallow	6.01	4.44	0.33	1.08	0.01	0.94	0.50	1.51	2.96	51	5.9	21	44	28	1.72
Bhengu	Hu 2200	Fallow	5.83	4.35	0.29	1.85	0.04	0.97	0.30	1.30	2.61	50	5.3	27	40	33	8.87
Mbili	Oa 1210	Vegetables	5.79	4.16	0.28	3.09	0.08	1.70	0.86	2.62	5.25	50	5.1	31	40	29	4.30
Bhengu	Hu 2200	Vegetables	5.51	4.35	0.33	3.16	0.07	1.00	0.51	1.57	3.15	50	6.1	16	45	39	4.13
F. Mkhize	Oa 1210	Vegetables	5.77	4.58	0.41	1.08	0.06	2.17	0.59	0.74	3.56	21	8.1	20	48	32	3.89
Mbili	Oa 1210	Taro	5.97	4.15	0.23	0.97	0.05	3.98	1.77	1.15	6.94	16	2.7	43	22	34	1.90
Mean			5.79	4.35	0.35	1.88	0.05	1.91	0.77	1.43	4.16	37	5.8	24	42	33	6.00
Median			5.79	4.35	0.33	1.86	0.05	1.80	0.68	1.37	3.86	44	5.9	23	43	33	4.13

*ECEC- effective cation exchange capacity (sum of bases + H).

Table 6. Soil chemical properties and particle size distribution of the B horizon of soils from Ezigeni village.

Homestead	Soil form and family	Land use	pH H₂O	pH KCl	N %	P mg kg⁻¹	K	Ca	Mg	H	ECEC*	Acid saturation	Organic Carbon %	Clay	Silt	Sand	Microbial activity µg g⁻¹ h⁻¹
							cmol$_c$ kg⁻¹							%			
F. Mkhize	Oa 1210	Veld	6.15	4.58	0.46	1.19	0.02	1.27	0.43	0.94	2.66	35	7.4	11	42	47	11.29
Mbili	Oa 1210	Fallow	6.25	4.49	0.33	1.04	0.01	1.14	0.57	0.98	2.70	36	3.9	19	39	42	0.87
Bhengu	Hu 2200	Fallow	6.27	4.53	0.26	1.09	0.02	1.34	0.62	0.62	2.60	24	3.0	43	40	17	2.51
Mbili	Oa 1210	Vegetables	6.00	4.25	0.29	2.08	0.06	1.58	0.81	2.25	4.69	48	4.5	35	36	29	4.01
Bhengu	Hu 2200	Vegetables	5.90	4.49	0.30	1.11	0.07	0.66	0.50	0.88	2.11	42	3.8	28	37	35	3.21
F. Mkhize	Oa 1210	Vegetables	6.13	4.62	0.43	1.11	0.06	1.76	0.51	0.66	2.98	22	6.3	18	43	39	2.72
Mbili	Oa 1210	Taro	5.80	4.08	0.18	0.97	0.04	2.89	1.54	1.77	6.24	28	2.5	56	14	30	0.19
Mean			6.07	4.43	0.32	1.23	0.04	1.52	0.71	1.16	3.43	34	4.5	30	36	34	3.54
Median			6.10	4.49	0.31	1.11	0.04	1.43	0.60	0.96	2.84	34	4.2	29	38	35	2.72

*ECEC- effective cation exchange capacity (sum of bases + H).

major role in whether farmers achieve a high yield rather than the village they reside in.

Conclusion

As expected, the farmers' soil indigenous know-ledge is rather abstract when compared to the more commonly obtained scientific knowledge. This is evident in farmers' soil classification which only takes into account the topsoil and extends to the way farmers perceive and assess soil fertility. Farmers' fertility indicators and soil taxonomy are based only on visible soil and crop properties and show that farmers are more concerned with soil productivity and food security. On the other hand, the scientific approach seeks to understand the processes of soil formation and has specific measured attributes that influence soil fertility (e.g. soil mineral elements). Despite many differences between the scientific and indigenous approaches,

Table 7. Soil chemical properties and particle size distribution of the A horizon of soils from Ogagwini village.

Homestead	Soil form and family	Land use	pH		N	P	K	Ca	Mg	H	ECEC*	Acid saturation	Organic carbon	Clay	Silt	Sand	Microbial activity
			H_2O	KCl	%	mg kg^{-1}				cmol$_c$ kg^{-1}				%			µg g^{-1} h^{-1}
Z. Mkhize	Hu 2200	Veld	5.75	4.24	0.25	2.83	0.05	2.55	1.83	1.12	5.56	20	3.6	33	22	45	13.90
Z. Mkhize	Hu 2200	Fallow	6.28	4.15	0.24	2.78	0.04	1.65	0.93	1.91	4.53	42	3.7	23	24	53	4.15
Gasa	Hu 2200	Fallow	5.92	4.40	0.34	3.30	0.06	2.60	1.30	1.09	5.05	22	4.9	19	23	58	10.59
Gasa	Hu 2200	Taro	5.98	4.45	0.22	1.06	0.02	2.37	1.24	0.97	4.61	21	3.8	42	34	24	2.77
Ngcamu	Hu 2200	Taro	5.73	4.08	0.29	6.25	0.06	0.50	0.22	4.91	5.69	86	4.1	19	23	58	2.77
		Mean	5.93	4.26	0.27	3.24	0.05	1.93	1.10	2.00	5.09	38	4.0	27	25	48	6.84
		Median	5.93	4.25	0.26	3.04	0.05	2.15	1.17	1.52	5.07	30	3.9	25	24	50	4.15

* ECE C- effective cation exchange capacity (sum of bases + H).

Table 8. Soil chemical properties and particle size distribution of the B horizon of soils from Ogagwini village.

Homestead	Soil form and family	Land use	pH		N	P	K	Ca	Mg	H	ECEC*	Acid saturation	Organic carbon	Clay	Silt	Sand	Microbial activity
			H_2O	KCl	%	mg kg^{-1}				cmol$_c$ kg^{-1}				%			µg g^{-1} h^{-1}
Z. Mkhize	Hu 2200	Veld	5.80	4.27	0.24	2.35	0.04	2.42	2.20	1.35	6.02	22	3.2	36	22	42	8.69
Z. Mkhize	Hu 2200	Fallow	6.08	4.22	0.20	4.44	0.03	2.11	1.43	1.77	5.33	33	3.3	38	16	46	1.32
Gasa	Hu 2200	Fallow	6.25	4.44	0.27	2.22	0.03	1.99	1.18	1.01	4.21	24	4.5	36	37	27	11.48
Gasa	Hu 2200	Taro	5.93	4.47	0.29	1.90	0.04	3.24	1.57	0.63	5.48	11	4.6	41	42	27	0.34
Ngcamu	Hu 2200	Taro	5.93	4.20	0.30	1.87	0.03	0.84	0.32	2.92	4.11	71	4.0	23	27	50	0.10
		Mean	6.00	4.32	0.26	2.56	0.03	2.12	1.34	1.54	5.03	32	3.9	35	29	38	4.39
		Median	5.96	4.30	0.27	2.29	0.03	2.12	1.39	1.44	5.18	28	4.0	36	28	40	2.85

* ECE C- effective cation exchange capacity (sum of bases + H).

results showed that there are many links between these two systems in terms of land evaluation, ranging from determination of land use to management issues which are critical components of sustainable agriculture. The farmers' soil classification and suitability evaluation, as well as their fertility assessment, correlate with the scientific evaluation. These significant agreements between the approaches imply that there are fundamental similarities between them. These similarities show that these two systems can complement each other to produce a hybrid approach that is highly contextual (an attribute of Indigenous knowledge) and that will help improve the relevance, adaptation and adoption of scientific interventions that provide the in-depth knowledge of soil processes. The challenge remains to achieve such integration between indigenous and scientific approaches at the local level to enable a better understanding of land management systems as applied by small-scale farmers and to ensure the protection of topsoil to maintain sustainable soil quality. This integration could be encouraged by production of an agricultural policy suitable at the farm level for small-scale farmers that utilizes their knowledge to support or perhaps even replace full scientific analysis, in instances where finances are not available. Such a hybrid approach could give direction to farmers and encourage them to play a

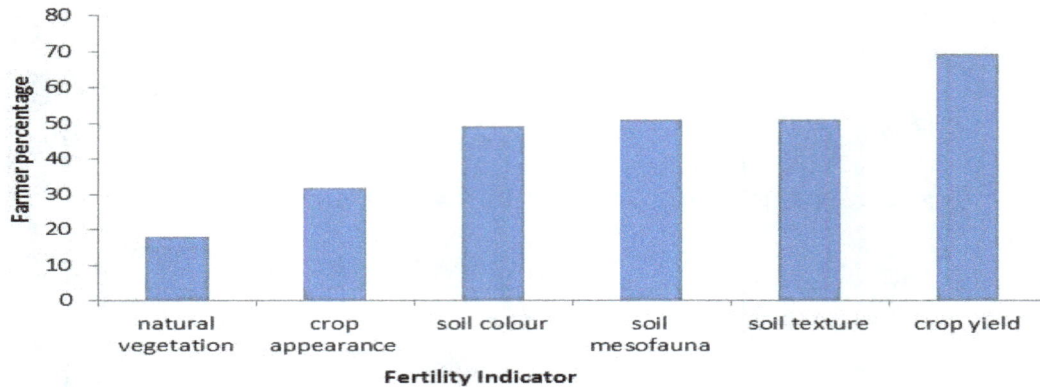

Figure 3. Local indicators identified by farmers for fertility assessment of Ezigeni and Ogagwini soils.

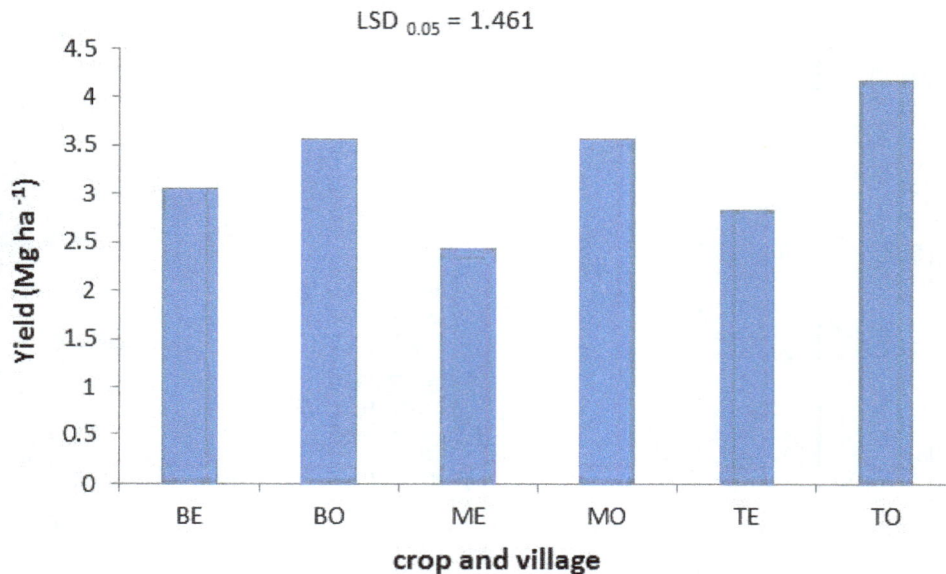

Figure 4. Means for crop yield across locations (p < 0.05). (B: Beans; M: Maize; T: Taro; E: Ezigeni; O: Ogagwini).

major role in the future sustainability of their land.

ACKNOWLEDGEMENTS

Financial support from SANPAD under Project 05/32 entitled: "The social agronomy of compound agriculture: A case study of the dynamics of commercialization", and the National Research Foundation of South Africa are gratefully acknowledge. Thanks goes to Ezemvelo Farmers Organisation members, Victor Bangamwabo, Tad Dorasamy, Charity Maphumulo, Karen Caister, Ncebo Zulu, Phesheya Dlamini, Jothan Buthelezi, Ayanda Mthalane, Simphiwe Ngcobo and Vusi Mbanjwa for their assistance.

REFERENCES

Agrawal A (1995). Dismantling the divide between indigenous and scientific knowledge. Dev. Change. 26: 413-439.

Asao T, Hasegawa K, Sueda Y, Tomita K, Taniguchi K, Hosoki T, Pramanik MHR, Matsu Y (2003). Autotoxicity of root exudates from taro. Sci. Hort. 97:389-396.

Barrera-Bassols N, Zinck JA (2003a). Ethnopedology: A worldwide view on the soil knowledge of local people. Geoderma 111:171-195.

Barrera-Bassols N, Zinck A (2003b). Land moves and behaves': Indigenous discourse on sustainable land management in Pichataro, Patzcuaro basin, Mexico. Swedish. Soc. Anthrophol. Geog. 85:229-245.

Barrera-Bassols N, Zinck JA, Van Ranst E (2006). Symbolism, knowledge and management of soil and land resources in indigenous communities: Ethnopedology at global, regional and local scales. Catena 65:118-137.

Barrera-Bassols N, Zinck JA, Van Ranst E (2009). Participatory soil survey: experience in working with a Mesoamerican indigenous community. Soil Use Manag. 25:43-56.

Barrios E, Delve RJ, Bekunda M, Mowo J, Agunda J, Ramisch J, Trejo MT, Thomas RJ (2006). Indicators of soil quality: A South-South development of a methodological guide for linking local and technical knowledge. Geoderma 135:248-259.

Birmingham DM (2003). Local knowledge of soils: The case of contrast in Cote d'Ivoire. Geoderma 111:481-502.

Braimoh AK (2002). Integrating indigenous knowledge and soil science to develop a national soil classification system for Nigeria. Agric. Human Values 19:75-80.

Briggs J (2005). The use of indigenous knowledge in development: problems and challenges. Progress Dev. Stud. 5:99-114.

Brinkman R, Smyth AJ (1972). Land evaluation for rural purposes. Publication ILRI, Wageningen, The Netherlands. p. 17.

Christanty L, Abdoellah OS, Marten GG, Iskandar J (1986). Traditional Agroforestry in West Java: The Pekarangan (Homegarden) and Kebun-Talun (Annual-Perennial Rotation) Cropping Systems. In Martin, G.G. (ed.) Traditional Agriculture in Southern Asia: A human ecology perspective, Westview Press, Boulder, Colorado. pp. 133-158.

Crevello S (2004). Land use systems and indigenous knowledge. J. Human Ecol. 16:69-73.

Davidson AD (1992). The Evaluation of Land Resources. Longman, Harlow, UK. pp. 57–59.

Ettema CH (1994). Indigenous Soil Classifications. What are their structure and function, and how do they compare with scientific soil classifications? Institute of Ecology, University of Georgia, Athens, GA, USA. [online] URL: http://www.nrel.colostate.edu:8080/simbobon/rkn.3b.SOIL.TEK.04.CHE.htm1#anchor1.

Foth HD, Ellis BG (1997). Soil fertility. Second edition. CRC Press Boca Raton, Florida. p. 290.

Gadgil M, Berkes F, Folke C (1993). Indigenous Knowledge for Biodiversity Conservation. Ambio 22:151-156.

Gagnon C, Berteaux D (2009). Integrating Traditional Ecological Knowledge and Ecological Science: a Question of Scale. Ecol. Soc. 14: 19. [online] URL: http://www.ecologyandsociety.org/vol14/iss2/art19/

Gee GW, Bauder JW (1986). Particle-size analysis. In Klute A (ed.) Methods of Soil Analysis: Part 1. Physical and Mineralogical Methods, ASA, CSSA and SSSA, Madison, WI, USA, pp. 312-383.

Gowing J, Payton R, Tenywa M (2004). Integrating indigenous and scientific knowledge on soils: recent experiences in Uganda and Tanzania and their relevance to participatory land use planning. Uganda J. Agric. Sci. 9:184-191.

Gruver JB, Weil RR (2006). Farmer perceptions of soil quality and their relationship to management-sensitive soil parameters. Renewable Agric. Food Syst. 22:271-281.

Habarurema E, Steiner KG (1997). Soil suitability classification by farmers in southern Rwanda. Geoderma 75:75-87.

Handayani IP, Prawito P (2010). Indigenous soil knowledge for sustainable agriculture. In Lichtfouse E (ed.) Sociology, Organic Farming, Climate Change and Soil Science, Sustainable Agricultural Reviews 3, Springer, London, UK, pp. 303-317.

Hart T, Vorste I (2006). Indigenous knowledge on the South African Landscape: Potentials for agricultural development. HSRC Press, Cape Town, South Africa, pp 18-34.

Hartemink AE, Johnston M, O'Sullivan JN, Poloma S (2000). Nitrogen use efficiency of taro and sweet potato in the humid lowlands of Papua New Guinea. Agric. Ecosyst. Environ. 79:271- 280.

He X, Xu Y, Zhang X (2007). Traditional farming system for soil conservation on slope farmland in south-western China. Soil Till. Res. 94:193-200.

Ingram J (2008). Are farmers in England equipped to meet the knowledge challenge of sustainable soil management? An analysis of farmer and advisor views. J. Environ. Manag. 86:214- 228.

Ingram J, Fry P, Mathieu A (2010). Revealing different understandings of soil held by scientists and farmers in the context of soil protection and management. Land Use Policy 27:51-60.

Kissing L, Pimentel A, Valido M (2009). Participatory soil improvement: A case study in fertility management. Cult. Trop. 30:43-52.

Krogh L, Paarup-Laursen B (1997). Indigenous soil knowledge among

the Fulani of northern Burkina Faso: linking soil science and anthropology in analysis of natural resource management. Geojournal 43:189-197.

Lynch AJJ, Fell DG, McIntyre-Tamwoy S (2010). Incorporating indigenous values with 'Western' conservation values in sustainable biodiversity management. Aust. J. Environ. Manag. 17:244-255.

Ma JF, Ryan PR, Delhaize E (2001). Aluminium tolerance in plants and the complexing role of organic acids. Trends in Plant Sci. 6: 273-278.

Mäder P, Fließbach A, Dubois D, Gunst L, Fried P, Niggli U (2002). Soil fertility and biodiversity in organic farming. Sci. New Series 296:1694-1697.

Mairura FS, Mugendi DN, Mwanje JI, Ramisch JJ, Mbugua PK, Chianu JN (2007). Integrating scientific and farmers' evaluation of soil quality indicators in Kenya. Geoderma 139:139-143.

Materechera SA (2008). Indigenous Knowledge and Approaches of Soil Fertility Management among Small Scale Farmers in Semi-Arid Areas of South Africa. International Meeting on Soil Fertility Land Management and Agroclimatology, Kusadasi, Turkey, pp. 627-646.

McGregor D (2004). Coming Full Circle: Indigenous Knowledge, Environment, and Our Future. Am. Ind. Q. 28:385-410.

Mutshinyalo TT, Siebert SJ (2010). Myth as a biodiversity conservation strategy for the Vhavenda, South Africa. Indilinga 9:151-171.

Nadasdy P (1999). The politics of TEK: Power and the "integration" of knowledge. Arct. Anthrophol. 36:1-18.

Neef A, Heidhues F, Stahr K (2006). Participatory and Integrated Research in Mountainous Regions of Thailand and Vietnam: Approaches and Lessons Learned. J. Mt. Sci. 3:305-324.

Nethononda LO, Odhiambo JJO (2011). Indigenous soil knowledge relevant to crop production of smallholder farmers at Rambuda irrigation scheme, Vhembe District, South Africa. Afr. J. Agric. Res. 6:2576-2581.

Newton J, Paci J, Ogden A (2005). Climate change and natural hazards in northern Canada: integrating indigenous perspectives with government policy. Mitigat. Adaptat. Strat. Glob. Change. 10:541-571.

Norton JB, Pawluk RR, Sandor JA (1998). Observation and experience linking science and indigenous knowledge at Zuni, New Mexico. J. Arid Environ. 39:331-340.

Ocholla D (2007). Marginalized Knowledge: An Agenda for Indigenous Knowledge Development and Integration with Other Forms of Knowledge. Int. Rev. Info. Ethics 7:1-9.

Pestalozzi H (2000). Sectoral fallow systems and the management of soil fertility: The rationality of indigenous knowledge in the High Andes of Bolivia. Mt. Res. Dev. 20: 64-71.

Phillips-Howard K, Oche C (2006). Indigenous knowledge and fertilizer uses in Transkei. In Normann H, Snaymann I, Cohen M (Eds), Indigenous knowledge and its uses in southern Africa, HSRC Press, Cape Town, South Africa, pp. 137-149.

Raji BA, Malgwi WB, Berding FR, Chude VO (2011). Integrating indigenous knowledge and soil science approaches to detailed soil survey in Kaduna State, Nigeria. J. Soil Sci. Environ. Manag. 2:66-73.

Riekert S, Bainbridge S (1998). Analytical methods of the Cedara Plant and Soil Laboratory. KwaZulu-Natal Department of Agricultural and Environmental Affairs, Pietermaritzburg.

Sandor J, Furbee L (1996). Indigenous knowledge and classification of soils in the Andes of Southern Peru. Soil Sci. Soc. Am. J. 60:1502–1512.

Schnürer J, Rosswall T (1982). Fluorescein diacetate hydrolysis as a measure of total microbial activity in soil and litter. Appl. Environ. Microbiol. 43:1256-1261.

Sileshi GW, Nyeko P, Nkunika POY, Sekematte BM, Akinnifesi FK, Ajayi OC (2009). Integrating Ethno-Ecological and Scientific Knowledge of Termites for Sustainable Termite Management and Human Welfare in Africa. Ecol. Soc. 14:48 [online] URL:http://www.ecologyandsociety.org/vol14/iss1/art48/.

Sillitoe P (1998). Knowing the land: soil and land resource evaluation and indigenous knowledge. Soil Use. Manag. 14:188-193.

Sillitoe P, Mazzano M (2009). Future of indigenous knowledge research in development. Futures 41: 13-23.

Smith CW (1995). Assessing the compaction susceptibility of South African forestry soils. PhD dissertation, University of Natal, Pietermaritzburg, South Africa.

Smith CW, Johnston MA, Lorentz SA (2001). The effect of soil compaction on water retention characteristics of soils in forest plantations. S. Afr. J. Plant Soil 18:87-97.

Soil Classification Working Group (1991). Soil Classification: a taxonomic system for South Africa. Department of Agricultural Development, Pretoria, South Africa.

Talawar S, Rhoades RE (1998). Scientific and local classification and management of soils. Agric. Human Values 15:3-14.

Tesfahunegn GB, Tamene L, Vlek PLG (2011). Evaluation of soil quality identified by local farmers in Mai-Negus catchment, northern Ethiopia. Geoderma 163:209-218.

Walkley A (1947). A critical examination of a rapid method for determining organic carbon in soils: effects of variations in digestion conditions and organic soil constituents. Soil Sci. 63:251-263.

7

In vitro evaluation of antitrypanosomal and cytotoxic activities of soil actinobacteria isolated from Malaysian forest

H. Lili-Sahira[1]*, K. Getha[1], A. Mohd Ilham[2], I. Norhayati I[1], M. M. Siti-Syarifah[1], A. Muhd Syamil[1], J. Muhd Haffiz[1] and G. Hema-Thopla[1]

[1]Natural Product Division, Forest Research Institute Malaysia, 52109, Kepong, Selangor, Malaysia.
[2]Malaysian Institute of Pharmaceutical and Nutraceutical, Bukit Gambir 11700, Penang, Malaysia.

A total of 83 actinobacteria isolates were successfully revived from cryovials stored in -80°C FRIM actinobacteria culture collection (FACC). Based on macromorphological observation, *Streptomyces*-like group is predominant among all the isolates. The representative actinobacteria isolates were grown in three different fermentation media (M1, M2 and M3) and a total of 249 culture broth extracts obtained were evaluated for *in vitro* antitrypanosomal activity against *Trypanosoma brucei brucei* strain BS221. Five extracts (2.0%) exhibited strong activity (score 3) with an IC_{50} value ≤ 1.56 µg ml^{-1} and 34 extracts (13.7%) exhibited moderate activity (score 2) of 1.56 µg ml^{-1} < IC_{50} ≤ 12.5 µg ml^{-1}. Extracts showing score 3 and score 2 activities were further tested for cytotoxicity. Eight extracts exhibited good selectivity with a SI (Selectivity Index) value ≥ 20. Among the isolates showing good selectivity, isolate FACC-A032 that belonged to the genus *Streptomyces* produced extract with the highest antitrypanosomal activity (IC_{50} = 0.23 µg ml^{-1}) and high selectivity (SI = 76.417). Growth profile study of isolates FACC-A032 in medium M3 exhibited maximum antitrypanosomal activity at day eight of fermentation with IC_{50} = 0.15 µg ml^{-1} and SI = 154.75. This is the first study of *in vitro* antitrypanosomal activity of soil actinobacteria isolated from Malaysian forest.

Key words: Antitrypanosomal, trypanosomiasis, actinobacteria, Alamar blue assay, *Streptomyces*.

INTRODUCTION

African trypanosomiasis causes sleeping sickness in humans and nagana in cattle. This disease is prominent on the World Health Organization (WHO) list of neglected tropical diseases and a major problem to the poorer countries in the world, especially throughout sub-Saharan Africa (WHO, 2010). Chemotherapy, jointly with vector control, remains one of the most important elements in the control of trypanosomatid disease, as there are currently no vaccines to prevent the trypanosome infection. In Malaysia, trypanosomiasis or surra has been reported in institutional farms of cattle (Cheah et al., 1999),

rhinocerous centre (Vellayan et al., 2004) and deer breeding centre (Nurulaini et al., 2007; Adrian et al., 2010). The disease was caused by *Trypanosoma evansi* infection. However, there are several cases reported that African trypanosomiasis has appeared in human caused by non-human pathogenic trypanosome species. These species are *T. brucei brucei*, *Trypanosoma congolense* and *Trypanosoma evansi* (Deborggraeve et al., 2008). Most pharmaceutical industry has declined their investment in drug development for trypanosomiasis because this disease affects populations who do not represent a profitable market. Thus, only few drugs are currently registered to treat this disease, however, the drugs is limited due to age, toxicity, difficulty to administer, cost and all current treatment suffer from significant drawbacks (Abdel Sattar et al., 2009). Hence,

*Corresponding author. E-mail: lili@frim.gov.my.

new strategies to treat sleeping sickness are urgently required.

Many of our best known and most valuable antibiotics are produced by actinobacteria. It is a prolific producer of structurally diverse bioactive metabolites, and has yielded some of the most important products of the drug industry, including penicillins, aminoglycosides, tetracyclines, cephalosporins and other classess of antibiotics. They have provided over two-thirds of the naturally occurring antibiotics discovered and continue to be a major source of novel and useful compounds (Berdy, 2005). Moreover, Woodruff (1966) proposed that all actinobacteria will produce antibiotics if provided the proper growth conditions.

The actinobacteria are a group of morphologically diverse and Gram-positive bacteria, which comprise a group of branching unicellular microorganisms. Actinobacteria can be isolated from soil and marine sediments. Soil, in particular is an intensively exploited ecological niche for isolation of actinobacteria that produce many useful biologically active natural products. Among actinobacteria, the genus *Streptomyces* are dominant (Balows et al., 1992). A large number of the commercially and medicinally useful antibiotics have been derived from this genus (Thakur et al., 2007). Several studies reported that actinobacteria have a promising antitrypanosomal activity *in vitro*. (Otoguro et al., 2008; Pimental-Elardo et al., 2010; Zin et al., 2011). Therefore, in the present study 83 isolates of action-bacteria from FACC which was isolated from Malaysian forest soil were studied for potential activity against the trypanosome parasite *T. brucei brucei* strain BS221.

MATERIALS AND METHODS

Soil sample collection, isolation and characterization of actinobacteria

Soil samples were collected from around roots of medicinal and forest plant species at Penang National Park, Malaysia. Soil samples were air-dried at room temperature for 5 to 7 days and were treated using chemical and physical pretreatment methods. Isolation of actinobacteria isolates were done according to the method described by Getha et al. (2004). Actinobacteria colonies were selected from isolates plates and maintained on yeast extract-malt extract (ISP2) agar. Pure cultures were stored as spore suspension in cryovials at -80°C within a FRIM Actinomycetes Culture Collection (FACC). An aliquot of 10 μl culture suspension of 83 actinobacteria isolates from cryovials was transferred to ISP2. Plates were incubated at 28 ± 2°C for seven to ten days until good growth was observed. After the incubation period, all isolates were assigned to the *Streptomyces*-like or non-*Streptomyces* groups based on colony macromorphological characteristics according to methods of Getha et al. (2004)

Fermentation and extract preparation

Three types of production media, M1, M2 and M3 (Getha and Vikineswary, 2002) were prepared and dispensed into many 125-ml Erlenmeyer flasks. Each flask containing 20 ml of medium was plugged with non-absorbent cotton wool and autoclaved at 121°C for 15 min. By using a sterile cork borer, a 5 mm diameter agar-plug was cut aseptically from rich aerial growth of seven to 10-day-old cultures and inoculated into duplicate flasks of each medium. Control flask contained only production medium. The flasks were then incubated at 28 ± 2°C in orbital shaker and shaken at 200 rpm. After seven days of incubation, the whole culture broth was extracted using 1:1 (v/v) butanol (BuOH) according to methods described by Getha et al. (2009). The crude extract was concentrated in a rotary evaporator and about 2 mg of the extracts were dispensed into 96-well microtiter plate and stored at -20°C before use.

Trypanosome parasite and culture medium

T. brucei brucei (*T. b. brucei*) BS221 strain was obtained from the Swiss Tropical Institute, Basel (Jean-Robert et al., 2009). The strain was cultured in minimum essential medium (MEM) with Earle's salts (powder, GIBCO), supplemented with 25 mM HEPES, 1 g L^{-1} additional glucose, 2.2 g l^{-1} NaHCO$_3$, and 10 ml l^{-1} MEM non-essential amino acids (100x). The medium was further supplemented with Balz supplement (Raz et al., 1997), 0.2 mM 2-mercaptoethanol and 15% heat inactivated fetal bovine serum. Fresh supplemented MEM (900 μl) was added into three to four wells of 24-well tissue culture plate and 1:10 dilution was prepared by adding 100 μl of the log phase trypanosome culture to the first well of 24-well plate and was mix well. 100 μl of trypanosome culture from the first well was removed to the second well containing another 900 μl supplemented MEM and the process was repeated to third and fourth well. The log phase of trypanosome culture was visually selected using an inverted microscope. The routine on sub-cultured was done once in a two days and the cultures were incubated at 37°C and 5% CO$_2$.

In vitro antitrypanosomal assay

The primary, secondary and tertiary *in vitro* antitrypanosomal screening methods were conducted in this study (Hitomi and Kazuhiko, 2005). In the primary assay, two concentrations of actinobacteria extracts (1.56 and 12.5 μg ml^{-1}l) were used for the testing. Extracts at each concentration dilution were tested in duplicate plates. Extracts that showed positive for antitrypanosomal activity in both or either one of concentrations well were selected for the secondary and tertiary assays, which consisted of seven final extract concentrations (12.5, 1.56, 0.78, 0.39, 0.19, 0.10 and 0.05 μg ml^{-1}). Extracts were tested in duplicate plates for secondary assay and triplicate plates for tertiary assay. All assays were performed in a flat bottom 96-well microtiter plate. *In vitro* antitrypanosomal activity of test sample was estimated by a dose response curve using Alamar Blue® sensitivity assay according to the method of Raz et al. (1997). Standard trypanocidal drugs, pentamidine isethionate (SIGMA) was dissolved in 5% dimethylsulfoxide (DMSO) and included in the assay as positive control. Negative control-solvent (5% DMSO and 25% ethanol) as well as negative control-blank (Sterile Milli-Q water) were also included in the experiment. Five micro liter of the pre-dilution of the extracts, standard drug and negative control were added to each well of a 96-well microtiter plate. Then, 95 μl of the trypanosomes suspension at a density of 20000 to 25000 trypanosomes ml^{-1} was inoculated to each well except control-blank well. After incubation for 72 h at 37°C under 5% carbon dioxide, 10 μl of fluorescent dye Alamar blue was added to each well and incubation was further continued for 3 to 6 h until colour change is observed. All tests were performed independently two to three times. Plates were analysed

Figure 1a. Culture of *Streptomyces*-like group growing on ISP2 agar.

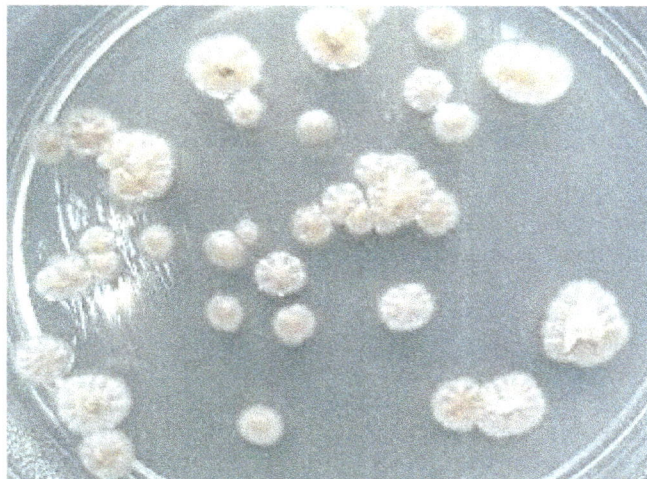

Figure 1b. Culture of non-*Streptomyces* group growing on ISP2 agar.

using Tecan Infinite M200 plate reader at an excitation wavelength of 530 nm and an emission wavelength of 590 nm. Absorbance data were transferred into a graphic program (Excel) and were evaluated to determine the IC_{50} values (Hitomi and Kazuhiko, 2005).

Cytotoxicity and selectivity index

In vitro cytotoxicity assay was carried out using normal kidney (Vero) cells according to the procedure described by Kohana and Otoguro (1999) and Malebo et al. (2009). Cells were grown in standard media according to the method reported by Siti Syarifah et al. (2011). Serial dilution of extracts was prepared to produce six final concentrations 0.1, 0.39, 1.56, 6.31, 25.0 and 100 µg ml^{-1} and 10 µl of pre-dilution of extracts, standard drugs, and negative control was added to each well of a 96-well microtiter plate containing of 40000 cells ml^{-1}. Standard drug pentamidine was used

as positive control, ethanol and DMSO as negative control-solvent and sterile Milli-Q water as negative control-blank.

The plates were incubated at 37°C under 5% carbon dioxide for 72 h and assayed using the alamar blue assay as described previously with shortened incubation time. After incubation, plates were read using an excitation wavelength of 530 nm and an emission wavelength of 590 nm using the fluorescent plate reader (Tecan Infinite M200) and Magellan software. IC_{50} was defined as the concentration of actinobacteria extract required to reduce a 50% of cell growth compared to control cultures. Based on the cytotoxicity results, calculation of selectivity index (SI) value was performed to select extracts that were very selective to trypanosome parasites and had low toxicity effects on normal cells by using the formula:

$$\text{Selectivity index (SI)} = \frac{IC_{50} \text{ value (cytotoxicity)}}{IC_{50} \text{ value (antitrypanosomal activity)}}$$

Growth profile study of potential actinobacteria isolate

The selected actinobacteria isolate was cultivated in medium M3 for ten days to study the maximum antitrypanosomal activity of active isolate during the fermentation period. Triplicate flasks for each isolate were harvested and analysed each day to observe the culture pH, wet weight of biomass and antitrypanosomal activity. The pH of the fermentation broth was measured with a stainless steel pH meter (IQ 150). Biomass production was expressed as wet weight of cells in 10 ml aliquots of well-mixed whole fermentation broth. The broth was centrifuged at 10,000 rpm for 20 min and the supernatant was discarded with the last drop was blot on a tissue paper and the pellet was then weighed. The remaining 10 ml aliquots of the fermentation broth at each day of fermentation started from day 0 were extracted using 1:1 (v/v) butanol. Crude extract was re-dissolved in ethanol and serially diluted of actinobacteria extracts from day 0 to day 10 of fermentation was then tested for second screening antitrypanosomal activity. Data on pH, wet weighed biomass and antitrypanosomal activity was recorded and plot into growth profile graph.

RESULTS

Macromorphology, fermentation and extraction of actinobacteria isolates

From the total of 83 isolates studied, 45 isolates (54%) belong to the group *Streptomyces*-like and 39 isolates (46%) to the group non-*Streptomyces*. The genus *Streptomyces* is common in actinobacteria population (Masayuki, 2008) and a large number of bioactive compounds are derived from *Streptomyces* species (Berdy, 2005). Isolates falling under *Streptomyces*-like group (Figure 1a) that are fast growers on ISP2 agar were observed after seven days of incubation. Whereas isolates under the non-*Streptomyces*, group (Figure 1b) grew slowly on ISP2 agar and can only be detected after 10 to 14 days incubation. Cultivation of these isolates in different production media was used to optimize expression of bioactive metabolites to produce high-quality extracts (Getha et al., 2009). This is also to increase the chance of isolating the active metabolites

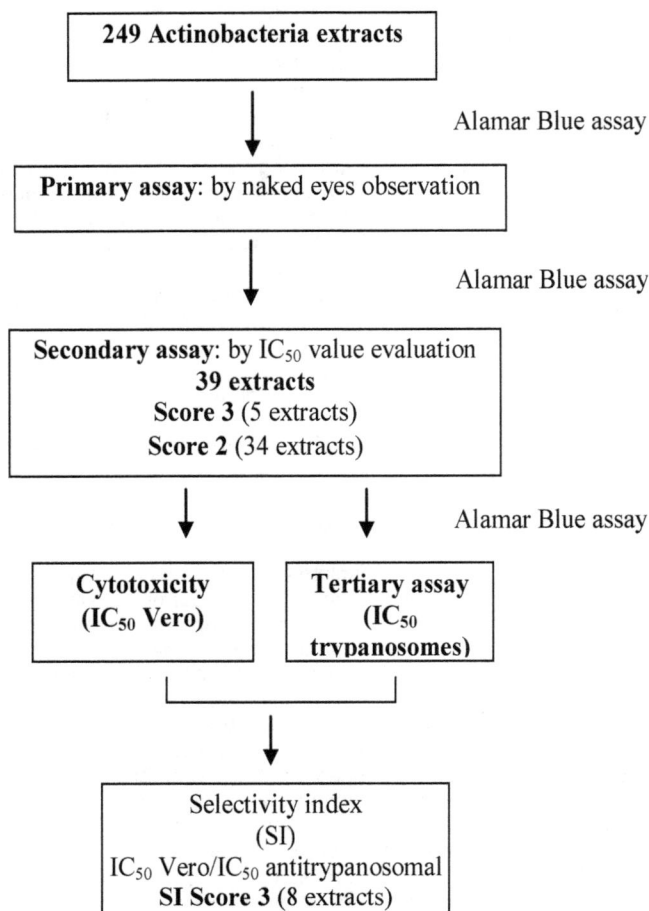

Figure 2. Flow chart showing screening strategy of actinobacteria extracts against trypanosomes and vero to select extracts with potential antitrypanosomal activity.

from actinobacteria. Butanol solvent was used in this study to extract the bioactive compounds from culture broth due to the solvent is most suitable for extraction according to Forar et al. (2007). An easy storage method of large number of extracts in 96-well microtiter plates at -20°C was used in this study before the extracts were tested in bioassay.

Antitrypanosomal activity, cytotoxicity and selectivity index

Antitrypanosomal activity was separated into three categories; $IC_{50} > 12.5$ µg ml^{-1}: score 1 (low activity); 1.56 µg ml$^{-1} < IC_{50} \leq 12.5$ µg ml^{-1}: score 2 (moderate activity) and $IC_{50} \leq 1.56$ µg ml^{-1}: score 3 (strong activity). During the primary assay, only two concentrations of extracts were used and the selections of potential extracts were evaluated using naked eye observation of the colour changes in Resazurin. This was a rapid selection method

for potential extracts. Rezasurin as a cell proliferation indicator dye (blue colour), will be reduced to a pink fluorescent dye in the medium by cell activity (O'Brien et al., 2000). If there is no cell activity, the blue colour will remain. Therefore when the blue colour appeared in both wells of an extract in the 96-well test plate, this indicates that the extract has a strong positive activity while moderate activity was observed if one of the well shows blue colour. Low activity is observed when both wells show pink colour. The primary and secondary assays showed that 5 extracts (2.0%) demonstrated strong antitrypanosomal activity, 34 extracts (13.7%) exhibited moderate activity and the other extracts showed low or no activity. Extracts that showed strong and moderate activity were selected for the tertiary and cytotoxicity assays to get the selectivity index value (Figure 2).An important criterion in the search of potential compounds against T. b. brucei is that they have no or very low toxic effects on mammalian host cell. For this purpose, out of 39 extracts selected after primary and secondary antitrypanosomal assays, the tertiary and cytotoxicity assays showed eight extracts (20.5%) with SI score 3 (SI ≥ 20), seven extracts (17.9%) exhibited SI score 2 (10 < SI < 20) and 24 extracts (61.5%) with SI score 1 (SI $\leq$ 10). The eight extracts (Table 1) that showed SI ≥ 20 were considered to have good selectivity for the parasite and will be studied for further bioassay guided isolation of the active antitrypanosomal compound/s. Among the eight isolates that produced extracts with good selectivity, one isolate, FACC-A032 exhibited the highest antitrypanosomal activity ($IC_{50} = 0.23$ µg ml^{-1}) and high selectivity (SI = 76.417) towards the trypanosomae parasite. This isolate was considered for characterization. The morphological, physiological and biochemical characterization indicated that isolate FACC-A032 was assigned to the genus *Streptomyces* (Lili Sahira et al., 2011).

The performance of fermentation media to produced bioactive metabolite in actinobacteria can be strongly influenced by the type of carbon and nitrogen sources (El-Refai et al., 2011). Among the three types of fermentation media used in this study, media M3 that contained the combination of cornstarch and corn steep solids as the carbon and nitrogen sources was found as the most suitable fermentation media for bioactive metabolite production in actinobacteria.

Growth profile of FACC-A032 in medium M3

The growth profile of isolates FACC-A032 in medium M3 is shown in Figure 3. The cell growth reached its maximum (217.59 wet weight/l) on day 3 and the mycelial wet weight of the actinobacteria slightly drop and remain almost constant after four to seven days. As reported by Junker et al. (2009), secondary metabolite fermentation are related with a period of rapid growth and followed by

Table 1. Antitrypanosomal activities of eight actinobacteria extract showing strong activity (Score 3) and moderate activity (Score 2) with high selectivity (SI $\geq$ 20).

| Extract code | Estimated group | IC$_{50}$ value | | SI (Vero / BS221) |
		Antitrypanosomal (BS221)	Cytotoxicity (Vero)	
A032-M3	STM	0.23 ± 0.014	17.34 ± 0.192	76.417
A026-M3	STM	0.86 ± 0.033	21.02 ± 0.955	24.478
A048-M3	n-STM	1.35 ± 0.046	73.84 ± 0.715	54.616
A049-M3	n-STM	1.40 ± 0.021	> 100	> 71.362
A085-M3	n-STM	4.55 ± 0.072	> 100	> 21.986
A016-M3	STM	3.75 ± 0.022	77.87 ± 0.137	20.7917
A039-M2	n-STM	4.17 ± 0.032	> 100	> 23.965
A052-M3	n-STM	4.69 ± 0.050	> 100	> 21.345
Positive control (Pentamidine)		0.00446 ± 0.00006	0.01883 ± 0.00075	4.257

BS221, *Trypanosoma brucei brucei*; Vero, normal kidney cell from African green monkey; STM, *Streptomyces*; n-STM, non-*Streptomyces*; n = 3. All data were recorded as ± standard deviation (SD).

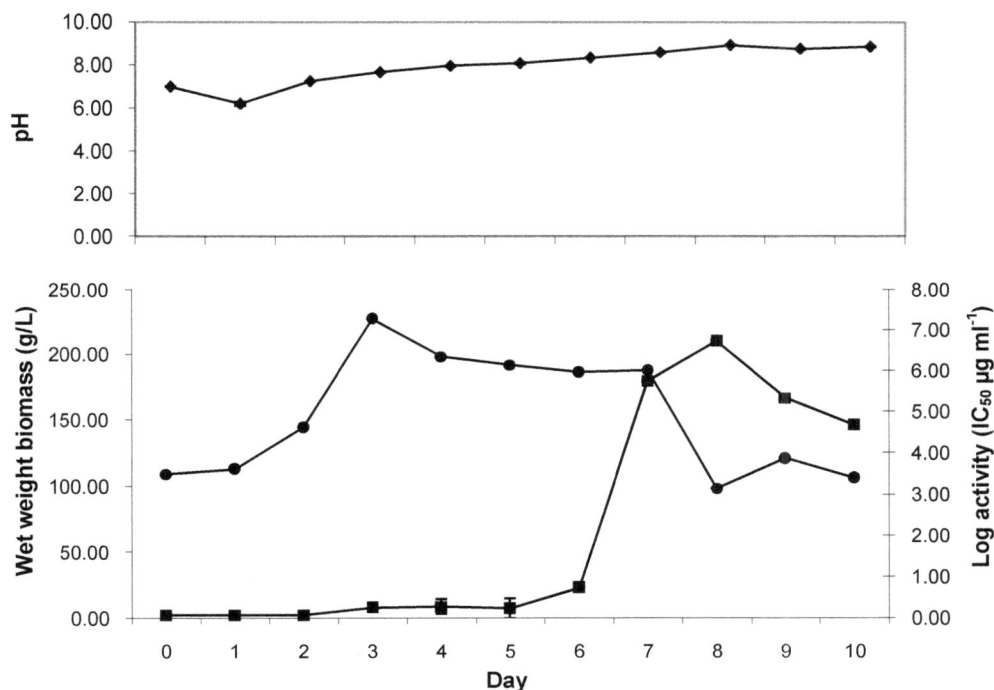

Figure 3. The growth profile of isolates FACC-A032 in medium M3 and their antitrypanosomal activity for ten days. The culture broth was analyzed for the wet weight biomass (•), antitrypanosomal activity (■), and pH (♦). The means ± standard deviations (SD) are display for all data points. Due to low SD, most error bars are not visible.

production phase of secondary metabolite, where production start at the same time with growth rate decline. Based on the previous reported, it was observed that after cell growth decrease on day seven; the antitrypanosomal activity becomes more strongly active. Antitrypanosomal activity was begins on day four of incubation and reach the maximum on day eight. The broth pH profile, which was initially neutral, became alkaline during the growth phase until the end day of fermentation. Results suggest that the maximal antitrypanosomal activity can be found after an optimization of the growth. On the other hand, results from growth profile study will be used as a basis for selection of harvest day (fermentation duration) during large scale fermentation for bioactive compounds purification.

DISCUSSION

It is known that secondary metabolites produced by actinobacteria possess a wide range of biological activities, and the majority of these compounds are derived from the genus *Streptomyces* (Solanki et al., 2008). Several compounds isolated from actinobacteria have been shown to exhibit antitrypanosomal activity *in vitro*. For example, compounds aureothin, cellocidin, destomycin A, echinomycin, hedamycin, irumamycin, LL-Z 1272β, O-methylnanaomycin A, venturicidin A and virustomycin A were isolated from soil microorganisms and shown to display potent antitrypanosomal activity (Otoguro et al., 2008). Trypanosome parasites GUTat 3.1 (*T.b.brucei*) was used for *in vitro* antitrypanosomal activity. Out of the ten compounds tested, virustomycin A and aureothin showed the highest antitrypanosomal activity, with an IC_{50} value around 0.001 μg ml^{-1}, however aureothin showed the highest SI value > 17,857. Ishiyama et al. (2008) discovered two compounds from soil microorganism KS-505a and alazopeptin, which exhibited antitrypanosomal activity with IC_{50} values of 1.03 and 0.51 μg ml^{-1} respectively, against *T. b. brucei* strain GUTat 3.1. KS-505a show high SI value >27.33 compared to alazopeptin with SI > 9.10. Furthermore, novel antitrypanosomal alkaloid spoxazomicin A-C were isolated from an endophytic actinobacteria, *Streptosporangium oxazolinicum* K07-0460^T as reported by Inahashi et al. (2011) and compounds isolated from *Streptomyces* sp. strains from Mediteranean sponges (valinomycin and staurosporine) show novel antitrypanosomal activity (Pimental-Elardo et al., 2010). Interestingly, isolates FACC-A032 at crude extract level resulted in much lower IC_{50} values of 0.15 μg ml^{-1} after growth profile study was conducted as compared to the KS-505a and alazopeptin. It also showed highest SI with values of 154.75. As reported by Badisa et al. (2009) the higher the SI value, the more selective to trypanosome parasite the extracts are. This paper only discusses antitrypanosomal activity of the crude extracts from actinobacteria, therefore, bioactivity guided isolation can be embarked upon to discover bioactive compound/s from *Streptomyces* sp. FACC-A032 and the other potential isolates from this study. Based on previous literature search, this is the first study to show potential antitrypanosomal activity from a Malaysian soil actinobacteria strain.

ACKNOWLEDGEMENTS

Authors would like to thank MOSTI for the grant (09-05-IFN-BPH-003) used in this study. We would also like to thank IPharm MOSTI, DNDi, Kitasato Institute, Japan and Swiss Tropical Institute, Switzerland for their valuable advice and assistance and MY Nur Fairuz, M. Faizulzaki and R. Ruzana for their technical assistance.

REFERENCES

Abdel Sattar E, Shehab NG, Ichino C, Kiyohara H, Ishiyama A, Otoguro K, Omura S, Yamada H (2009). Antitrypanosomal activity of some pregnane glycosides isolated from *Caralluma* species. Phytomedicine 16:659-664.

Adrian MS, Rehana AS, Latiffah H, Wong MT (2010). Outbreaks of trypanosomiasis and the seroprevalence of *T. evansi* in a deer breeding centre in Perak, Malaysia. Trop. Anim. Health. Prod. 42:145-150

Badisa RB, Darling-Reed, SF, Joseph P, Cooperwood JS, Latinwo LM, Goodman CB (2009). Selective cytotoxic activities of two novel synthetic drugs on human breast carcinoma MCF-7 cells. Anticancer Res. 29:2993-2996.)

Balows A, Trupper HG, Dworkin M, Harder W, Schleifer KH (1992). The prokaryotes: a handbook on the biology of bacteria: ecophysiology, isolation, identification, applications. Springer-verlag 1:811-815.

Berdy J (2005). Bioactive microbial metabolites: Review article. J. Antibiot. 58:1-26.

Cheah TS, Sani RA, Chandrawathani P, Sansul B, Dahlan I (1999). Epidemiologi of Trypanosoma evansi infection in crossbred dairy cattle in Malaysia. Trop. Anim. Health. Prod. 31:25-31.

Deborggraeve S, Koffi M, Jamonneau V, Bonsu FA, Queyson, Richard, Simarro PP, Herdewijn P, Buscher P (2008). Molecular analysis of archived blood slides reveals an atypical human *Trypanosoma*. Diagnostic Microbiol. Infect. Dis. 61:428-433.

El-Refai HA, AbdElRahman HY, Abdulla H, Hanna AG, Hashem AH, El-Refai AH, Ahmed EM (2011). Studies on the production of actinomycin by *Nocardioides luteus*, a novel source. Curr. Trends Biotechnol. Pharm. 3:1282-1297.

Forar LR, Ali E, Mahmoud S, Bengra C, Hacene H (2007). Screening, isolation and characterization of a novel antimicrobil producing actinomycete, strain RAF10. Biotechnology 6:489-496.

Getha K, Hatsu M, Wong HJ, Lee SS (2009). Submerge cultivation of basidiomycete fungi associated with root diseases for production of valuable bioactive metabolites. J. Trop. Forest Sci. 21:1-7.

Getha K., Vikineswary S (2002). Antagonistic effects of *Streptomyces violaceusniger* strain G10 on *Fusarium oxysporum* f.sp. *Cubense* race 4: Indirect evidence for the role of antibiosis in the antagonistic process. J. Ind. Microbiol. Biotechnol. 28:303-310.

Getha K, Vikineswary S, Wong WH, Seki T, Ward A, Goodfellow M (2004). Characterization of selected isolates of indigenous *Streptomyces* species and evaluation of their antifungal activity against selected plant pathogenic fungi. Malaysian J. Sci. 23:37-47.

Hitomi S, Kazuhiko O (2005). *Trypanosoma brucei brucei* In vitro Screening Model. Standard Operating Procedure. p. 11.

Inahashi Y, Iwatsuki M, Ishiyama A, Namatame M, Nishihara-Tsukashima A, Matsumoto A, Hirose T, Sunazuka T, Yamada H, Otoguro K, Takahashi Y, Omura S, Shiomi K (2011). Spoxazomicins A-C, novel antitrypanosomal alkaloids produced by an endophytic actinomycete, Streptosporangium oxazolinicum K07-0460T. J. Antibiot. 64:303-307.

Ishiyama A, Otoguro K, Namatame M, Nishihara A, Furusawa T, Masuma R, Shiomi K, Takahashi Y, Ichimura M, Yamada H, Omura S (2008). *In vitro* and *in vivo* antitrypanosomal activity of two microbial metabolites, KS-505a and Alazopeptin. J. Antibiotics 61:627-632.

Jean-robert I, Reto B, Tanja W, Marcel K, Vanessa Y (2009). Drug screening for kinetoplastids diseases. A training manual for screening in Neglected Diseases. pp. 74.

Junker B, Walker A, Hesse M, Lester M, Christensen J, Connors N (2009). Actinomycetes scale-up for the production of the antibacterial, nocathiacin. Biotechnol. Prog. 25:176-188.

Kohana A, Otoguro K (1999). MRC-5 cells Cytotoxicity evaluation. Standard Operating Procedure. p. 6.

Lili Sahira H, Getha K, Mohd Ilham A, Hema Thopla G, Norhayati I, Muhd Syamil A, Muhd Haffiz J, Siti Syarifah MM, Nur Fairuz MY (2011). Antitrypanosomal activity and characterization of potential actinobacteria isolate. In Raha *et al.* (Eds.) *Leveraging on MicrobialDiversity for a Sustainable Future Proceedings of International Congress of Malaysian Society for Microbiology, 8-11 December 2011, Penang. pp. 413-416

Malebo HM, Tanja W, Cal M, Swaleh SAM, Omolo MO, Hassanali A, Sequin U, Hamburger M, Brun R, Indiege IO (2009). Antiplasmodial, anti-trypanosomal, anti-leishmanial and cytotoxicity activity of selected Tanzanian medicinal plants. Tanzanian J. Health Res. 11:226-234.

Masayuki H (2008). Studies on the isolation and distribution of rare actinomycetes in soil. Actinomycetologica 22:12-19.

Nurulaini R, Jamnah O, Adnan M, Zaini CM, Khadijah S, Rafiah A, Chandrawathani P (2007). Mortality of domesticated java deer attributed to Surra. Trop. Biomed. 24:67-70.

O'Brien J, Wilson I, Orton T, Pognan F (2000). Investigation of the Alamar Blue (Resazurin) fluorescent dye for the assessment of mammalian cell cytotoxicity. Eur. J. Biochem. 267:5421-5426.

Otoguro K, Ishiyama A, Namatame M, Nishihara A, Furusawa T, Masuma R, Shiomi K, Takahashi Y, Yamada H, Omura S (2008). Selective and potent *in vitro* antitrypanosomal activities of ten microbial metabolites. J. Antibiotics 61:372-378.

Pimental-elardo SM, Kozytska S, Bugni TS, Ireland CM, Moll H, Hentschel U (2010). Anti-parasitic compounds from *Streptomyces* sp. strains isolated from Mediterranean sponges. Marine drugs 8:373-380.

Raz B, Iten M, Brun YGR (1997). The Alamar Blue® assay to determine drug sensitivity of African trypanosomes (*T.b. rhodesiense* and *T.b. gambiense*) *in vitro*. Acta Tropica. 68:139-147.

Siti Syarifah MM, Nurhanan MY, Muhd Haffiz J, Mohd Ilham A, Getha K, Asiah O, Norhayati I, Lili Sahira H, Anee Suryani S (2011).Potential anticancer compound from *Cerbera Odollam*. J. Trop. Forest. Sci. 23:89-96.

Solanki R, Khanna M, Lal R (2008). Bioactive compounds from marine actinomycetes. Indian J. Microbiol. 48:410-431.

Thakur D, Yadav A, Gogoi B, Bora TC (2007). Isolation and scereening of streptomyces in soil of protected forest areas from the states of assam and tripura, India, for antimicrobial metabolites. J. Med. Mycol. 17:242-249.

Vellayan S, Aidi M, Robin WR, Linda JL, Epstein J, Simon AR, Donald EP, Rolfe MR, Terri LR, Thomas JF, Khan M, Vijaya J, Reza S, Abraham M (2004). Trypanosomiasis (Surra) in the captive Sumatran Rhinoceros (*Dicerorhinus sumatrensis sumatrensis*) in Peninsular Malaysia. In Iskandar *et al.* (Eds.) *Animal health: a* breakpoint in economic development? Proceedings of the 16th veterinary Association Malaysia Congress and the 11th International Conference of the Association of Institutions for tropical veterinary medicine, 23-27 August 2004, Selangor. pp. 187-189

WHO (2010). African trypanosomiasis or sleeping sickness. Fact sheet, p. 259.

Woodruff HB (1966). The physiology of antibiotic production: the role of the producing organism. Symp.Soc.Gen.Microbiol. 16:22-46.

Zin NM, Ng KT, Sarmin NM, Getha K, Tan GY (2011). Anti-trypanosomal activity of endophytic Streptomycte. Current. Res. Bacteriol. 4:1-8.

8

Spatial variability in soil properties of a continuously cultivated land

Denton Oluwabunmi Aderonke[1]* and Ganiyu Adeniyi Gbadegesin[2]

[1]Institute of Agricultural Research and Training, Obafemi Awolowo University, Moor plantation Ibadan, Nigeria.
[2]Department of Geography, University of Ibadan, Nigeria.

Poor knowledge of soil suitability for agricultural production constitutes a major problem to land users. For proper assessment of the distribution soil properties, the use of geographic information system (GIS) has been considered a very effective tool to achieve this. A study on an Alfisol at the Institute of Agricultural Research and Training, Ibadan (7°23'N, 3° 51'E), Nigeria, was carried out to measure the spatial variation of soil properties of a continuously cultivated land under rain-fed and irrigation systems. The study involved a systematic grid mapping of about 3 ha of an experimental plot subjected to maize cultivation for more than 15 years. This study area was divided into 20 m by 20 m grids, and samples were collected at each grid point for laboratory analysis while the coordinates of each sampling point was taken for interpolation in ArcGIS. The results obtained from the analysis of various elements were imported into GIS environment and then presented in form of digital maps that shows the spatial distribution of the soil properties, which can be used for precision agriculture. The results obtained showed that the area had medium acid in majority of the area covered with a pH range of 5.5 to 5.9. The %N was majorly low at < 0.08%, the organic matter content ranges between 0.4 and 3.0%, the ECEC was found to be low at < 4 cmol/kg, potassium was medium at 0.2 to 0.5 while the phosphorus content was also low having < 7 ppm. The results showed that the fertility of the area is not so high with majority of the nutrients having low to medium amounts.

Key words: Geographic information system (GIS), spatial variation, soil properties, precision agriculture.

INTRODUCTION

Soil is an essential part of any terrestrial ecosystem. It is defined as the product of interactions between parent materials, biota, topography and climate through time. Because of human activities, the soil is also one of the most affected parts of the ecosystem (Flechsig et al., 1995; Rapaport et al., 1995; Schlesinger, 1991). Human activities have however resulted in soil degradation and reduction in soil functions. For sustainable crop production, reliable soil data are the most important prerequisite for the design of appropriate land-use systems and soil management practices as well as for a better understanding of the environment.

Though soil classification and mapping are necessary and very useful for general land use planning, what is of utmost importance to the farmer is knowing how profitable it is to grow a particular crop or series of crops on a given plot of land, and what amendments are necessary to optimize the productivity of the soil for specific crops. In the recent past, the ill effects of land use on the environment and environmental sustainability of agricultural production systems have become an issue of concern. The problems of declining soil fertility, low crop yield and accelerated soil erosion are associated

---

*Corresponding author. E-mail: bunmidenton@gmail.com.

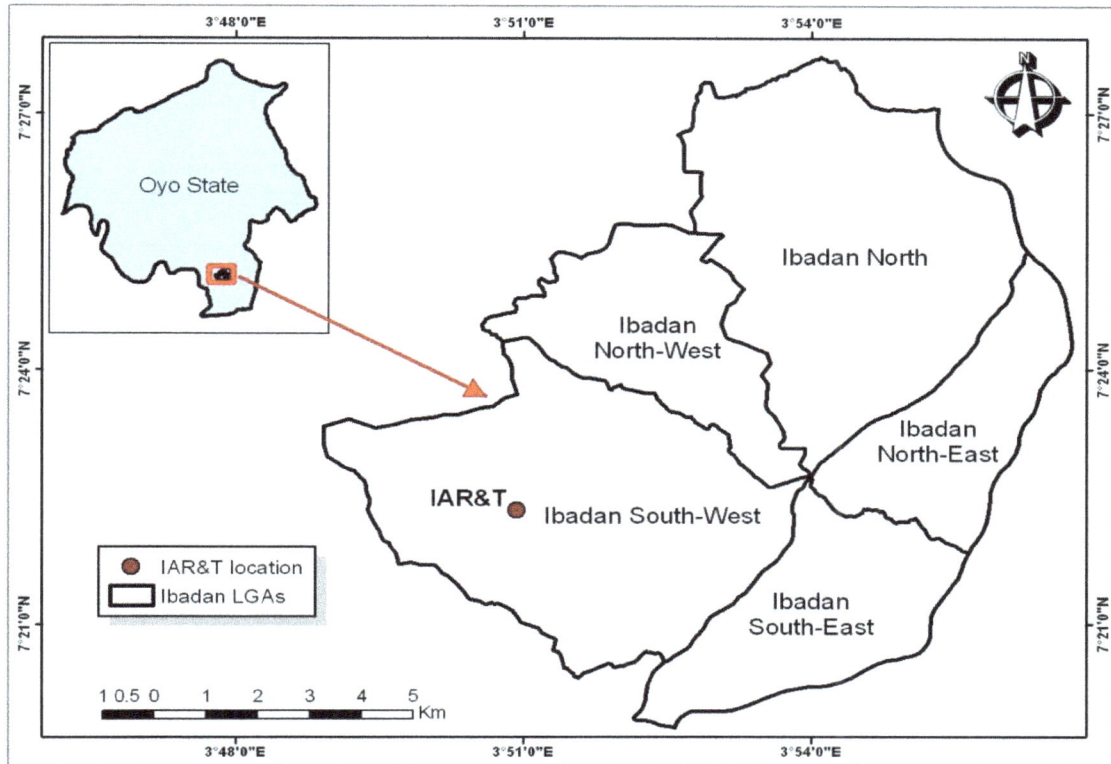

Figure 1. Oyo state showing core LGAs and location of study area.

with intensive and mechanized cultivation, while over-exploitation of natural resources and incessant use chemical fertilizers denote intensive agriculture in the developing areas. Soil survey evaluation is a tool to assess, manage and induce changes in the soil and to link existing resource concerns to environmentally sound land management practices and soil survey assessment, which can then be used to evaluate the effects of management on the soil (FAO, 1988). An intimate knowledge about the types of soils and their spatial distribution is a prerequisite in developing rational land use plan for agriculture, forestry, irrigation and drainage.

Technological advances in geographical information systems (GIS) have recently given land use planners as well as agriculturists a more efficient and effective way of handling large amounts of spatial data. The use of the global positioning system (GPS) and remote sensing in agriculture offers at least four advantages: (1) provision of data cheaply and quickly at a variety of resolutions; (2) use of repeatable methods; (3) provision of improved diagnostics for error detection and accuracy determinations; and (4) generation of information that can be used with the visualization tools in GIS to develop customized as well as tabular summaries.

A major problem of agricultural development in Nigeria is poor knowledge and appraisal of suitability of parcels of land for agricultural production. This has adverse

implications for agricultural development since the bulk of agricultural production takes place under traditional systems where soil fertility is a key component. The result is poor farm management practices, low yield and an unnecessary high cost of production.

The objective of this study was to assess the spatial variation of soil properties of a continuously cultivated land under rain-fed and irrigation systems using the GIS approach to be able to determine agricultural suitability of the site and make suggestions for improvement in crop production.

MATERIALS AND METHODS

This study was carried out at the experimental farm of the Institute of Agricultural Research and Training (IAR&T), Ibadan (7° 23' N; 3° 51'E and 160 m above mean sea level), Nigeria (Figure 1). The area is characterized by a tropical climate marked with wet and dry seasons. It is characterized by a bimodal rainfall pattern with rainfall peaks occur mostly in June and September. Annual temperature ranges from 21.3 to 31.2°C. There are two cropping seasons: Early (March/April to early August) and late (mid-August to October/November) seasons. The study area covered a 3 ha of land that has been under continuous cultivation for more than 10 years. During the dry season, the land is being cultivated under irrigation system. Several crops has been grown on the piece of land such as maize, upland rice, okra, kenaf, e.t.c without adequate knowledge as to which area of the land is most suitable for which crop.

Table 1. Critical limits of nutrients

S/N	Criteria	High	Medium	Low
1	Nitrogen (%)	>0.15	0.08-0.15	<0.08
2	Phosphorus (ppm)	>22	7-22	<7
3	Potassium (Cmol/kg)	>0.5	0.2-0.5	<0.2
4	Organic matter (%)	>3	0.4-0.3	<0.4
5	ECEC (Cmol/kg)	>15	4-15	<4

Source: Okalebo et al. (1993).

The surface soil is coarse gravelly soil ranges from sandy loam to loamy sand. It belongs to Alfisol, classified as Typic Kanhaplustalf according to USDA classification, and locally classified as Iwo association (Smyth and Montgomery, 1962).

Field work

Reconnaissance survey of the area was carried out to delineate the boundary and points taken with the use of global positioning system (GPS). The area was then gridded at an interval of 20 m. 9 transects and 40 points were identified. The GPS coordinates of the grid points were taken and imported into ArcGIS to generate the map of the area. Soil samples were taken at each grid point at the varying distances with the use of a Dutch auger. Depths of collection were at 0 to 25 cm and 25 to 50 cm so as to observe variation on surface soil and at depths. 80 samples were collected for onward transition to the laboratory for analysis.

Laboratory analysis

The samples were air-dried, crushed and allowed to pass through a 2 mm sieve. The gravel content (materials >2 mm) was determined and expressed as a percentage of the total weight of the soil. Soil samples were analysed for soil pH in both water and 0.01 M potassium chloride solution (1:1) using glass electrode pH meter (Mclean, 1965). Total nitrogen was determined by the macro-kjeldahl digestion method as described by Jackson (1962). Bray-1 P was determined by molybdenum blue colorimetry (Bray and Kurtz, 1945) while exchangeable cations were extracted with 1 M NH4OAC (pH 7.0) to determine K and Na using flame photometer and exchangeable Mg and Ca by atomic absorption spectrophotometer (Sparks, 1996). Exchangeable acidity was determined by the KCl extraction method (Mclean, 1965) and organic carbon was after dichromate wet oxidation method (Walkey and Black, 1934). Conversions between values of organic carbon and organic matter was made using Van Bemmelen factor of 1.724 on the assumption that, on average, SOM contains 58% of organic C. Cation Exchange Capacity (CEC) was calculated from the sum of all exchangeable cations. Particle size distribution was determined using hydrometer method (Day, 1965). All the data analysed were imported into GIS environment.

GIS datasets and analysis

The criteria's listed in Table 1 formed the basis for the GIS datasets that were analysed. The data were inputted into ArcGIS and interpolated using Inverse distance Weighted (IDW) technique, this is a technique used to interpolate a surface from points. The GPS coordinates of each grid point, the corresponding nutrient values were interpolated, and a raster image derived for each. After interpolation the raster data obtained was then reclassified using

the reclassify module of a spatial analyst tool in ArcGIS to group into three classes that is, low, medium and high based on the values in the raster data set.

The nutrient status of the soil was assessed based on the concentration of the macronutrients (N, P, K), organic matter and the ECEC. These criteria's were used to classify the fertility of the area into three major classes low, medium and high (Okalebo et al., 1993)

RESULTS AND DISCUSSION

Particle size distribution

The spatial variability of particle size distribution plays an important role in crop production as they impact the soil texture, soil quality and soil erosion. The textural classifications of 0 to 25 cm depth of the study area are presented in Figure 2. The result showed that the surface soil texture ranges from sandy loam to loamy sand. Sandy loam dominated the study site with pockets of loamy sand, which accounted for about 8% of the whole area. This textural class is particularly noted for its high infiltration rate making it very good for cultivation.

Chemical properties of the soil

The soil reaction in terms of soil pH as presented in Figure 3 was tested in water; this showed that the surface soil (0 to 25 cm depth) ranges from strongly acid, medium acid, slightly acid and very slightly acid. Most of the study area had medium acid with a range of 5.5 to 5.9. This could be attributed to the nature of the area as it is been intensively cultivated throughout the year with the use of chemical additives in form of fertilizer could also be a factor.

The soil nitrogen and phosphorus of the soil were low based on the classification scale for maize production as shown in Figures 4 and 5. This could be due to the continuous cropping of the area both in the raining and dry season. Only a small spot shows a high concentration of nitrogen and it is towards the tail end of the area, this could be as a result of rainfall eroding the soil and thus concentrating the nutrients there. Areas having a medium range of nitrogen are also not much and these are the places that are least cultivated and have been left to

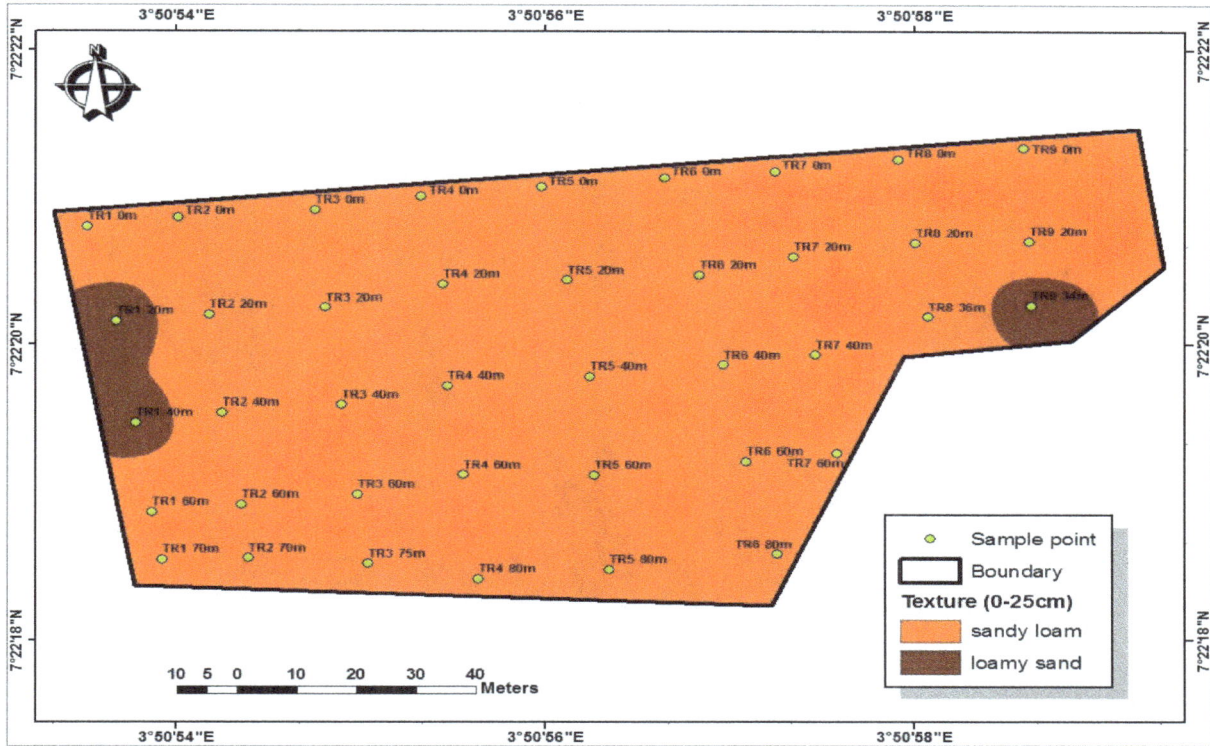

Figure 2. Map showing soil textural classification of the area at depth 0 to 25 cm.

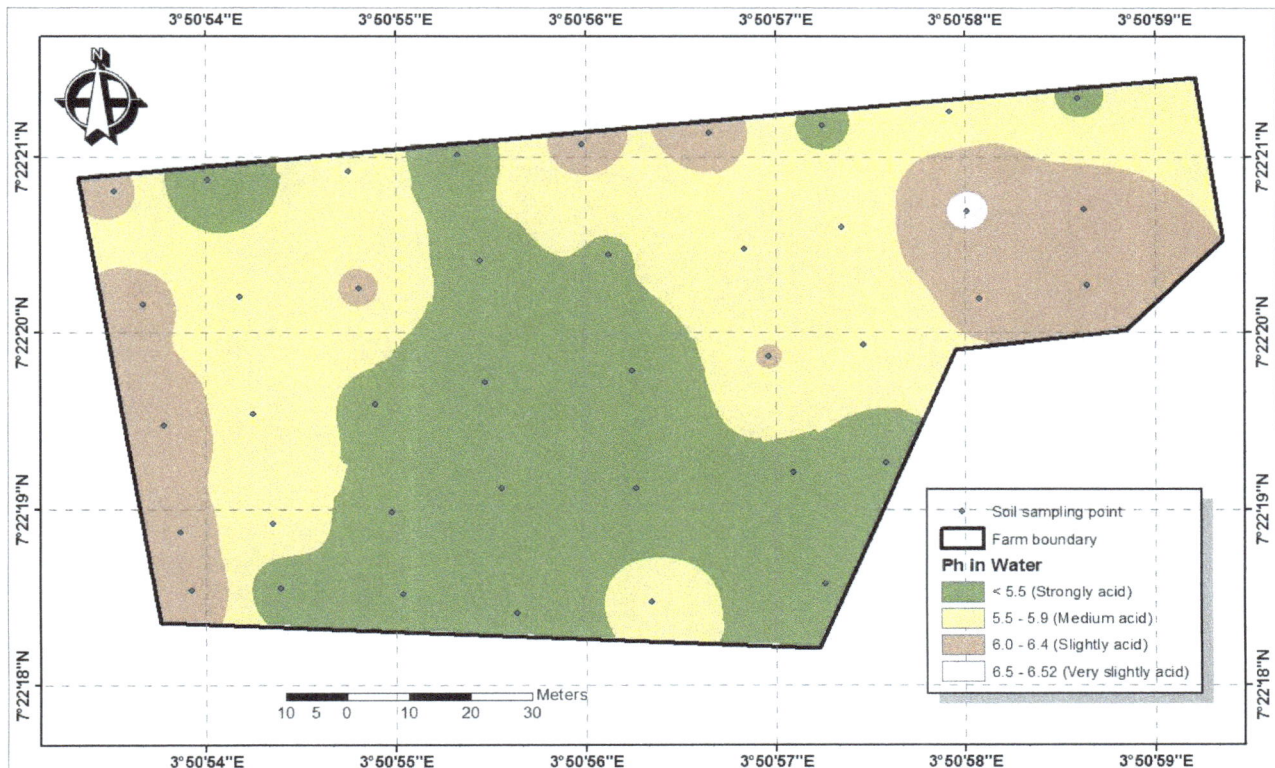

Figure 3. Soil pH rating.

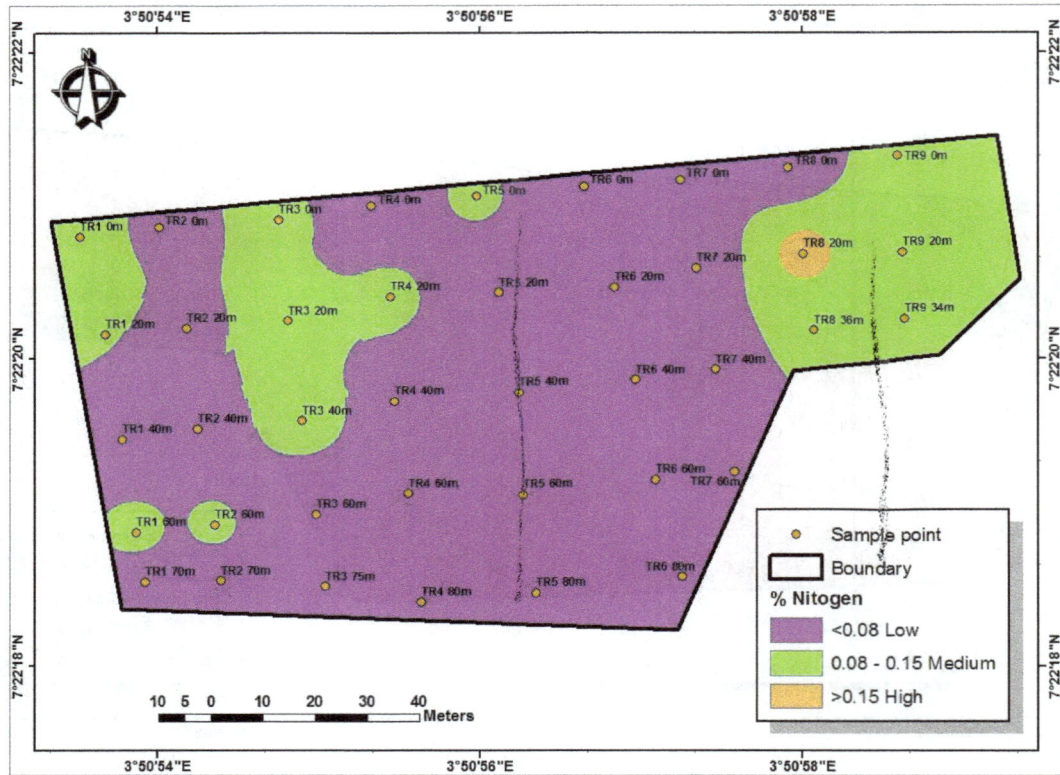

Figure 4. Spatial distribution of Nitrogen in this study area.

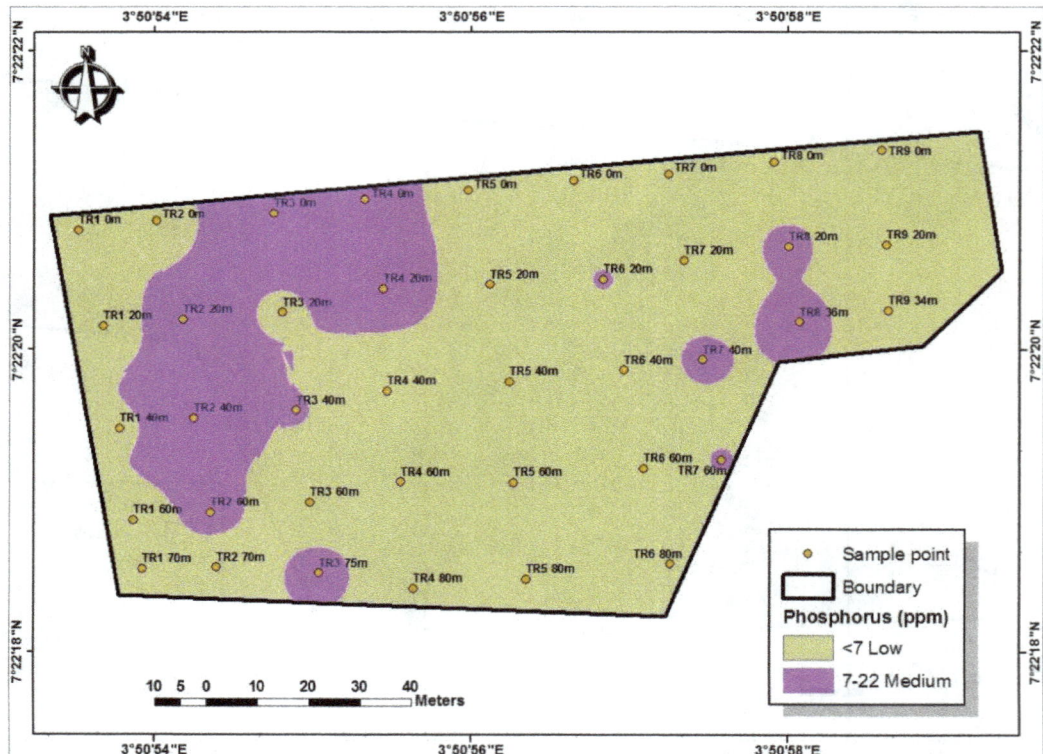

Figure 5. Spatial distribution of Phosphorus in this study area.

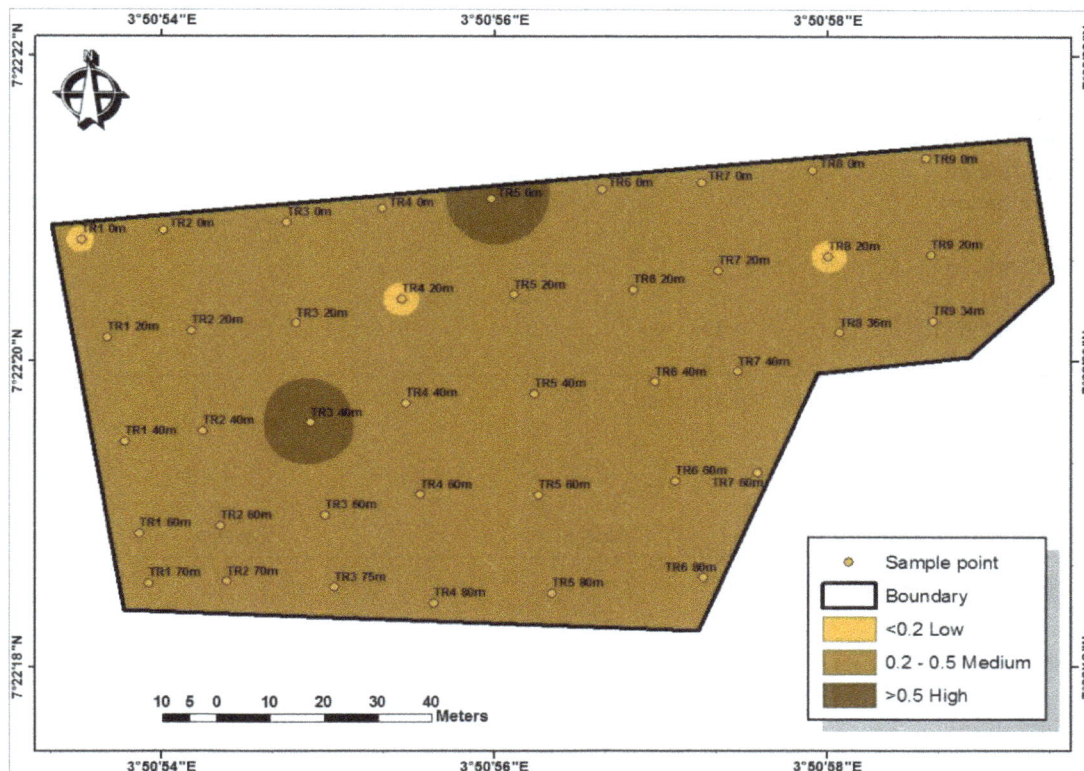

Figure 6. Spatial distribution of potassium in the study area.

fallow over time Figure 6 shows that the potassium content in the area is of the medium limit, it is evenly spread across the entire field with just a few spots having a high concentration. The ECEC which was calculated from the combination of the exchangeable bases (Ca, Mg and Na, K and EA) in Figure 8 is generally low in the study area. This has implication on the overall productivity of the soil because soils with CEC less than 5 meq/100 g generally have a low clay and organic matter content, have a low water holding capacity, requires more frequent lime and fertilizer additions, and is subject to leaching of NO_3, B, NH_4, K and perhaps Mg. Such soils will have lower yield potential than soils with higher CEC under the same level of management, but high productivity can be maintained by intensive management.

Land suitability classification

This study area was evaluated based on the physical soil characteristics that is, the texture and gravel content and the fertility characteristics that is, effective CEC, base saturation and organic matter content. Suggested land characteristics and scores for suitability evaluation for maize that was used for the evaluation and suitability classification of the area are listed in Table 2. Land characteristics (LC) are simple attributes of the land that

can be directly measured or estimated in routine surveys, including remote sensing as well as resource survey (Rossiter, 1996; FAO, 1976; Dent and Young, 1981).

The mean values of the two sampling depths for these land characteristics was obtained and used for the classification. Based on the data, two capability classes were identified as S1 and S2, this being highly suitable and marginally suitable respectively (Figure 7).

Conclusion

The outcome of this research reveals the effectiveness and usefulness of GIS especially in the analysis and interpolation of soil data which is then used in the production of thematic maps. However GIS cannot stand alone in the absence of conventional soil survey if accurate results are to be obtained, they are meant to complement each other so as to yield better results in the shortest possible time thereby reducing cost. Other limitations of GIS include the lack of adequate and up to date maps and data; most of the legacy soil maps available are old and should have been updated. GIS might enable us to make better use of information but it cannot work without data. Though the combination of field inspection and GIS tools can be used in the production of soil maps, follow-up by agronomic

Table 2. Land suitability classes.

Land characteristics	Land classes				
	S1	S2	S3	N1	N2
Physical soil characteristics					
Texture/structure	C+60s to SCL	C+60v to LS	C+60v to fS	C+60v to fS	Cm to cS
Gravel content (%)	<15	<35	<55	<55	any
Fertility characteristics					
Apparent CEC	>3	1-3	<1	-	-
Base saturation (%)	>35	>20	Any		
Organic matter (%)	>1.2	>0.8	Any		

SCL, Sandy clay loam; LS, loamy sand; FS, fine sand.

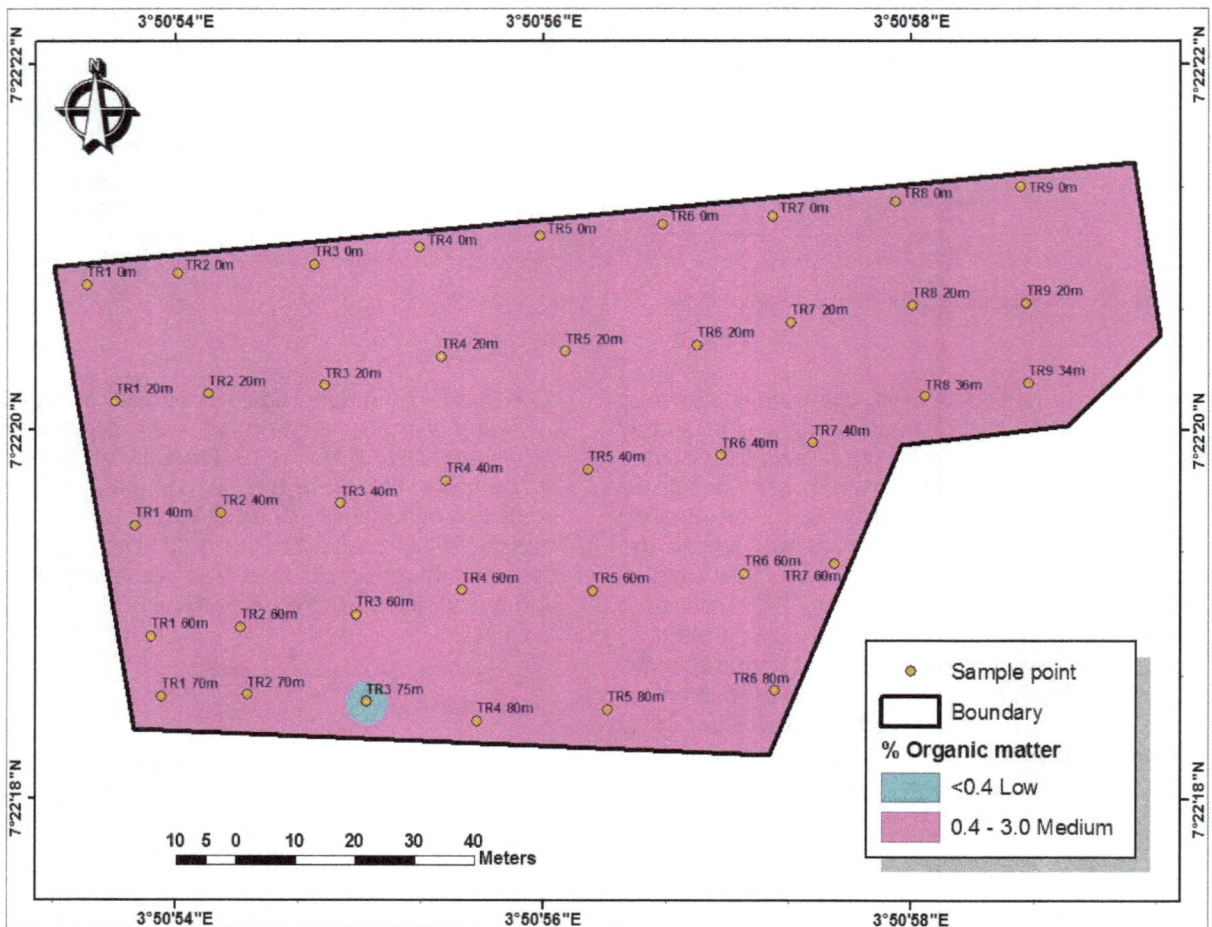

Figure 7. Spatial distribution of organic matter content in this study area.

experimentation are vital. The result reflect the soil properties of this study area, there is need to add appropriate fertilizer N, P, K to the soil in order to increase the fertility status. Also areas that have a high concentration of the nutrients needed for maize production should be cultivated more as this would save cost and

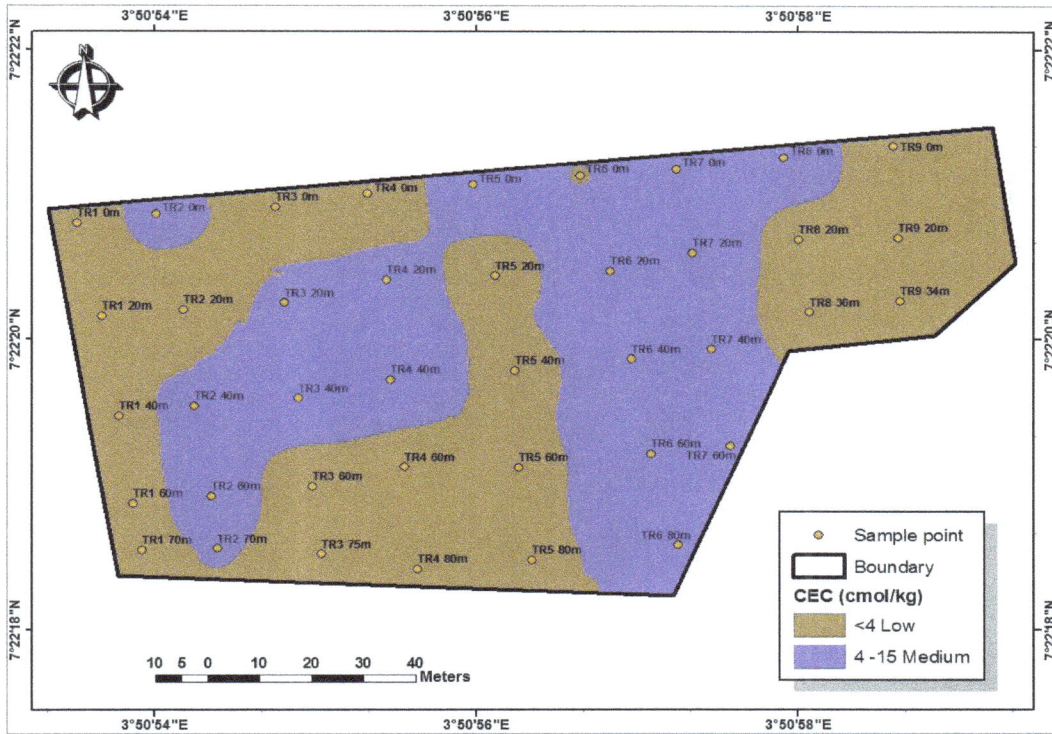

Figure 8. Spatial distribution of ECEC in the area.

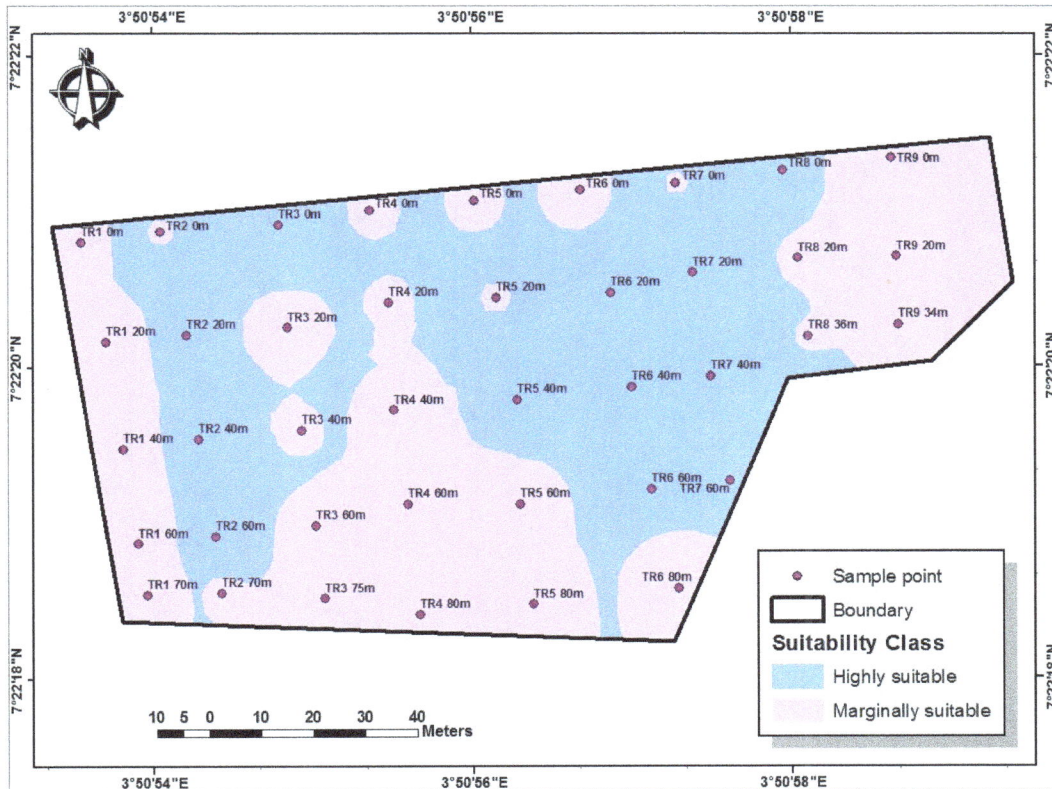

Figure 9. Spatial distribution of land suitability classes.

increase production.

REFERENCES

Bray RH, Kurtz LT (1945). Determination of total organic and available forms of phosphorus in soils. Soil Sci. 59:22-229.

Day PR (1965). Particle fraction and particle size analysis. In: Black CA et al. (Eds). Methods of soil analysis. Part 2. American Society of Agronomy, Madison. pp. 545 -567.

Dent D, Young A. (1981). Soil survey and land evaluation. George Allen and Unwin, London.

FAO (1976). A Framework for land evaluation. Rome, FAO Soils Bulletin. p. 32.

Flechsig M, Erhard M, Grote R (1995). Landscape Models For the evaluation of ecosystem stability under environmental change: The Duebener Heide case study. In: M, Heit HD, Parker and A Shortreid (eds.). GIS applications in natural resources 2. GIS World. Vancouver, British Columbia, Canada. pp. 493-500.

Jackson ML (1962). Soil chemical analysis. Prentice Hall New York. pp. 263-268.

Mclean EO (1965). Aluminum: In methods of soil analysis (ed. C. A. Black) Agronomy No. 9 Part 2, Am. Soc. Agronomy, Madison, Wisconsin. pp. 978-998.

Okalebo JR, Gathua KW, Woomer PL (1993). Laboratory methods of soil and plant analysis: A working manual. p. 88.

Rapaport DJ, Withford WG, Korporal K (1995). Evaluating ecosystem health: opportunities for GIS. In: M. Heit HD, Parker and A Shortreid (eds.). GIS applications in natural resources 2. GIS World. Vancouver, British Columbia, Canada. pp. 408-413.

Rossiter DG (1996). A theoretical framework for land evaluation. Geoderma 72:165-190

Schlesinger WH (1991). Biogeochemistry: An analysis of global change. Academic Press New York.

Smyth AJ, Montgomery RF (1962). Soil and land use in central western Nigeria. Govt. Printer, Ibadan, Western Nigeria. pp. 264.

Sparks DL (1996). Methods of Soil Analysis. Part 3. Chemical Methods. SSSA and ASA. Madison, 1:551-574.

Walkey A, Black IA (1934). Determination of organic matter in soil. Soil. Sci. 37:549-556.

9

Evaluation of soil biological properties of 9- and 15-year-old stands in the oil palm plantation in Perak, Malaysia

Daljit Singh Karam[1], A. Arifin[1,2], O. Radziah[3,4], J. Shamshuddin[3], Hazandy Abdul-Hamid[1,2], Nik M. Majid[1], I. Zahari[5], Nor Halizah Ab. Halim[5] and Cheng Kah Yen[1]

[1]Department of Forest Production, Faculty of Forestry, Institute of Tropical Forestry and Forest Products, Universiti Putra Malaysia, 43400 UPM Serdang, Selangor, Malaysia.
[2]Laboratory of Sustainable Bioresource Management, Institute of Tropical Forestry and Forest Products, Universiti Putra Malaysia, 43400 UPM Serdang, Selangor, Malaysia.
[3]Department of Land Management, Faculty of Agriculture, Universiti Putra Malaysia, 43400 UPM Serdang, Selangor, Malaysia.
[4]Laboratory of Food Crops and Floriculture, Institute of Tropical Agriculture, Universiti Putra Malaysia, 43400 UPM Serdang, Selangor, Malaysia.
[5]Forestry Department Peninsular Malaysia, Universiti Putra Malaysia, 43400 UPM Serdang, Selangor, Malaysia.

Opening land for oil palm cultivation provokes many debates around the world regarding on the fate of biodiversity. A study was conducted to evaluate and compare soil biological properties of 9-year-old (P1) and 15-year-old (P2) stands of an oil palm plantation in Bikam, Perak, Malaysia. Composite samples were collected at depths of 0-15 cm (topsoil) and 15-30 cm (subsoil) located within six subplots (20 m × 20 m). The microbial population count was estimated using a spread-plate technique, and the Fluorescein diacetate (FDA) hydrolysis assay was used to measure microbial enzymatic activity. A rapid ethanol-free chloroform fumigation extraction technique was used for microbial biomass extraction, and the extracts were respectively analyzed by wet dichromate oxidation and Kjeldahl digestion for biomass carbon (C) and nitrogen (N). At the 0-15 cm depth, the microbial biomass C and N contents in the soils from both plots were significantly different (P<0.05). At the 15-30 cm depth, only microbial enzymatic activity was significantly different between plots. Although the addition of fertilizers to the soil is believed to be a predisposing factor, no significant differences in P1 and P2 plots for the biomass C and N in soils at the 15-30 cm depth were observed. Variations in the MBC/MBN ratio in soils of the P1 and P2 plots indicate that changes occurring in the soil microbial composition are due to the availability of soil organic substrates and N. Thus establishment of an oil palm plantation does contribute to changes in soil biological properties.

Key words: Oil palm, different ages, microbial population, enzymatic activity, biomass C and N, rapid chloroform-fumigation extraction technique.

INTRODUCTION

Oil palm production has increased in many Association of Southeast Asian Nations (ASEAN) countries, especially in Malaysia and Indonesia, because it is an important commercial product for domestic and international use

(Laurance, 2007). As such, clearing of forest areas for oil palm cultivation has resulted in massive deforestation. In Malaysia, the area of oil palm plantations was about 4 million ha in 2005 and had an annual growth of 10.06% (Basiron, 2007). Butler and Laurance (2009) reported that expansion of oil palm cultivation contributes to the loss of biodiversity in lowland and peat swamp forests. Meanwhile, the members of The European Commission were trying to uphold a ban on the import of fuel crops, which includes oil palm, that are cultivated in areas on the endangered list, such as natural forests (Koh and Wilcove, 2008). In contrast, Basiron (2007) stated that Malaysia's oil palm industry is the country's best-organized sector, and many programs are in place to educate communities about the management, conservation and community services provided for the environment and people.

Therefore, it is crucial for scientists working on oil palm to provide essential information on the land management of the estate rather than just focusing on the yield of oil palm. Through such practice, we can predict the degree of disturbance caused by the plantation to the environment. We all are aware of the high levels of fertilizer application to this plant; hence, we need to quantify the effect of it on soil degradation and biodiversity. Lalfakzuala et al. (2008) suggested that heavily cultivated agricultural areas disrupt soil productivity due to depletion of soil organic matter. However, Henson (1999) believed that oil palm cultivation contributes little to environmental damage. He also clarified that negative effects only occur at the beginning of forest clearing operations.

One of the approaches used to determine the soil quality of a particular plantation is to analyze soil biological properties. Sánchez-Monedero et al. (2008) found that microbial enzymatic activity is an important aspect of the decomposition of organic materials, such as humus; in the degradation of pollutants; and in the transformation of nutrients suitable for plant uptake. In addition, Lalfakzuala et al. (2008) suggested that it is important to study biological properties as well because they are a sensitive indicator that can be used to quantify soil fertility and quality. Moreover, examination of microbial enzymatic activity gives instant view of the organic matter turnover in soil (Joergensen and Emmerling, 2006). Ajwa et al. (1999) also stated that soil microbial activity is a sensitive indicator that proportionally changes with the other changes or disturbances occurring in the soil of particular area. The fertility of the soil is proportional to the amount and the number of times fertilizer is applied to enhance oil palm growth (Phosri et al., 2010).

Gaspar et al. (2001) reported that soil microbial biomass contributes 1 to 4% of organic carbon and 2 to 6% of organic nitrogen in the soil. Furthermore, microbial biomass C and N are important constituents in soil organic matter and are the main nutrients stored for plant uptake. Furthermore, Ajwa et al. (1999) and Rice et al. (1996) found that environmental changes (e.g., weather changes, physical disturbances and chemical toxicity) influence the activities of microbial biomass in forest soils. All of these factors make soil microbial biomass a good, sensitive indicator of soil quality and fertility (Islam and Weil, 2000).

In this study, the Fluorescein diacetate (FDA) hydrolysis assay described by Sánchez-Monedero et al. (2008) was chosen for the evaluation of microbial enzymatic activity because of its rapid estimation of overall enzymatic activity. In addition, Schnürer and Rosswall (1982) claimed that FDA possesses the ability to rapidly hydrolyze a wide range of enzymes including esterases, lipases and proteases.

Soil microbial biomass C and N are two important components of soil organic matter because plant or tree species in a particular area obtain nutrients from nutrient storage in organic matter (Barbhuiya et al., 2004). A higher level of organic matter in the soil indicates that a larger proportion of C and N are available. Studies of the relationship between soil microbial biomass and selected soil properties, such as organic matter and total nitrogen, will give us a clear view of the current fertility and quality of cultivated land. Soil acidity and moisture content can affect the distribution of microorganisms in the soil. Behera and Sahani (2003) reported that most bacteria cannot withstand acidic or dry soil conditions. Highly acidic soil conditions inhibit microbial activity in the soil, which subsequently affects nutrient cycling in the soil (González-Pérez et al., 2006). However, Shamshuddin and Che Fauziah (2010) reported that oil palm trees were able to achieve good growth performance in highly acidic soil.

To our knowledge, studies of the soil biological properties of oil palm plantations are lacking. Therefore, the objective of the current study was to evaluate and compare the soil biological properties of 9-year-old and 15-year-old oil palm stands in Bikam Plantation, Perak, Malaysia.

MATERIALS AND METHODS

Description of study site

A 9-year-old (N 03°.99'862° E 101°.24'220°, approximately 12 m above sea level) (P1) and a 15-year-old (N 03°.99'811° E 101°.24'110°, ± 9 m a.s.l) (P2) oil palm stand in a plantation in Bikam, Perak were selected as study plots. The mean annual rainfall and temperature are 2,417 mm and 24.5°C, respectively. The soils in this study area are classified as Ultisols; these soils are characterized as highly weathered because they contain a large amount of low-activity clays associated with high Al saturation (Arifin et al., 2008). Both plots were managed independently by local farmers. The basic fertilizers annually applied to these crops were N:P:K (15:15:15) fertilizers. The distance of the palm stand from each other was 9 m × 9 m.

Soil sampling

At each plot, six subplots were established that were each 20 m

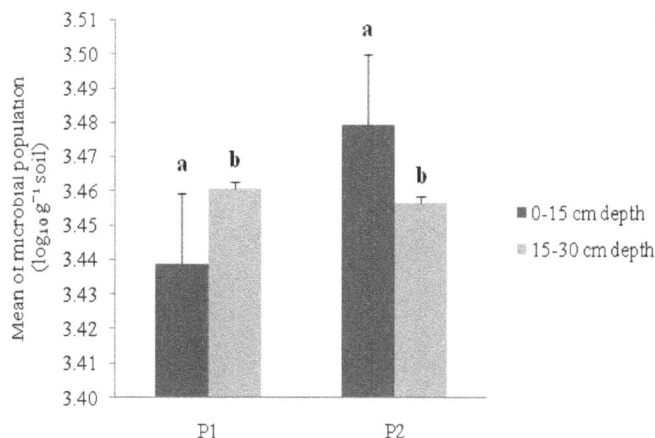

Figure 1. Means of the soil microbial populations at 9-year-old (P1) and 15-year-old (P2) oil palm plots. Different letters indicate significant differences between the means of the same soil depths at the 9-year-old (P1) and 15-year-old (P2) oil palm plots using the Student's t-test ($P<0.05$).

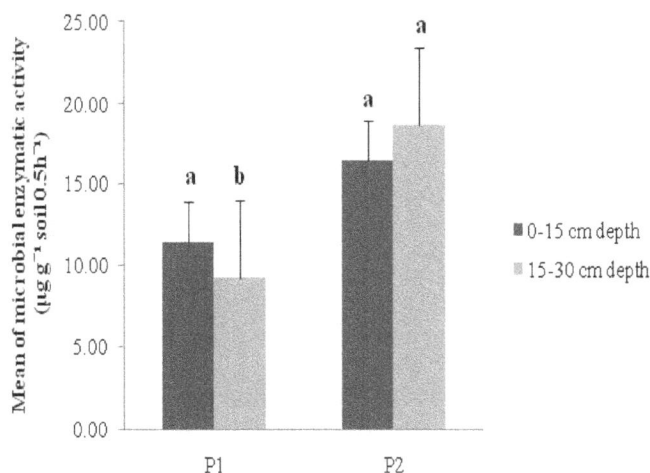

Figure 2. Means of the microbial enzymatic activity at 9-year-old (P1) and 15-year-old (P2) oil palm plots. Different letters indicate significant differences between the means of the same soil depths at the 9-year-old (P1) and 15-year-old (P2) oil palm plots using the Student's t-test ($P<0.05$).

× 20 m in size. At each plot, a composite sample was obtained after mixing six soil samples that were collected randomly from 0-15 cm and 15-30 cm depths. A total of 24 composite samples were obtained from the plots. All composite samples collected were wrapped in UV-sterilized polyethylene bags and stored in ice-filled polystyrene boxes before being analyzed in the laboratory.

Soil analyses

A number of parameters were selected for determining the soil biological properties of P1 and P2. The microbial population count was estimated via the spread-plate technique (Sleytr et al., 2007).

FDA was used for microbial enzymatic activity analysis. The rapid ethanol-free chloroform fumigation extraction approach was used for the extraction of soil microbial biomass (Witt et al., 2002).The extract was subjected to wet dichromate oxidation to determine microbial biomass C (Vasquez-Murrieta et al., 2007; Vance et al., 1987) and to Kjeldahl digestion to determine microbial biomass N (Simmone et al., 1997; Brookes et al., 1985). Selected soil physico-chemical analyses were also performed. Organic matter and organic C were elucidated via the loss-on-ignition technique (Ahmadpour et al., 2010); total N was determined using Kjeldahl digestion. Soil pH was determined using a glass electrode and a soil: distilled water ratio of 1:2.5 (w/w) (Akbar et al., 2010). Bulk density was determined using disturbed soil sample technique as described by Gupta (2007). Gravimetric method was performed to elucidate soil moisture content for each sample.

Data analyses

The mean values obtained from the same soil depths in P1 and P2 plots were analyzed using the Student's t-test. Pearson correlation analyses were performed using SPSS (version 16.0) to detect any linear relationships between microbial biomass C and organic matter and between microbial biomass N and total N from the same soil depths at both sites.

RESULTS

Microbial population

For both plots, no significant differences ($P<0.05$) were observed in microbial population counts of soil at the same soil depths (Figure 1). The means of the microbial population counts for the soils at 0-15 cm and 15-30 cm depths in P1 were 3.44 ± 1.97 $\log_{10}$ g^{-1} soil and 3.46 ± 0.05 $\log_{10}$ g^{-1} soil, respectively; for the P2 plot, the means were 3.48 ± 0.03 log10g^{-1} soil and 3.46 ± 0.04 $\log_{10}$ g^{-1} soil at the 0-15 cm and 15-30 cm depths, respectively.

Microbial enzymatic activity

The microbial enzymatic activity in the soils of P2 (9.27 ± 1.71 µg g^{-1} soil 0.5h^{-1}) was significantly higher ($P<0.05$) than that in the soils of P1 (18.69 ± 1.97 µg g^{-1} soil 0.5h^{-1}) at the 15-30 cm depth (Figure 2). In contrast, no significant difference ($P<0.05$) in microbial enzymatic activity was observed at the 0 to 15 cm depth for the P1 (11.39 ± 3.01 µg g^{-1} soil 0.5h^{-1}) and P2 (16.44 ± 1.28 µg g^{-1} soil 0.5 h^{-1}) plots.

Microbial biomass C (MBC)

The MBC was found to be significantly higher ($P<0.05$) in P1 (409 ± 142 µg g^{-1} soil) compared to that in P2 (82 ± 33 µg g^{-1} soil) at the 0-15 cm depth (Figure 3). However, no significant difference ($P<0.05$) in MBC was found in the soil at the 15-30 cm depth for either plot. The means of the soil MBC for the P1 and P2 plots at the 15 to 30 cm depth were 330 ± 78 and 421 ± 312 µg g^{-1} soil, respectively.

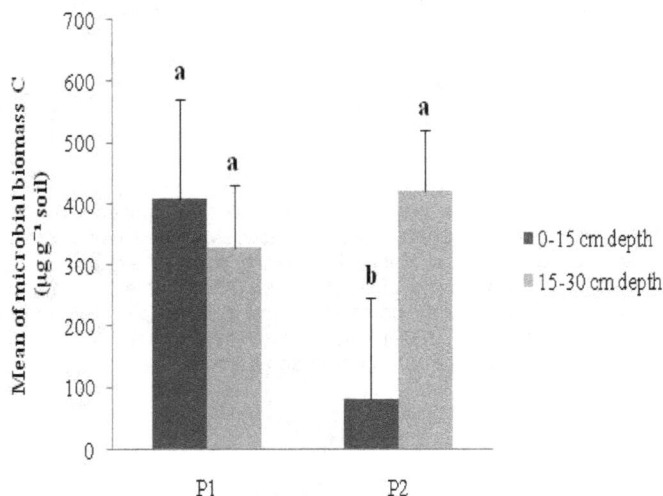

Figure 3. Means of the soil microbial biomass C at the 9-year-old (P1) and 15-year-old (P2) oil palm plots. Different letters indicate significant differences between the means of the same soil depths at the 9-year-old (P1) and 15-year-old (P2) oil palm plots using the Student's t-test (P<0.05).

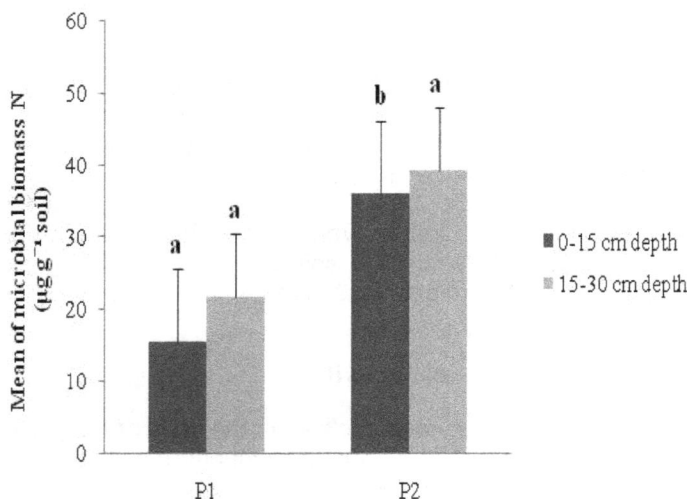

Figure 4. Means of the soil microbial biomass N at the 9-year-old (P1) and 15-year-old (P2) oil palm plots. Different letters indicate significant differences between the means of the same soil depths at the 9-year-old (P1) and 15-year-old (P2) plots using the Student's t-test (P<0.05).

Microbial biomass N (MBN)

The mean of the soil MBN was found to be significantly lower (P<0.05) in P1 (15 ± 3 µg g^{-1} soil) than in P2 (22 ± 4 µg g^{-1} soil) at the 0-15 cm depth (Figure 4). On the other hand, the means of the soil MBN at the 15-30 cm depth for both the P1 (36 ± 8 µg g^{-1} soil) and P2 (39 ± 12 µg g^{-1} soil) plots were not significantly different (P<0.05).

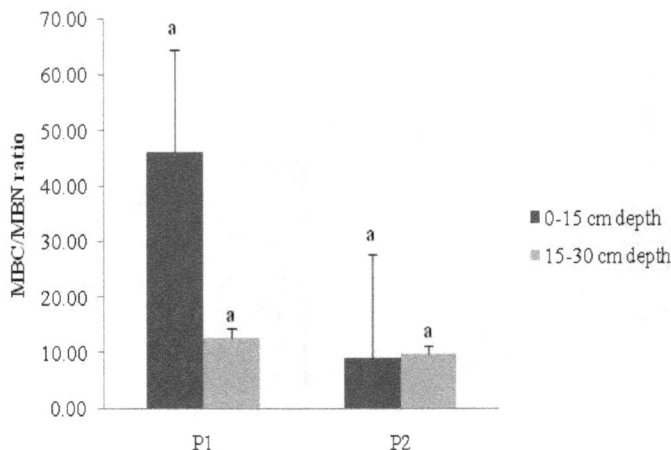

Figure 5. Ratios of microbial biomass C to microbial biomass N (MBC/MBN) at the 9-year-old (P1) and 15-year-old (P2) oil palm plots.

Microbial biomass C to microbial biomass N ratio (MBC/MBN)

The ratios of MBC/MBN in P1 at the 0-15 cm and 15-30 cm depths were higher compared to the P2 ratios (Figure 5). The values of the MBC/MBN ratio for P1 at the 0-15 cm and 15-30 cm depths were 46.03 ± 2.99 and 12.77 ± 2.95, respectively. In contrast, the values of the MBC/MBN ratio for P2 at the 0-15 cm and 15-30 cm depths were 9.08 ± 2.28 and 9.71 ± 2.23, respectively.

Organic matter, organic C, total N and soil acidity

Table 1 shows the soil organic matter, organic C, total N and pH of the P1 and P2 plots. The organic matter and organic C were significantly higher (P<0.05) in the soils of P1 compared to those of P2. However, no significant difference in total N (P>0.05) in soil was found between the plots. The soil of P1 exhibited a significantly different pH value compared to the soil of P2 (P<0.05). P2 possessed higher (P<0.05) bulk density compared to P1 for both soil depths. In contrast, P1 showed significantly higher (P<0.05) moisture content compared to P2.

Pearson correlation analysis

Pearson correlation analysis indicated that no strong relationship exists between biomass C and organic matter or between biomass N and total N for the same soil depths in either of the plots (Table 2). Correlation analysis done between organic matters with MBC/MBN ratio showed no strong relationship. Hence, these data showed that organic matter and total N did not directly impact the microbial biomass C and N availability in the soil.

Table 1. Selected soil physico-chemical properties of the 9-year-old (P1) and 15-year-old (P2) plots.

Parameter	P1	P2	P value
0-15 cm depth			
Organic matter (%)	5.18 ± 0.49^b	3.19 ± 0.11^c	0.002644
Organic carbon (%)	3.00 ± 0.28^b	1.85 ± 0.06^c	0.002644
Total nitrogen (%)	0.54 ± 0.04^b	0.54 ± 0.02^b	0.888409
pH-H_2O	4.47 ± 0.09^b	4.80 ± 0.04^c	0.006851
Bulk density (g cm^{-3})	1.43 ± 0.03^b	1.52 ± 0.02^c	0.016433
Moisture content (%)	19.50 ± 0.72^c	17.83 ± 2.39^c	0.518794
15-30 cm depth			
Organic matter (%)	6.30 ± 0.73^b	11.27 ± 0.78^c	0.007515
Organic carbon (%)	3.66 ± 0.43^b	6.54 ± 0.45^c	0.003105
Total nitrogen (%)	0.64 ± 0.17^a	0.49 ± 0.05^a	0.392245
pH-H_2O	4.47 ± 0.11^b	4.84 ± 0.07^c	0.015383
Bulk density (g cm^{-3})	1.45 ± 0.02^a	1.54 ± 0.01^b	0.005173
Moisture content (%)	22.50 ± 1.96^a	15.33 ± 1.58^b	0.017503

Different letters within each row indicate significant differences between the means of soil properties at both depths at the 9-year-old (P1) or 15-year-old (P2) oil palm plots using the Student's t-test ($P<0.05$).

Table 2. Pearson correlation analysis results comparing microbial biomass C (MBC) with organic matter (OM), microbial biomass N (MBN) with total N (TN) and OM with MBC/MBN ratio for both plots at the same soil depths.

Soil depth (cm)	MBC versus OM		MBN versus TN		OM versus MBC/MBN ratio	
	P value	r^2	P value	r^2	P value	r^2
P1 (0-15)	0.193	0.616	0.346	0.470	0.168	0.644
P1 (15-30)	0.476	-0.365	0.868	0.088	0.627	-0.254
P2 (0-15)	0.248	-0.565	0.059	0.794	0.324	-0.490
P2 (15-30)	0.322	-0.492	0.373	0.448	0.526	-0.335

P1, 9-year-old oil palm stand; P2, 15-year-old oil palm stand.

DISCUSSION

Significant differences in microbial population counts for both P1 and P2 plots were detected; this result could be due to the adequate application of fertilizer. An adequate supply of nutrients helps microbes to enhance nutrient cycling within soils. It has been found that differences in the slope gradient contribute to the uneven distribution of soil nutrients that leads to the uneven distribution of soil microorganisms. However, both the P1 and P2 plots possess the same type of topography, which partly explains the similarities in the distributions of the microbial populations in these plots. The high rate of microbial enzymatic activity in the soils from the 15-30 cm depth of P1 could be due to the effects of fertilizer addition to the soil. Addition of fertilizer causes certain soil bacteria to become inactive due to the excessive supply of fertilizer in the soil (Klose et al., 1999; Deng and Tabatabai, 1997). After 12 years of planting, farmers normally apply less fertilizer. Hence, the reduced nutrient availability in the soil triggers or forces soil bacteria,

especially nitrogen-fixing bacteria, to reactivate nutrient cycling. In addition, Klose et al. (1999) stated that soil enzymes are vital tools for assessing soil quality because soil biota rapidly respond to changes in the soil environment.

In the topsoil, the MBC was higher in P1 compared to P2; this phenomenon was enhanced by the availability of high levels of organic C. Hu et al. (1997) illustrated that the turnover and changes occurring in soil organic C and biomass C of agricultural soils were influenced by management practices. Organic matter content in the oil palm soil triggers soil microbial activities because it is one of the vital nutrient sources for microbes. In addition, large amounts of organic matter and C in the soil provide an ideal medium for the growth of soil microorganisms. Moreover, the decomposition of soil organic matter by soil microbes contributes the high levels of humus and organic C to the soil; oil palm can then take up these nutrients. In contrast, low biomass C in P2 could be due to the lack of monitoring of the ground cover availability. The P2 area was monitored less than P1 because of the age of the oil palm; this stand is 15 years old and does not require as much attention as the younger palm stand. On the other hand, the MBN in the soil at the 15 to 30 cm depth was higher in P2 than in P1. Low N availability in the soils of P1 could be due to the decomposition of litter (Barbhuiya et al., 2004). Besides low N availability, Omay et al. (1997) claimed that N fertilization and cropping systems used for long-term production affect certain soil physico-chemical properties and nutrient cycles. Murwira and Kirchmann (1993) found that immobilization of N in soil was due to the addition of certain types of N fertilizerand helped to increase crop yield.

Behera and Sahani (2003) reported that variations in the MBC/MBN ratio indicate a qualitative change in the microbial composition of soil. The MBC/MBN ratios at both soil depths were higher for the P1 than for the P2 soils. This result could be due to the reduced soil accumulation of N in P1 compared to P2. Furthermore, most of the total N will be absorbed through plant roots for oil palm growth; hence, this process results in lower N availability in the soil. In addition, Arunachalam and Pandey (2003) reported that fungal domination of microbial biomass in the soil of a disturbed forest potentially results in soil nutrient retention and conservation. The higher MBC/MBN ratio found in P1 soil is also due to the reduced microbial N availability that results from the slow rate of organic matter accumulation in fine roots. Arunachalam and Pandey (2003) also stated that deforestation or land opening for agricultural cultivation affects the restoration of total N.

From the Pearson correlation analysis, no strong relationship was detected between microbial biomass C and organic matter available in the soil of the oil palm plantation. This condition was probably due to the fact that organic matter content at oil palm plantation is relatively low due to no litter or forest ground (Haron et al., 1998) at the respective plot. In addition low contribution of organic matter on microbial biomass C through the study done by Lalfakzuala et al. (2008) by which the high rate of fertilizer application to soils of oil palm farms subsequently over took the function of organic matter as the nutrient reservoir in the soil. Furthermore, Patzel et al. (1999) stated that there was little organic matter available in agricultural land. Total N was also uncorrelated with microbial biomass N at both plots in the current study. This finding proves that the soil conditions were altered by the high or excessive amounts of nitrogen fertilizers in the soil, provided that excessive fertilizer application or available nitrogen in the oil palm plantation is the crucial factor that influences soil microbial biomass N. Large amounts of nitrogen in the soil reduce the ability of microbes to convert organic nitrogen into inorganic nitrogen (Murwira and Kirchmann, 1993; Omay et al., 1997).

Conclusion

Land opening for an oil palm plantation affects the distribution of selected soil biological properties, including microbial enzymatic activity and biomass C and N in different age groups of oil palm plots. Land management practices for oil palm cultivation include fertilizer applications; these applications result in soil alterations that enhance or negatively impact soil microbial components. Hence, it is important to include soil biological properties in the evaluation of plantation soils. Soil biological properties, such as microbial enzymatic activity and biomass, are important for describing nutrient sustainability and regeneration that are used to assess land management of the oil palm plantation.

ACKNOWLEDGEMENTS

The authors wish to thank Perak South District, Department of Forestry, Perak who allowed us to carry out the research project. This study was financially supported by the Fundamental Research Grant Scheme (FRGS-5523723) and Research University Grant Scheme (RUGS 91709) from the Ministry of Higher Education of Malaysia (MOHE) through the Universiti Putra Malaysia (UPM), Malaysia. They also would like to express their gratitude to Forestry Department Peninsular Malaysia and Perak Forestry Department staff that helped us with the fieldwork.

REFERENCES

Akbar MH, Jamaluddin AS, Majid NM, Nik Ab, Abdul-Hamid H, Jusop S, Hassan A, Yusof KH, Abdu A (2010). Differences in soil physical and chemical properties of rehabilitated and secondary forests. Am. J. Appl. Sci. 7:1200-1209.

Arunachalam A, Pandey HN (2003). Ecosystem restoration of jhum fallows in Northeast India: microbial C and N along altitudinal and successional gradients. Restor. Ecol. 11:1–6.

Ahmadpour P, Nawi AM, Abdu A, Abdul-Hamid H, Singh DK, Hassan A, Majid NM, Jusop S (2010). Uptake of heavy metals by Jatropha curcas L. planted in soils containing sewage sludge. Am. J. Appl. Sci. 7:1291-1299.

Ajwa HA, Dell CJ, Rice CW (1999). Changes in enzyme activities and microbial biomass of tall grass prairie soil as related to burning and nitrogen fertilization. Soil Biol. Biochem. 31:769-777.

Arifin A, Tanaka S, Jusop S, Majid NM, Ibrahim Z, Wasli ME, Sakurai K (2008). Assessment on soil fertility status and growth performance of planted dipterocarp species in Perak, Peninsular Malaysia. J. Appl. Sci. 8(21):3795-3805.

Basiron Y (2007). Palm oil production through sustainable plantations. Eur. J. Lipid Sci. Technol. 109:289-295.

Barbhuiya AR, Arunachalam A, Pandey HN, Arunachalam K, Khan ML, Nath PC (2004). Dynamics of soil microbial biomass C,N and P in disturbed and undisturbed stands of a tropical wet-evergreen forest. Eur. J. Soil Biol. 40:113-21.

Behera N, Sahani U (2003). Soil microbial biomass activity in response to Eucalyptus plantation and natural regeneration on tropical soil. Forest Ecol. Manag. 174:1-11.

Brookes PC, Landman A, Pruden G, Jenkinson DS (1985). Chloroform fumigation and the release of soil nitrogen: A rapid direct extraction method to measure microbial biomass nitrogen in soil. Soil Biol. Biochem. 17:837-842.

Butler RA, Laurance WF (2009). Is oil palm the next emerging threat to the Amazon? Tropical Conservation Sci. 2:1-10.

Deng SP, Tabatabai MA (1997). Effect of tillage and residue

management on enzyme activities in soils. III Phosphatases and arysulphatase. Biol. Fert. Soils 22:208-213.

Joergensen RG, Emmerling C (2006). Methods for evaluating human impact on soil microorganisms based on their activity, biomass, and diversity in agricultural soils. J. Plant Nutr. Soil Sci. 169:295-309.

González-Pérez M, Martin-Neto L, Colnago LA, MIlori DMBP, de Camargo OA, Berton R, Bettiol W (2006). Characterization of humic acid extracted from sewage sludge-amended oxisols by electron paramagnetic resonance. Soil and Tillage Research. 91:95-100.

Gupta PK (2007). Soil, Plant, Water and Fertilizer Analysis. 2nd Edition. India: Agrobios.

Haron K, Brookes PC, Anderson M, Zakaria ZZ (1998). Microbial biomass and soil organic matter dynamics in oil palm (Elaeis guineensis Jacq.) Soil Biol. Biochem. 30:547-552.

Henson IE (1999). Comparative ecophysiology of oil palm and tropical rain forest. In: Oil Palm and the Environment – A Malaysian Perspective. Eds. G. Singh, L.K. Huan, T. Leng and D.L. Kow, Malaysian Oil Palm Growers Council, Kuala Lumpur, Malaysia. P. 9-39.

Hu S, Coleman DC, Carroll CR, Hendrix PF, Beare MH (1997). Labile soil carbon pools in subtropical forest and agricultural ecosystems as influenced by management practices and vegetation types. Agric. Ecos. Environ. 65:69-78.

Islam KR, Weil RR (2000). Land use effects on soil quality in a tropical forest ecosystem of Bangladesh. Agric. Ecosys. Environ. 79:9-16.

Klose S, Moore JM, Tabatabai MA (1999). Arylsuphatase activity of microbial biomass in soils as affected by cropping systems. Biol. Fert. Soils 29:46-54l.

Koh LP, Wilcove DS (2008). Is oil palm agriculture really destroying tropical biodiversity? Conserv. Lett. 1:60-64.

Lalfakzuala R, Kayang, Dkhar MS (2008). The effect of fertilizers on soil microbial components and chemical properties under leguminous cultivation. Am-Eurasian J. Agric. Environ. Sci. 3:314-324.

Laurance WF (2007). Forest destruction in tropical Asia. Curr. Sci. 93:1544-1550.

Murwira H, Kirchmann H (1993). Carbon and nitrogen mineralization of cattle manures, subject to different treatments, in Zimbabwe and Swedish soils. In: Mulongoy, K. and R. Merckx (Eds.). Soil Organic Matter Dynamics and Sustainability of Tropical Agriculture, John Wiley and Sons, New York, pp. 189-198.

Omay AB, Rice CW, Maddux LD, Gordon WB (1997). Changes in soil microbial and chemical properties under long-term crop rotation and fertilization. Soil Sci. Soc. Amsterdam J. 61:1672-1678.

Patzel N, Sticher H, Karlen DL (1999). Soil fertility-phenomenon and concept. J. Plant Nutr. Soil Sci. 163:129-142.

Phosri C, Rodriguez A, Sanders IR, Jeffries P (2010). The role of mycorrhizas in more sustainable oil palm cultivation. Agric. Ecosyst. Environ. 135:187-193.

Sánchez-Monedero MA, Mondini C, Cayuela ML, Roig A, Contin M, Nobili M De (2008). Fluorescein diacetate hydrolysis, respiration and microbial biomass in freshly amended soils. Biol. Fertil. Soils 44:885-890.

Shamshuddin J, Fauziah IC (2010). Alleviating acid soil infertility constraints using basalt, ground magnesium limestone and gypsum in a tropical environment. Malaysian J. Soil Sci. 14:1-14.

Schnürer J, Rosswall T (1982).Fluorescein diacetate hydrolysis as a measure of total microbial activity in soil and litter. App. Environ. Microbiol. 43:1256–1261.

Simmone AH, Simmone EH, Eitenmiller RR, Mills HA, Cresman III CP (1997). Could the Dumas method replace the Kjeldahl digestion for nitrogen and crude protein determination in foods? J. Sci. Food Agric. 73:39-45.

Sleytr K, Tietz A, Langergraber G, Haberl R (2007). Investigation of bacterial removal during the infiltration process in constructed wetlands. Sci. Total Environ. 380:173-180.

Vance ED, Brookes PC, Jenkinson DS (1987). An extraction method for soil measuring soil microbial biomass C. Soil. Biol. Biochem. 19:703-707.

Vasquez-Murrieta MS, Govarts B, Dendooven L (2007). Microbial biomass C measurements in soil of the central highlands of Mexico. Appl. Soil Ecol. 35:432-440.

Witt C, Gaunt JL, Galicia CC, Ottow JCG, Neue HU (2000). A rapid chloroform-fumigation extraction method for measuring soil microbial biomass carbon and nitrogen in flooded rice soils. Biol. Fertil. Soils 30:510-519.

10

Impact of effective and indigenous microorganisms manures on *Colocassia esculenta* and enzymes activities

Hermann Désiré Mbouobda[1,2], Fotso[1,2], Carole Astride Djeuani[2], Kilovis Fai[1] and Ndoumou Denis Omokolo[2]

[1]Department of Biology, Higher Teachers' Training College (HTTC), University of Bamenda, P.O. Box 39, Bamenda, Cameroon.
[2]Laboratory of Plant Biology, Department of Biological Sciences, Ecole Normale Supérieure (ENS), University of Yaoundé 1, P.O. Box 47, Yaoundé, Cameroon.

This study deals with the evaluation of the effect of effective microorganisms (EM) and indigenous microorganisms (IMO) manure on *Colocasia esculenta* (Taro) in Bambili-Cameroon. A randomized complete block design (RCBD) with three treatments (EM manure, IMO manure and control) and six replications were conducted. Investigations were performed taking into account morphological and agronomical parameters as well as disease incidence, total phenol contents, peroxidase (Pox) and polyphenoloxidase (PPO) activities. There was a significant difference ($p < 0.05$) in the height of plants and number of leaves throughout the period of research in plants treated with EM manure. Plants treated with EM manure gave the heaviest corms and cormels (15.549 ± 2.17 tons/ha) followed by plants treated with IMO manure (12.335 ± 1.69 tons/ha) and then the control plants (10.539 ± 2.24 tons/ha). Both EM and IMO manures were ineffective in controlling taro leaf blight disease that emerged in the field. Total phenolic content as well as Pox and PPO activities increased significantly during the first 5 months of development with EM manures producing the highest quantities followed by that of IMO manure. This is due to the microbial diversity of the manures which in turn improves soil quality and enhances the growth and yield of *C. esculenta*. These results suggest that EM and IMO manures can be used to ameliorated taro productivity but cannot be used to combat disease.

Key words: Taro, yields, phenolic content, peroxidase, polyphenoloxidase.

INTRODUCTION

Taro (*Colocasia esculenta* L.) is an herbaceous perennial plant widely cultivated in West and Central Africa. It is the third most important staple root/tuber crop after yam and cassava in Nigeria and second after cassava in Cameroon and first in Ghana (Echebiri, 2004). The main nutrient supplied by taro is carbohydrate (Jirarat et al., 2006), and it also contains proteins, vitamins and minerals (Duru and Uma, 2002). The prevalence of taro leaf blight caused by *Phytophthora infestans* (Brunt et al., 2001) and the declining soil fertility have had a negative impact on the yield of taro. Changes in dietary habits and preferences for exotic foods, and introduction of other crop species with better comparative advantages such as sweet potato and cocoyam have hindered taro production

(Joughin and Kalit, 1986). During growth processes, many enzymes play physiological roles in plants. Phenolic compounds are essential for the growth and productivity of plants; they act as metal chelators by directly scavenging molecular species of active oxygen or by inhibiting lipid peroxidation by trapping the lipid alkaxyl radical (Michalak, 2006). Peroxidases (Pox) are a family of isoenzymes found in all plants and they are involved in the scavenging of reactive oxygen species (ROS), produced throughout plant development in response to biotic and abiotic stresses. Polyphenoloxidase (PPO) is important in the oxidation of phenolic compounds to quinines and the reinforcement of physical barriers of cells through the process of lignification (Campos-Vargas and Saltveit, 2002).

The increased use of chemical fertilizers and some organic fertilizers in agriculture helped the country in achieving self sufficiency in food production. However, it has also polluted the environment and caused slow deterioration of soil health. The chemical residues in the food product are also causing injury to human beings and animal production. To combat these problems and in the light of sustainable agriculture, green technology is now used by most farmers (Piqueres et al., 2005).

It has led to increased research efforts on the biological components of dynamic soil microorganisms that are beneficial for plant growth, resulting in rapid minerali-zation of organic matter, suppression of soil-borne pathogens and increased crop yield and quality (Higa, 1996). Such products include indigenous microorganisms (IMO) and effective microorganisms (EM) (Helen et al., 2006). IMO are beneficial members of the soil including filamentous fungi, yeast and bacteria collected from non-cultivated soils near the area where they are applied. IMO is collected from the environment surrounding the farm and its use is aimed at protecting life and integrity of the natural world. However, a high degree of farm management needs to be put in place if maximum benefits from IMO need to be obtained (Helen et al., 2006). EM consists of mixed culture of beneficial and naturally occurring microorganisms that can be applied as inoculants to increase the microbial density of soils and plants (Zimmermann and Kamukuendjie, 2008). Research has shown that the inoculation of EM cultures to the soil/plant ecosystem can improve soil quality, soil health and the growth, yield and quality of crops (Daly and Steward, 1999).

However, the use of mixed cultures of beneficial microorganisms as soil inoculants to enhance growth, health and quality of crops is still questionable by researchers since the claim lacks conclusive scientific proof (Szymanski and Patterson, 2003). In fact, data to justify the use of EM and IMO in cultivation and production of crops in Cameroon are lacking. Therefore, the aim of this study was to evaluate the effect of two organic manures (EM and IMO) on field cultivation of C. esculenta through morphological, agronomic and biochemical markers.

MATERIALS AND METHODS

Location

The experiment was carried out at the research farm of Higher Teacher Training College (HTTC) Bambili, University of Bamenda (Cameroon). It is located at Latitude 5°99'0" north and longitude 10°15'00" east. Bamenda is in the North West Cameroon highlands, having a mean temperature of about 24°C, a humid tropical climate with an average rainfall of 2000 mm and an altitude of 900 m above sea level (Gwaabe, 2000).

Land preparation

The land for planting was cleared and raked, and then ridges (70 to 100 cm) were formed. The experimental design was the randomized complete block design (RCBD) with three treatments and six replications: EM manure, IMO manure and control.

Manure preparation and application

The EM manure was prepared according to the method of Higa (1991), whereas IMO manure was prepared according to the method of Helen et al. (2006), using local farming field material. The treatments were applied 1 week before planting and 3 months after planting at a concentration of 20 g per hole. Cormels of C. esculenta (100 to 125 g) with 3 to 4 buds were planted at 1 m interval. Weeds were controlled manually and by mulching done 3 months after planting.

Measurement of morphological parameters

Plant height was measured with a tape at 2 weeks intervals across 5 months. Plant height was taken from the plant base to the tip of the top most leaf. The mean number of leaves for each ridge and treatment was also recorded.

Harvest

The taro cormels were harvested 9 months after planting when all the leaves had turned yellow. The number and weight of cormels for each treatment were recorded.

Disease incidence

Disease incidence was evaluated by counting number of leaves in each treatment showing yellowish-brown-soaked lesions. This was done at monthly intervals for 6 months.

Phenol analysis

Fresh leaves (0.5 g) were ground at 4°C with 80% methanol, and then centrifuged at 6 000 g for 15 min after 30 min incubation. The total phenolic compounds were determined using Folin-Ciocalteu reagent according to the method described by Macheix et al. (1990). Results were expressed in μg equivalent of chlorogenic acid by reference to the standard.

Extraction of proteins and analysis of peroxidase and polyphenoloxidase

Fresh leaves (200 mg) were extracted with 3 ml Tris-maleate buffer

Table 1. Variation of the height of *C. esculenta* plants under different treatments with time.

Duration (weeks)	Control	EM manure	IMO manure
2	3.500 ± 0.985^a	3.778 ± 0.732^a	3.278 ± 0.958^a
4	6.278 ± 1.127^a	6.611 ± 1.037^a	6.389 ± 1.037^a
6	9.389 ± 0.698^a	9.944 ± 1.552^a	9.667 ± 0.686^a
8	11.389 ± 0.979^a	11.944 ± 1.162^a	11.611 ± 0.979^a
10	14.556 ± 1.042^a	15.222 ± 1.801^a	14.556 ± 1.042^a
12	18.111 ± 1.451^a	19.944 ± 2.014^b	18.667 ± 0.907^b
14	20.000 ± 1.237^a	21.667 ± 0.970^b	21.000 ± 0.840^a
16	21.222 ± 1.166^a	22.500 ± 1.200^b	21.667 ± 0.907^a
18	20.389 ± 0.850^a	22.667 ± 1.138^b	20.722 ± 1.127^a
20	20.500 ± 1.618^a	22.278 ± 1.776^b	20.389 ± 1.145^a

Means with same letter in the same line are not significantly different at $P < 0.05$ (Student Newman and Keuls test).

(0.1 M, pH 6.5) containing Triton X-100 (0.1 g.L^{-1}) and centrifuged for 15 min at 6 000 g after 1 h incubation at 4°C. The supernatant obtained was used as the crude protein extract. Pox activity was assayed spectrophotometrically at 470 nm using guaiacol as a substrate. Twenty-five microliters of enzyme extracts was added to 2.5 ml of reaction mixture containing a solution of 0.1M Tris-maleate buffer (pH 6.5) and 25 mM guaiacol. Reactions were initiated with 20 µl of H_2O_2 (10%) and ascorbic acid (0.25 mM) and stopped after 2 min. PPO activity was determined by measuring the oxidation of 0.2 M catechol at 420 nm. Enzyme activity was expressed as enzymatic unit.mg^{-1} fresh weight (EU.mg^{-1} FW)

Data analysis

The data obtained were expressed as means ± SD and were statistically analyzed using the SPSS statistical software Version 17.0 (SPSS Inc., Chicago). The significant difference between mean values was determined using analysis of variance (ANOVA). Student Newman-Keuls test was used to compare means at 0.05 level of significance.

RESULTS

Variation of the height of *C. esculenta* plants

Results showed a gradual increase in height of plants for all treatments up to week 16 with control having the longest plants. From week 18 till the end of the experiment, EM manure produced the longest plants. No significant difference was detected for plant treated with IMO manure and that of control, while a significant difference (P < 0.05) was noted for plants treated with EM manure (Table 1).

Variation of the number of leaves of *C. esculenta* plants

The number of leaves per plant increased in all treatments till the 12th week of the experimental period. Plants treated with EM manure and IMO manure showed

a significant difference in number of leaves during the early stages of experiment (up to week 8) when compared with control plants. Between weeks 16 and 18, there was a significant difference (P < 0.05) in number of leaves in plants treated with both EM and those treated with IMO manure compared with control plants. During the last week, there was a significant difference in the number of leaves for plants treated with EM manure only (3.222 ± 0.732) (Table 2).

Evaluation of productivity

Significant differences at P < 0.05 were observed for mean weight of cormels produced per treatment. Plants treated with EM manure recorded cormels that weighed the most (15.549 ± 2.17 tons/ha) followed by plants treated with IMO manure (12.335 ± 1.69 tons/ha) and then control plants (10.539 ± 2.24 tons/ha) (Figure 1).

Evaluation of the disease incidence

The disease incidence increased for all treatments up to the 4th month and started decreasing. After 1 month, control plants had the highest disease incidence (2.5 ± 1.25) followed by plants treated with IMO manure (2.16 ± 1.72) and then plants treated with EM manure (1.83 ± 0.87). At the 4th month, this disease increased in all plants, with plants treated with IMO manure having the highest value (8.66 ± 3.74) over plants treated with EM manure (7.83 ± 1.47) and control plants (7.66 ± 3.47). These values decreased at 6 months with EM manure plants recording the highest and significant value at p < 0.05 (6.00±1.26) followed by control plants (4.02 ± 1.78) and IMO manure plants (4.83 ± 2.71) (Figure 2).

Evaluation of the total phenol content

After 2 months of growth, EM manure treated plants

Table 2. Variation of the number of leaves of *C. esculenta* plants under different treatments with time.

Duration	Control	EM manure	IMO manure
2	3.111 ± 0.963[a]	2.778 ± 0.943[a]	3.333 ± 0.907[a]
4	3.944 ± 0.873[a]	4.556 ± 1.199[b]	4.611 ± 0.778[b]
6	4.444 ± 0.922[a]	4.756 ± 1.097[a]	4.944 ± 0.873[a]
8	4.944 ± 1.162[a]	5.644 ± 0.984[b]	5.889 ± 0.758[b]
10	5.778 ± 1.114[a]	5.722 ± 0.826[a]	5.833 ± 0.707[a]
12	6.056 ± 1.110[a]	6.156 ± 0.938[a]	6.111 ± 1.132[a]
14	5.389 ± 0.850[a]	5.278 ± 1.127[a]	5.222 ± 1.309[a]
16	4.222 ± 0.943[a]	4.611 ± 1.145[b]	4.611 ± 1.092[b]
18	3.244 ± 0.938[a]	3.922 ± 0.575[b]	3.833 ± 0.857[b]
20	2.667 ± 0.594[a]	3.222 ± 0.732[b]	2.778 ± 0.808[a]

Means with same letter in the same line are not significantly different at P < 0.05 (Student Newman and Keuls test).

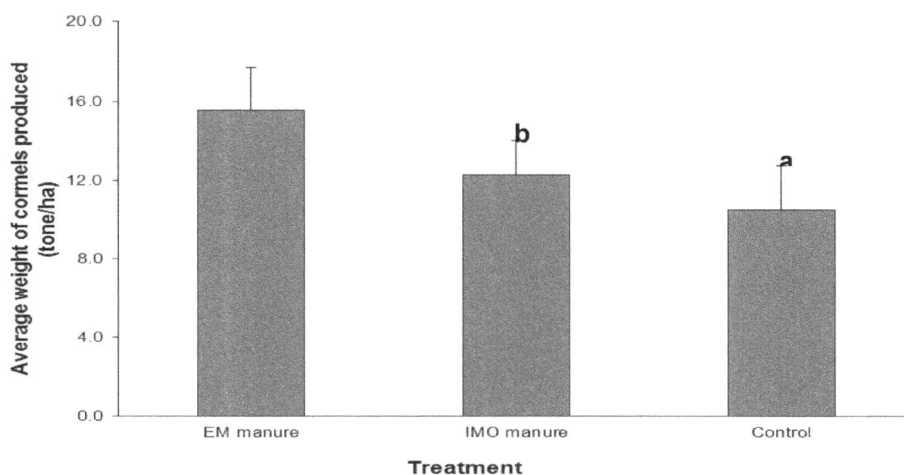

Figure 1. Average weight of cormels of *C. esculenta* produced (tons.ha^{-1}) per treatment. Histograms with same letters are not significantly different at P<0.05 (Student Newman and Keuls test).

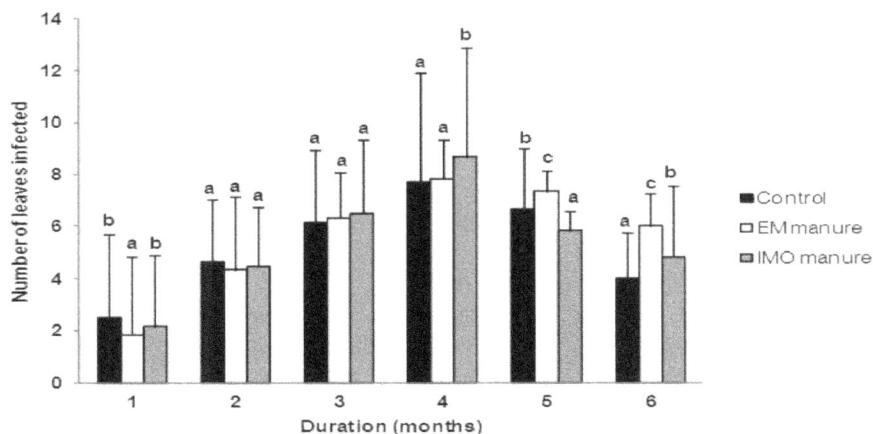

Figure 2. Evaluation of the disease incidence in *C. esculenta* plants under different treatments with time. Histograms with same letters are not significantly different at P<0.05 (Student Newman and Keuls test).

Table 3. Variation of the total phenol (μg of CA.mg^{-1} FW) content in the leaves of *C. esculenta* plants under different treatments with time.

Duration	Control	EM manure	IMO manure
2	3.929 ± 1.396^a	6.346 ± 0.779^b	4.596 ± 0.618^a
3	13.717 ± 0.926^a	15.408 ± 1.280^b	13.996 ± 0.591^a
4	18.772 ± 0.704^a	23.116 ± 0.484^b	25.670 ± 1.306^b
5	16.743 ± 3.957^a	19.732 ± 0.453^{ab}	23.791 ± 2.294^b

Means with same letter in the same line are not significantly different at P < 0.05 (Student Newman and Keuls test).

Table 4. Variation of POX activity (EU.mg^{-1} FW) in leaves of *C. esculenta* under different treatments with time.

Duration	Control	EM manure	IMO manure
2	4.680 ± 0.195^{ab}	5.617 ± 0.168^b	4.172 ± 0.231^a
3	4.595 ± 0.168^a	6.468 ± 0.132^b	5.01 ± 0.384^a
4	7.411 ± 0.502^a	7.517 ± 0.162^a	7.303 ± 0.768^a
5	9.184 ± 0.524^{ab}	9.34 ± 0.539^b	8.581 ± 0.475^a

Means with same letter in the same line are not significantly different at P < 0.05 (Student Newman and Keuls test).

Table 5. Variation of PPO activity (EU.mg^{-1} FW) in leaves of *C. esculenta* under different treatments with time.

Duration	Control	EM manure	IMO manure
2	0.936 ± 0.12^a	1.163 ± 0.189^b	0.814 ± 0.165^a
3	1.212 ± 0.32^a	1.356 ± 0.22^b	1.109 ± 0.097^a
4	1.39 ± 0.245^a	1.691 ± 0.479^{ab}	1.814 ± 0.506^b
5	1.498 ± 0.174^a	1.976 ± 0.169^b	1.773 ± 0.186^{ab}

Means with same letter in the same line are not significantly different at P < 0.05 (Student Newman and Keuls test).

produced the highest phenolic content (6.346 ± 0.779 μg of CA. mg^{-1} of FW) followed by plants treated with IMO manure (4.596 ± 0.618 μg of CA. mg^{-1} of FW) and that of control plants (3.929 ± 1.396 μg of CA. mg^{-1} of FW). This content increased with time and peak was achieved after 4 months after which it decreased in all treatments. At this time, the total phenolic content noticed was highest in plants treated with IMO manure (25.670 ± 1.306 μg of CA. mg^{-1} of FW) followed by that of EM manure (23.116 ± 0.484 μg of CA. mg^{-1} of FW) and lowest in control (18.772 ± 0.704 μg of CA. mg^{-1} of FW). At 5 months, the total phenolic content decreased in all treatments, but at this time, IMO manure plants produced the highest quantity of phenolic content followed by that of EM manure, then control plants (Table 3).

Evaluation of Pox activity

The evaluation of Pox activity showed that after 2 months of development, this activity was significantly higher (P <

0.05) in plants treated under EM manure (5.617 ± 0.168 EU.mg^{-1} FW) than that of IMO manure-treated plants (4.172 ± 0.231 EU.mg^{-1} FW) and control plants (4.680 ± 0.195 EU.mg^{-1} FW). This activity increased significantly during growth independent of treatment. After 4 months of growth, there were no significant differences of the activities of plants treated with EM manure and control plants (Table 4).

Evaluation of PPO activity

The PPO activity increased significantly during the development of plants independent of the treatment. This activity was highest in plants treated with EM manure along the process followed by plants treated with IMO manure and with control plants. After 5 months, this activity was 1.976 ± 0.169 EU.mg^{-1} FW; 1.773 ± 0.186 EU.mg^{-1} FW and 1.498 ± 0.174 EU.mg^{-1} FW for EM manure plants, IMO manure plants and control plants, respectively (Table 5).

DISCUSSION

Results obtained in this study showed that plant height increased gradually up to week 16 for control and plants treated with IMO and then week 18 for plants treated with EM manure. This result of differences in plant heights may be attributed to the gradual release of essential nutrient elements as required by C. esculenta plants. EM and IMO manure continuously supply nutrients to plants which enhance growth. The microorganisms associated with these amendments enhance the production of plant growth regulators (Arshad and Frankenberger, 1992). The longest plants obtained with EM manure could be attributed to the effect of EM stimulating mineralization of organic matter, with subsequent release of more nutrients into the soil-plant system (Daly and Stewart, 1999).

The production of more leaves by plants treated with manures than control plants was because EM and IMO manure had a steady supply of nutrients resulting from the gradual breakdown of the manure components. Similar results were obtained by Xu et al. (2000), where application of EM increased fruit yield and plant growth of a tomato crop. Apparently, the application of EM manure into the soil is usually associated with an increase in soil microbial biomass, which increases the rate of symbiotic biological nitrogen fixation through increases in Azotobacter bacteria (Hussain et al., 1994). The EM contains phytohormones and other biologically active substances delay senescence in plants (Yamada and Xu, 2000). Depletion of nutrients during the cultivation period explains why control plants had the least number of leaves.

The application of EM and IMO manure was ineffective in controlling taro leaf blight that emerged in the field. EM and IMO manures in the soil served as a good substrate for microorganisms, some of them pathogenic, like those causing taro leaf blight. The results are similar to those obtained by Ncube et al. (2010), where application of EM had no effect on the control of tomato leaf blight. It is possible that in other instances where EM has been found to have positive effects on leaf blights, the weather was not favorable for blight attack compared with our environment where the weather is at times very conducive for the development of blight and the results clearly indicate that EM may not be effective in controlling it. The IMO manure was also ineffective in controlling taro leaf blight, showing very high disease incidence especially in the 3rd and 4th months. The overall population dynamics of indigenous microbial communities during composting is influenced by a number of factors such as available nutrients, moisture, temperature, oxygen and pH (Epstein, 1997). Consequently, factors such as different bulking materials, composting and storage conditions are expected to influence the population dynamics of IMO and ultimately pathogen growth potential. Therefore the increased occurrence of disease was attributed to the inability of IMO to compete with harmful microorganisms in the presence of limited nutrients and unfavorable conditions, inability to produce inhibitory compounds or secondary metabolites and the absence of highly active IMO to lyse harmful microorganisms.

Total phenol contents are widespread in plant and tissue and could more or less accumulate depending on the physiological condition of plants (Mbouobda et al., 2010). General tendency indicates relevant activation of phenol contents, which increases with time and dependent on the manure used. Phenol content as well as Pox and PPO activities were high in plants treated with EM and IMO manures during growth development (5 months). These biochemical activities played an important role in the growth process of plants. Poxs are related to manure detoxification presumably by catalyzing the phenol oxidation at the expense of hydrogen peroxide (Wang and Balligton, 2007). They are also involved in lignin biosynthesis as a physical barrier against several stresses (Belaqziz et al., 2008). The PPOs are highly reactive intermediates whose secondary reactions are believed to be responsible for the oxidative browning which accompanies plant senescence, wounding, and responses to pathogens (Friedman and Baker, 2007). Bioactive substances such as hormones and enzymes produced by yeasts promote active cell and root division. Their secretions are useful substrates for EM and IMO such as lactic acid bacteria and actinomycetes which are intermediate to that of bacteria and fungi producing antimicrobial substances from amino acids secreted by photosynthetic bacteria and organic matter.

Conclusion

Application of EM and IMO manures increased growth parameters, yields as well as biochemical parameters of C. esculenta. However, they did not show any potential for controlling taro blight disease that affected the C. esculenta crops during growth season.

ACKNOWLEDGEMENTS

Authors are indebted to IMO student family of the 2nd batch of the Biology Department, HTTC Bambili, University of Bamenda, Cameroon for their participation in this research.

REFERENCES

Arshad M, Frankenberger WT (1992). Microbial production of plant growth regulators. In: Xu (eds). Effects of organic fertilizers and a microbial inoculants on leaf photosynthesis and fruit yield and quality of tomato plants. Crop Prod. 3:173-182.
Belaqziz M, Lakhal EK, Mbouobda HD, El Hadrami I (2008). Land spreading of Olive Wastewater: effect on maize (Zea mays) crop. J. Agronomy. 7(4): 297-305.
Brunt J, Hunter D, Delp C (2001). A bibliography of taro leaf blight. AUSAID. Taro genetic resources: conservation and utilization, Secretariat of the pacific community, Suva, Fiji. p. 96

Campos-Vargas R, Saltveit ME (2002). Involvement of putative chemical wound signal in the induction of phenolic metabolism in wounded lettuce. Physiologia Plantarum 114:73-84.

Daly MJ, Stewart DPC (1999). Influence of effective microorganisms (EM) on vegetable production and carbon mineralization: A preliminary investigation. J. Sustain. Agric.14:222-232.

Duru CC, Uma NU (2002). Post harvest spoilage of cormels of Xanthosoma sagitttifolium (L).Schott. Biosci. Res Commun. 14:277-283.

Echebiri RN (2004). Socio-economic factors and resource allocation in cocoyam production in Abia State, Nigeria: A case study. J. Sustainable Trop. Agric. Res. 9:69-73.

Epstein E. (1997). The Science of Composting. Technologic Publishing Company Inc., Lancaster.

Friedman AR, Baker BJ (2007). The evolution of resistance genes in multi-protein plant resistance systems. Curr. Opin. Genet. Dev. 17:493-499.

Gwaabe GK (2000). A comparative study in the use of inorganic and organic fertilizer in Bambili village, DIPES I, ENSAB. University of Yaounde I, Cameroon. p. 38.

Helen J, Leopold G, Gerry G (2006). A handbook of preparations, techniques and organic amendments inspired by nature farming and adapted to locally available materials and needs in the Western Visayas region of the Philippines. Natural farming manual. pp. 1-37.

Higa T (1991). Effective microorganisms: A biotechnology for mankind. Sunmark Publishing Inc, Tokyo. pp. 8-14.

Higa T (1996). Effective microorganisms - Their role in Kyusei Nature Farming. Proceedings of the 3rd International Nature Farming Conference. USDA; Washington. pp. 20-23.

Hussain T, Ahmad R, Jilani G, Higa T (1994). Applied EM technology. Nature Farming Research Center. University of Agriculture, Faisalabad. pp 1- 6.

Jirarat T, Sukruedee A, Pasawadee P (2006). Chemical and Physical Properties of flour extracted from Taro (Colocasia esculenta (L)) Schott grown in different regions of Thailand. Sci. Asia 32:279-284.

Joughin J, Kalit K (1986). The changing cost of food in Papua New Guinea-an analysis of five urban markets. Technical Report N 14. Department of Primary Industry, Port Moresby. p. 56.

Macheix JJ, Billot J, Fleuriet A (1990). Fruits Phenolics. 1st Edn, CR Press, Inc., Boca. Raton, FL. Pp. 149-237.

Mbouobda HD, Fotso, Djocgoue PF, Omokolo ND, El Hadrami I, Boudjeko T (2010). Benzo-(1,2,3)-thiadiazole-7-carbothionic S-methyl ester (BTH) stimulates defense reactions in Xanthosoma sagittifolium. Phytoparasitica 38:71-79.

Michalak A. (2006). Phenolic compounds and their antioxidant activity in plants. Pol. J. Environ. Stud. 15:347-362.

Ncube L, Pearson NMS, Brutsch MO (2010). Agronomic suitability of effective micro-organisms for tomato production, Full Length Research Paper, South Africa.

Piqueres PA, Hermann EV, Alabouvette C, Steinberg C (2005). Response of soil microbial communities to compost amendments. Soil Biol. Biochem. 38:460-470.

Szymanski N, Patterson RN (2003). Effective microorganisms (EM) and waste water systems in future directions for on-site systems. Best management practice. pp. 259-262.

Wang SY, Ballington JR (2007). Free radical scavenging capacity and antioxidant enzyme activity in deerberry (Vaccinium stamineum L.). Food-Sci. Technol. 140:1352-1361.

Xu HL, Wang RA, Mridha MAU (2000). Effects of organic fertilizers and microbial inoculants on leaf photosynthesis and fruit yield and quality of tomato plants. Crop Prod. 3:173-182.

Yamada K, Xu H (2000). Properties and Applications of an Organic Fertilizer Inoculated with Effective Microorganisms. J. Crop Prod. 3(1):255-268.

Zimmermann I, Kamukuenjandje R (2008). Overview of a variety of trials on agricultural application of effective microorganisms (EM). Agricola. pp. 17-26.

11

Irrigation with saline-sodic water: Effects on soil chemical-physical properties

G. Cucci*, G. Lacolla and P. Rubino

Department of Agricultural and Environmental Science, University of Bari Via G. Amendola 165/A 70126 Bari, Italy.

The results of a two-year research, aimed at studying the effect of irrigation with saline and sodic water on soil physical and chemical properties, are reported. Bean and capsicum were grown in pots filled with two different clay-loam soils, irrigated with 9 types of water obtained from the factorial combination of three salt concentration levels (0.001 – 0.01 – 0.1 M for bean, and 0.01-0.032- 0.1 M for capsicum) with three sodium adsorption ratio (SAR) levels (5, 15 and 45) and were subjected to two leaching fractions (10 and 20%). The results did not show any significant effect of irrigation water's salinity and sodicity, and of the leaching fraction, on soil type. The use of irrigation water with 0.1 M salt concentration caused an increase in electrical conductivity (ECe) from an initial average value of 0.71 dS m^{-1} to 13.9 and 19.5 dS m^{-1}, at the end of the first and the second irrigation season, respectively; small variations were, instead, observed, for soil pH. Despite the use of leaching fractions, any increase in the salt concentration and SAR of irrigation water resulted in an increase in the exchangeable Na percentage and a decrease in the exchangeable K, Ca and Mg.

Key words: Soil type, sodic-saline water, leaching, exchangeable sodium percent (ESP), soil aggregates stability.

INTRODUCTION

Soil salinization and sodification have been identified as major causes of land degradation. Postel (1996) reports that salt-affected areas increase at a high rate, by about 2 million hectares per year. Secondary salinization is the consequence of a not optimal irrigation water management and of the use of saline water for irrigation. This problem is particularly critical in arid and semi-arid regions where total water availability is limited and good quality water is addressed to high-valued uses, and thus poor quality waters, including wastewaters (Minhas et al., 2007; UNESCO, 2003), is often used for irrigation (Richards et al., 1954; Szabolcs, 1989; So and Aylmore, 1993; Tedeschi and Dell'Aquila, 2005). Particularly, if this concerns domestic wastewater, it may have high Na-concentrations resulting from the salt content of human

food (Tedeschi and Menenti, 2002; van der Zee et al., 2010).

The problems related to the use of highly saline water affect above all the desert areas in Southern America; some states of the U.S.A: such as California, Arizona (Ayers and Westcot, 1985); many Asian regions, including Pakistan, India, Bangladesh, China, Japan (Levy et al., 1988); the Middle East, the areas between Tigris and Euphrates (Iraq), Bahrain (Ayers and Westcot, 1985); the Negev Desert (Israele) (Pasternak and De Malach, 1987) and the Mediterranean region (Levy et al., 1988). The problem does also exist in Italy, especially in Central-Southern areas, where groundwater is the main source of water supply. As a consequence of the water table drawdown, caused both by its continuous and intensive exploitation and by the low rainfall recorded over the last decades, a significant increase in the salt content of water occurs especially in coastal areas, thus causing a further deterioration of the existing conditions (Graifenberg et al., 1993; Postiglione et al., 1994).

*Corresponding author. E-mail: giovanna.cucci@agr.uniba.it.

The major problem of irrigating with saline water is not actually the crop response to irrigation (which is a basically short-term effect) but rather the long-term changes on soil properties that might seriously alter its fertility.

The risk of soil fertility degradation depends both on the total salt content of irrigation water and on the salt composition, especially in relation to Na concentration.

Soil sodification refers to the accumulation of Na, in relation to divalent cations (mainly Ca and Mg), in the soil solution and at the cation exchange complex and may induce severe structural degradation in loamy and clayey soils that contain swelling minerals (Bresler et al., 1982), as consequence of clay particle dispersion. Among physical parameters, the soil aggregates stability index can be considered a good indicator of the soil quality status as function of the agronomic practices adopted (Manachini et al., 2009). Soil fertility degradation depends on irrigation water quality and on soil physical properties as well, with particular regard to clay mineral characteristics (Cavazza et al., 2002).

The de-flocculating effect of sodium ion on clay increases with the concentration of adsorbed sodium; the critical level of exchangeable sodium percentage (ESP) is usually taken as 15% of the cation exchange capacity, although in some soils the sodification characteristics occur at much lower values (Murray and Quirk, 1990).

If ESP becomes too large (e.g. over 15%), the hazard of organic and inorganic colloid dispersion upon introducing good quality water (such as rainwater) increases. Swelling, compression of larger pores, and a severe and often irreversible reduction of hydraulic conductivity can be the result (So and Aylmore, 1993; Halliwell et al., 2001). Since the development of soil sodicity is gradual, and often irreversible within limits imposed by reasonable time scales and costs, it is essential to anticipate its onset. Unfortunately, relatively simple conceptual tools such as the leaching requirement for salinity control (Richards et al., 1954; Howell, 1988; Corwin et al., 2007) are not available for sodicity control. The analysis of the evolution of soil chemical and physical properties, as consequence of supplying water characterized by different total salt concentrations and type, could aid to a better understanding of processes involved and to the definition of tools and land managements able to reduce the soil degradation risk.

Therefore, to provide additional insight into this issue, a two-year research has been performed at Bari University (Italy) with the aim of assessing the effects of irrigation with saline and sodic water on physical and chemical properties of two contrasting soils.

MATERIALS AND METHODS

The research was carried two-year period at the Campus of the Agricultural Faculty of Bari University (Italy) on bean (*Phaseolus vulgaris* L.) and capsicum (*Capsicum annuum* L.) successive crops, grown in cylindrical pots of 20 and 100 cm respectively in diameter and height, supplied with a bottom valve to collect drained water, and located under shed to prevent the leaching action of rainfall.

In both years, thirty-six treatments obtained from the factorial combination of two not saline soil types with nine types of water and two leaching fractions (10 to 20%) were compared.

The soils were clay loam; the first (T1) was poor in iron and aluminium sesquioxides, non calcareous, taken from the AP horizon of a *Udertic Ustochrept* fine, mixed mesic, series Montefalcone on the Emilia-Romagna soil map, northern Italy; the second (T2) contained more kaolin, it was calcareous and rich in sesquioxides, taken from the AP horizon of a *Pachic Haploxeroll*, fine mixed, thermic, series Cutino on the Apulia soil map, southern Italyn (Table 1).

The nine types of water were obtained by dissolving adequate amounts of NaCl and CaCl$_2$ in de-ionised water, and from the factorial combination of three salt concentration levels (0.001- 0.01 - 0.1 M for bean irrigation in the first year; 0.01 – 0.032 – 0.1 M for capsicum irrigation in the second year) with 3 SAR (sodium adsorption ratio) levels (5 - 15 - 45) (Table 2).

A split plot design with two replicates was used, with the two soil types in plots (18 pots), the two leaching fractions in sub-plots (9 pots) and the nine types of water in sub-sub-plots (single pots). As concerns, the fat round bean, cv. Taylor's Horticultural, just after sowing several water applications were effected using the compared different types of water to favour seed germination and plantlet emergence, which occurred 15 to 20 days after sowing. For capsicum (cv. Argo) just after transplantation, plantlets were instead irrigated with de-ionised water till rooting. Just after bean emergence and capsicum plantlet rooting, irrigation was applied whenever 30% of the maximum available moisture was lost by evapotranspiration. The amount of irrigation water corresponded to the volume required to restore the field capacity in the whole soil mass contained in each pot, plus the leaching fraction.

Through the cropping cycles, the water drained from each pot was collected and analysed for the leached solutes.

To characterise the soil at the end of the cropping cycle both for fat round bean and capsicum, for each pot (when it had lost 30% of the maximum available moisture) soil samples were taken from the upper layer (0 to 30 cm) for the following determinations:

1. Electrical conductivity and pH of the saturation extract, soluble bases (Na, Ca, Mg), SAR;
2. Exchangeable cations (Na - K - Ca - Mg), exchangeable sodium percentage.

After the two years of irrigation with saline-sodic waters, the soil stability has been evaluated on average samples collected along the whole profile, after separating the soil aggregates ranging between 1 and 2 mm diameter and using wet sieving with vertical oscillation, with or without alcohol pretreatment (Hénin et al., 1969; Kemper and Rosenau, 1986). The tests of stability of the aggregates allow differentiating soils according to their physical properties, but little is known about the relationship between indicators of aggregate stability and soil response to specific destabilizing factors.

All data were then submitted to analysis of variance using the SAS software (S.A.S.INSTITUTEINC.-USA), and the differences between the means were assessed by the Student-Newman-Keuls test; the most significant ones are reported in Figures 1 to 7.

RESULTS

In the adopted watering regime, the irrigation variables were equal for both soil types being compared but varied

Table 1. Main properties of the soils being tested.

Parameter	T1	T2
Chemical properties		
Total nitrogen (Kjeldahl method) (g kg^{-1})	0.79	1.65
Available phosphorus (Olsen method) (mg kg^{-1})	31.50	52.50
Exchangeable potassium (BaCl$_2$ method) (mg kg^{-1})	160.00	352.00
Organic matter (Walkley Black method) (g 100 g^{-1})	1.21	3.13
Total limestone (g 100 g^{-1})	0.47	2.58
Active limestone (g 100 g^{-1})	0.05	1.40
pH (pH in H$_2$O)	6.80	6.90
ECe (dS m^{-1})	0.65	0.78
ESP	0.70	0.80
CEC (BaCl$_2$ method) (meq 100 g^{-1} of dry soil)	29.54	31.61
Particle-size analysis		
Total sand: 2 > $\varnothing$ > 0.02 mm (g 100 g^{-1})	30.27	20.94
Silt (%): 0.02 > $\varnothing$ > 0.002 mm (g 100 g^{-1})	33.10	44.00
Clay (%): $\varnothing$ < 0.002 mm (g 100 g^{-1})	33.63	35.06
Hydrologic properties		
Field capacity (field determ.) (g 100 g^{-1} of soil dry mass)	34.50	35.80
Wilting point (-1.5 MPa) (g 100 g^{-1} of soil dry mass)	14.70	18.40
Bulk density (t m^{-3})	1.20	1.20

Table 2. Salt concentration, sodium adsorption ratio (SAR) and electrical conductivity (ECw) of the irrigation water compared.

Bean			Capsicum		
Salt concentration (M)	SAR	ECw (dS m^{-1})	Salt concentration (M)	SAR	ECw (dS m^{-1})
0.001	5	0.13	0.01	5	1.47
0.001	15	0.12	0.01	15	1.24
0.001	45	0.12	0.01	45	1.19
0.01	5	1.47	0.032	5	4.65
0.01	15	1.24	0.032	15	3.86
0.01	45	1.19	0.032	45	3.59
0.1	5	13.55	0.1	5	13.55
0.1	15	11.18	0.1	15	11.18
0.1	45	10.20	0.1	45	10.20

according to the quality of applied water. The number of waterings and the seasonal irrigation volumes supplied to each pot based on evapotranspiration, decreased as the irrigation water salinity and SAR increased, due to the smaller plant development. On average, the seasonal irrigation volumes per pot decreased as the salt concentration of the irrigation water increased, respectively with leaching fractions of 10 and 20% of the watering volume, from 35 to 13 L and from 40 to 16 L for bean, from 67 to 56 L and from 75 to 64 L for capsicum. Even the solutes supplied to each pot varied as a function of the quality and quantity of applied water (Figure 1).

In particular, with 10% leaching fraction, applied solutes were, 10.2 and 204 mg/100 g of soil dry mass respectively, for bean and capsicum when irrigated with water of lower salt concentration, and 484 and 1476 mg/100 g of soil d. m. when irrigated with water of higher salt concentration; with 20% leaching fraction, instead, the applied solutes were 11.9 and 228 and 554 and 1595 mg/100 g of soil dry mass (Figure 1), respectively, for the waters of lower and higher salt concentrations.

The amounts of water drained from each pot were

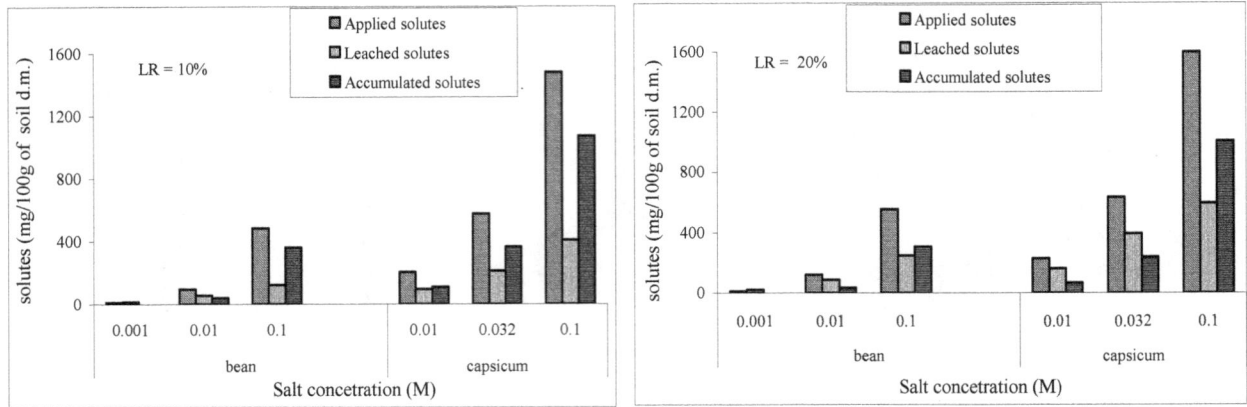

Figure 1. Applied, leached and accumulated solutes in the soil irrigated with water of different salt concentrations and subject to two leaching fractions in t bean and capsicum end of the cropping cycle. For each effect considered, the values followed by the same letter are not significantly different, according to the SNK test at P ≤ 0.01.

proportional to the applied leaching fractions; the amount of leached solutes, in turn, varied as influenced by the quantity and concentration of drainage water. As a consequence, the leached solutes changed, on average, from the lowest to the highest salt concentration, with LR = 10%, from 14 to 123 mg/100 g of soil dry mass for bean and from 94.5 to 407 mg/100 g of dry mass for capsicum; instead with LR = 20%, it varied from 21 to 248 mg/100 g of soil dry mass for the crop and from 162 to 592 for the crop (Figure 1). From the balance between applied and leached salts it was found that, irrigating bean with water of lower salt concentration (0.001 M) for both leaching requirements (10 and 20% of the watering volume), leached solutes were lower than solutes applied through irrigation water; on average, the loss was 6.6 mg/100 g of soil dry mass. For the pots irrigated with water of higher salt concentrations (0.01 and 0.1 M for bean and 0.032 and 0,1 M for capsicum), instead, accumulated salts were, on average, 36 and 334 mg/100 g of soil dry mass after bean and 300 and 1035 mg/100 g of soil dry mass after capsicum (Figure 1). In both years, no marked differences were observed in terms of amounts of leached solutes varying the leaching fraction from 10 to 20% of the watering volume. As a result of the salt balance, the EC values of the saturation of the top layer soil, 30 cm deep, extract at the end of the 1st and 2nd irrigation seasons respectively, increased, on average, from 1 and 2.2 dS·m^{-1} for the soils irrigated with water of lower salinity level to 2.1 and 4.3 dS m^{-1} and to 13.9 and 19.5 dS m^{-1} for the soils irrigated with waters of intermediate and higher salt concentrations (Figure 2).

With the different SAR values of irrigation water (5 - 15 - 45), the electrical conductivity of the saturation extract (ECe) varied respectively from 5.9 to 5.5 and 4.2 dS · m^{-1} in the 1st year, and from 8.7 to 8.4 and 7.9 in the 2nd year (Figure 2).

Slight modifications were, instead, recorded for the pH of the soil saturation extract, which ranged between 7.4

and 8.1, respectively, using waters of increasing salinity levels and SAR (Figure 3).

With the increase in irrigation water salt concentration, the exchangeable Na gradually increased whereas the exchangeable K, Ca and Mg did not vary significantly, as shown in Figure 4

With the increase in the SAR of irrigation water, the exchangeable Na and, to a lower extent, the exchangeable Mg increased, whereas the amounts of exchangeable K did not vary significantly (Figure 5).

As a result, the exchangeable sodium percentage increased with the increase in the salt concentration and SAR of irrigation water (Figure 6). The SAR of the soil saturation extract was closely correlated with the exchangeable sodium percentage.

Specifically, exchangeable sodium percentage (ESP), as average of the whole soil profile, consistently increased to the increase of the salinity of the water used and, in any case, to the raise of the SAR value. Increasing ESP values worsen soil structure; in the red soil (T2), rich in organic matter, soil structure remains more stable (stability index = 38.8%); this behavior is not observed in the gray calcareous soil (T1) (stability index = 38.8%) (Figure 7).

Illite and caolinite play an important role in stabilizing soil structure especially in soils rich in sesquioxides (Cavazza et al., 2002). In both soils, to the increase of salt concentration of the irrigation water, the trend of the soil aggregates stability indices reflected the ESP level. The structural aggregates stability index decreased on average of more than 12% passing from soils previously irrigated with water having a salt concentration of 0.01 M to those irrigated with water characterized by a salt concentration of 0.1 M (Figure 7). To the low structural stability of not pretreated samples corresponded a higher stability after alcohol pretreatment (Figure 7). The first result represents the conditions at soil surface (effect of rain), while the second better indicates the effect under

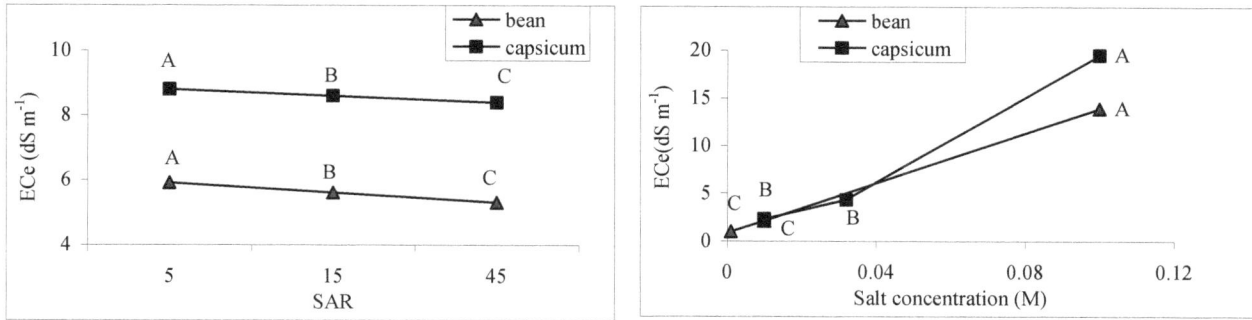

Figure 2. Electrical conductivity of the saturation extract (ECe) of the layer soil, 30 cm deep, versus the salt concentration and sodium adsorption ratio (SAR) of the solutions used for irrigation in bean and capsicum end of the cropping cycle. For each effect considered, the values followed by the same letter are not significantly different, according to the SNK test at P ≤ 0.01.

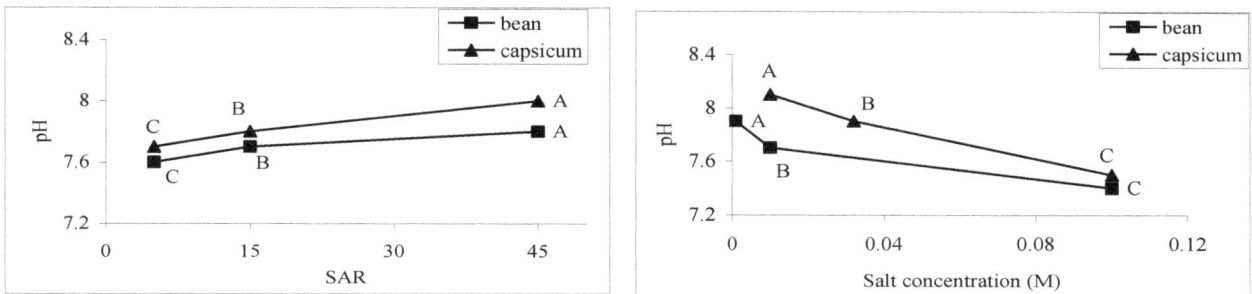

Figure 3. pH of the saturation extract of the layer soil, 30 cm deep, versus the salt concentration and the SAR of the solutions used for irrigation in bean and capsicum end of the cropping cycle. For each effect considered, the values followed by the same letter are not significantly different, according to the SNK test at P ≤ 0.01.

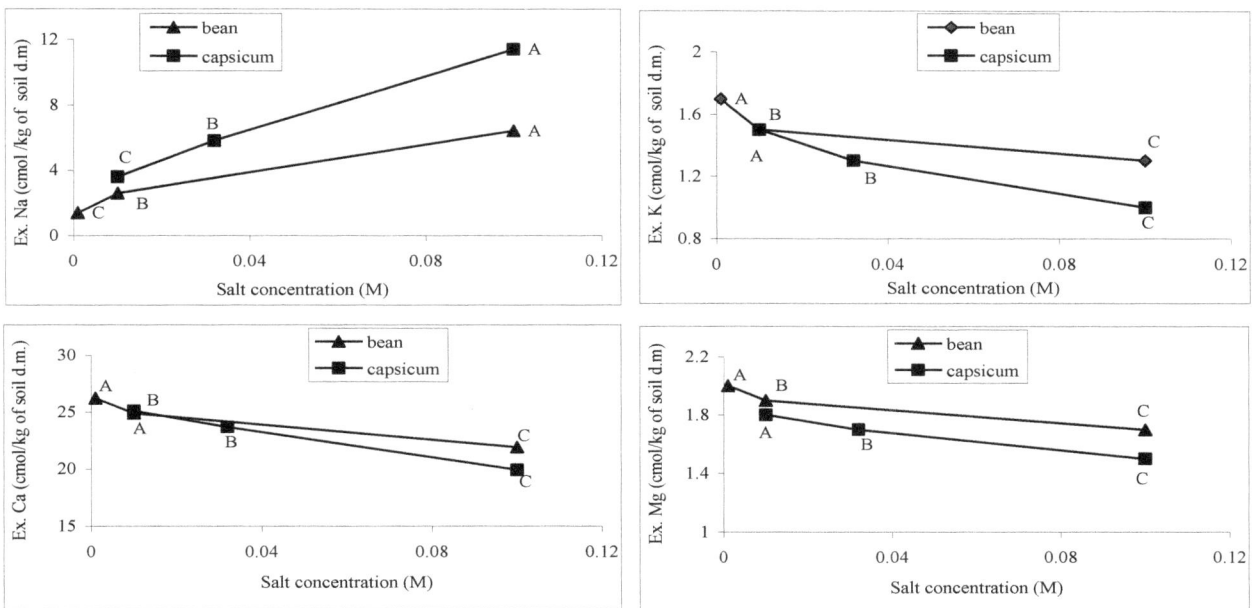

Figure 4. Variations of the soil exchangeable bases versus the concentration of the solutions used for irrigation in bean and capsicum end of the cropping cycle. For each effect considered, the values followed by the same letter are not significantly different, according to the SNK test at P ≤ 0.01.

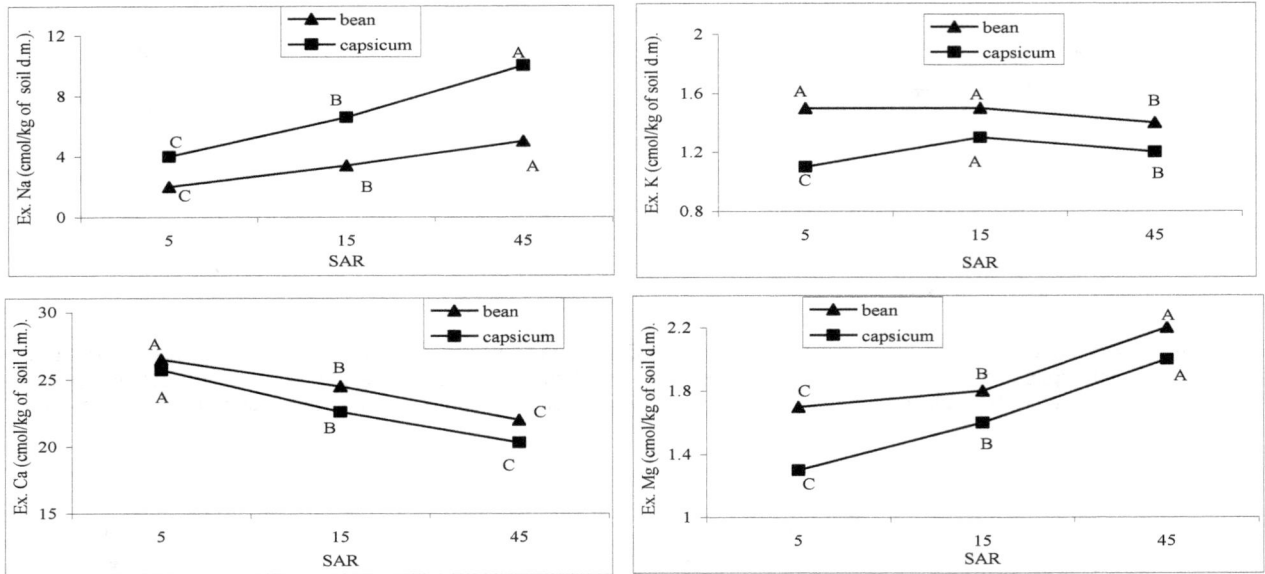

Figure 5. Variations of the soil exchangeable bases versus the SAR (Sodium Adsorption Ratio) of the solutions used for irrigation in bean and capsicum end of the cropping cycle. For each effect considered, the values followed by the same letter are not significantly different, according to the SNK test at P ≤ 0.01.

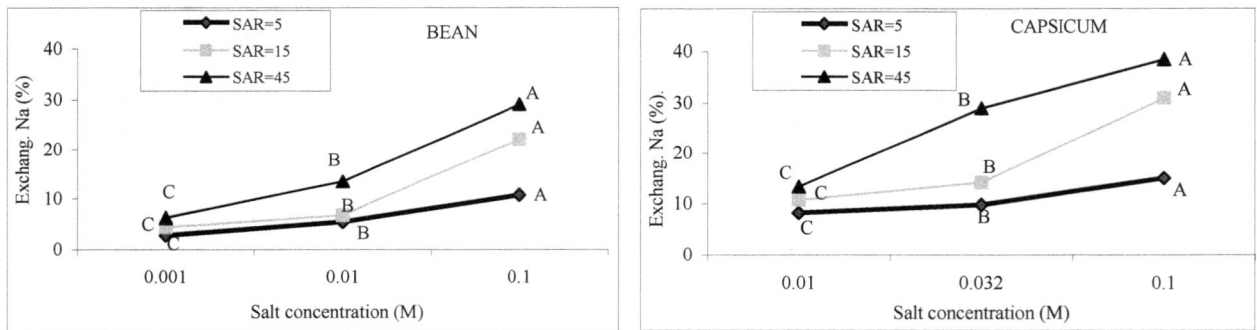

Figure 6. Soil exchangeable sodium percentage (ESP) as influenced by the salt concentration and the SAR (Sodium Adsorption Ratio) of the solutions used for irrigation in bean and capsicum end of the cropping cycle. For each effect considered, the values followed by the same letter are not significantly different, according to the SNK test at P ≤ 0.01.

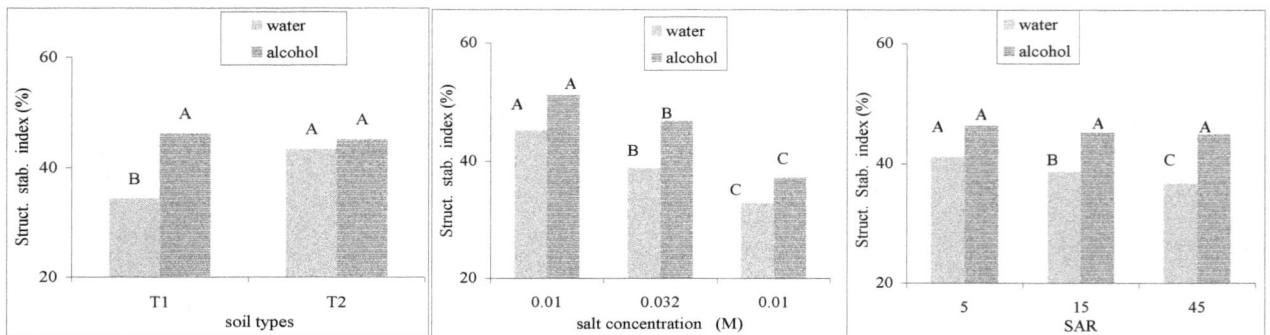

Figure 7. Variation of the soil aggregate stability index, with or without alcohol pretreatment as function of soil type, salt concentration and irrigation water SAR. For each effect considered, the values followed by the same letter are not significantly different, according to the SNK test at P≤ 0.01.

the soil surface (Cavazza et al., 2002).

DISCUSSION

A two-year research was conducted on two soil types, packed in cylindrical pots located under shed, in which bean and capsicum were grown in succession and irrigated with nine types of water, with different salt concentrations and SAR values, and subject to two different leaching requirement levels. The following conclusions may be drawn: The seasonal irrigation volume increased when the leaching requirement (LR) was doubled from 10 to 20% of the watering volume; it decreased, instead, as the salinity of the irrigation water increased. This is due to the fact that salts induced less crop growth and reduced evapotranspiration.

The soil applied solutes increased proportionately to the applied water volume and its salinity. The drainage water volumes were different in relation to the applied leaching requirements. In the first year, because of the soil pore-size reduction, due to soil compaction, and water salinity, drainage water volumes were low, when low and medium salinity waters were used; they were higher, instead, when higher salinity water was used. In the second year, when the salt concentrations of low and medium salinity waters were higher than those of the first year, the drainage water volumes were higher as compared to the applied leaching requirement. The amounts of leached solutes varied with the amount and salt concentration of drainage water. However, with the same amounts of leaching requirements, the leached solute percentage, as compared to those supplied with irrigation water, decreased considerably as the irrigation water salinity increased, with a subsequent reduction of the leaching efficiency of applied water. Therefore, the amount of solutes accumulated in the soil increased as the salt concentration of irrigation water increased; on the other hand, there was a slight variation with the higher leaching requirement and between the two compared soils. The results show that in the Mediterranean areas, where the long-term average yearly rainfall is not less than 450 to 500 mm, winter rainfall could effectively leach the solutes applied with saline water, thus reducing the amounts of irrigation water and the solutes applied to the soil.

As a result of the balance between applied and leached solutes, at the end of the irrigation season of the first and second years, the electrical conductivity of saturation extract (ECe) of the 0.3 m top soil layer, irrigated with the lowest and the highest saline waters, resulted, respectively, equal to 1 and 2.2 dSm^{-1} and to 13.9 and 19.5 dSm^{-1}, against an average value of 0.71 $dS\ m^{-1}$ recorded before starting the research. As to the characteristics of the saturation extract by increasing the salt concentration of irrigation water, the pH decreased slightly, while by increasing the SAR, the ECe varied a little and the pH increased. The exchangeable sodium percentage (ESP) increased gradually while the exchangeable calcium percentages decreased gradually as the increasing salinity and SAR of irrigation water increased. The exchangeable potassium percentage did not vary appreciably, and the exchangeable magnesium percentage increased with the rise of the irrigation water SAR.

Soil structure stability index progressively decreased to the increase in soil salinization and sodication in the two soil types. In the red Locorotondo soil, with the clay fraction rich in illite and caolinite, and also in organic matter and iron and aluminum oxides, the structure aggregates remained more stable.

ACKNOWLEDGEMENTS

This work is part of the co-operative research programme n. 9807038430 co-financed by the *Ministero dell'Università e della Ricerca Scientifica e Tecnologica* and the University of Bari.

REFERENCES

Ayers RS, Westcot DW (1985). Water quality for agriculture. Irrigation and Drainage. Paper No 29, FAO, p. 174.

Bresler E, MacNeal BL, Carter DL (1982). Saline a nd sodic soils: PrinciplesDynamics-Modeling. Springer-Verlag, Berlin/Heidelberg/New York. p. 236.

Cavazza L, Patruno A, Cirillo E (2002). Soil trait and structure stability in artificial sodicated soil. Italian J. Agron. 6(1):15-25.

Corwin DL, Rhoades JD, Simunek J (2007). Leaching requirement for soil salinity control: steady-state versus transient models. Agric. Water Manage. 90:165-180.

Graifenberg A, Lipucci Di Paola M, Giustiniani L (1993). Yield and growth of globe artichoke under saline-sodic conditions. HortSci, 8:791-793.

Halliwell DJ, Barlow KM, Nash DM (2001). A review of the effects of wastewater sodium on soil physical properties and their implications for irrigation systems. Australian J. Soil Res. CSIRO publishing 39:1259–1267.

Hénin S, Gras R, Monnier G (1969). Le profile cultural. Masson et Cie, Paris. p.332.

Howell TA (1988). Irrigation Efficiencies: Handbook of Engineering in Agriculture. CRC Press, Boca Raton, Florida. pp.173–184.

Kemper WD, Rosenau RC (1986). Aggregate stability and size distribution. In: Klute A. (ed.): Methods of Soil Analysis. Parti. Physical and mineralogical methods. Agronomy 9:425-442. ASA and SSSA, Madison, WI, USA.

Levy GJ, Van der Watt HVH, Shainberg I, Du Plessis HM (1988). Potassium-calcium and sodium-calcium exchange on kaolinite and Kaolinitic soils. Soil Sci. Soc. Am. J. 52:1259-1264.

Manachini B, Corsini A, Bocchi S (2009). Soil quality indicator as affected by a long term barley-maize an cropping systems. Italian J. Agron. 1(15):15-22.

Minhas PS, Dubey SK, Sharma DR (2007). Comparative affects of blending, intera/inter-seasonal cyclic uses of alkali and good quality waters on soil properties and yields of paddy and wheat. Agric. Water Manage. 87:83–90.

Murray RS, Quirk JP (1990). Interparticle force in relation to the stability aggregates. In M.F. De Boodt, M.H.B. Hayes, A. Herbillon, Editors, Soil colloids and their Associations in aggregates, Plenum Press, New York. pp. 439-461.

Pasternak D, De Malach Y (1987). Saline water irrigation in the Negev desert. Conference on Agriculture and Food Production in the Middle East. Athens, Greece, January 21-26, p. 24.

Postel S (1996). Forging a sustainable water strategy. In: State of the world 1996. Ed.: Brown R.L., London. pp. 40-59.

Postiglione L, Barbieri G, Tedeschi A (1994). Long-term effects of irrigation with saline water on some characteristics of a clay loam soil. ESSC Conference on Problems and Management of soil salinization-alkalinization in Europe. Budapest, 26-30 April.

Richards LA, Allison LE, Bernstein L, Bower CA, Brown JW, Fireman M, Hatcher JT, Hayward HE, Pearson GA, Reeve RC, Wilcox LV (1954). Diagnosis and Improvement of Saline and Alkali Soils. U.S. Dept.

Agr. Handbook p. 60, U.S. Govt. Printing Office, Washington D.C. p. 160.

So HB, Aylmore LAG (1993). How d o sodic soils behaved the effects of sodicity on soil behavior. Australian J. Soil Res. 31:761-778.

Szabolcs I (1989). Salt Affected Soils. CRC Press Inc, Boca Raton, Florida. p. 274.

Tedeschi A, Dell'Aquila R (2005). Effects of irrigation with saline waters, at different concentrations, on soil physical and chemical characteristics. Agric. Water Manage 77:308-322.

Tedeschi A, Menenti M (2002). Simulation studies of long-term saline water use: model validation and evaluation of schedules. Agric. Water Manage. 54:123-157.

UNESCO (2003). World Water Development Report. Water for People, Water for Life. UNESCO Publishing, Barghahn Books.

van der Zee SEATM, Shah SHH, van Uffelen CGRPAC, Raats dal Ferro N (2010). Soil sodicity as a result of periodic al drought. Agric. Water Manage. 97:41-49.

12

Soil pH, electric conductivity and organic matter after three years of consecutive amendment of composted tannery sludge

Ademir Sérgio Ferreira Araújo[1], Maria Dorotéia Marçal Silva[1], Luiz Fernando Carvalho Leite[2], Fabio Fernando de Araujo[3] and Nildo da Silva Dias[4]

[1]Federal University of Piauí, Agricultural Science Center, Soil Quality Laboratory, Campus da Socopo, CEP 64000-000, Teresina, PI, Brazil.
[2]Embrapa Mid-North, Av. Duque de Caxias, Teresina, PI, Brazil.
[3]UNOESTE, Campus II, Presidente Prudente, SP, Brazil.
[4]Federal University of Semi-Arid, UFERSA, Mossoró, RN, Brazil.

Experiments were carried out under field conditions to evaluate the effects of composted tannery sludge (CTS) on soil pH, electric conductivity (EC) and soil organic matter (SOM) content after three years of amendment. The CTS was applied at 0, 2.5, 5.0, 10.0 and 20.0 ton ha^{-1}. In each year, soil samples were collected 60 day after CTS amendment. After three years, the CTS increased about 5, 46 and 69% of the soil pH, EC and SOM content as compared with the control. Consecutive amendment of crescent rates of CTS promotes linear increases in soil pH, electric conductivity and organic matter content.

Key words: Composting, industrial wastes, salinity soil.

INTRODUCTION

Brazil is one of the biggest leather producers in the world with processing 45 million units per year. In addition, tannery industry in Brazil exports about 28 million units of leather with asserts of 21 billion dollars per year (Santos et al., 2011). Annually, tannery industries produce high volumes of solid wastes, commonly known as tannery sludge (TS), which presents high amounts of organic and inorganic elements (Santos et al., 2011). Therefore, the agricultural use of TS is one alternative for TS recycling (Silva et al., 2010), once that it may promote plant growth supported by its nutrients supply, and increase soil pH with consequent reduction in exchangeable aluminum availability (Teixeira et al., 2006).

Moreover, TS organic matter may improve the soil properties (Araújo et al., 2008). Adversely, this sludge contains high sodium, hydroxides and carbonates which causing soil alkalinity and salinity in long-term (Ferreira et al., 2003; Teixeira et al., 2006). Although, TS improves soil properties in the last years, the composting has been suggested as a suitable method for TS treatment before amendment on soil (Araújo et al., 2007; Santos et al., 2011). This process may improve TS characteristics improving its quality for soil use (Santos et al., 2011).

The agricultural use of composted tannery sludge (CTS) for agricultural purposes implicates in the knowledge of its characteristics, mainly alkalinity, salinity

Table 1. Chemical properties of composted tannery sludge (CTS).

Properties	CTS			Limits of heavy metals permitted[a]
	2009	2010	2011	
pH	7.8	7.2	7.5	
C (g kg^{-1})	187.5	195.3	201.2	
N (g kg^{-1})	1.28	1.39	1.51	
P (g kg^{-1})	4.02	3.83	4.91	
K (g kg^{-1})	3.25	3.51	2.90	
Ca (g kg^{-1})	95.33	84.28	121.18	
Mg (g kg^{-1})	6.80	5.71	7.21	
S (g kg^{-1})	9.39	8.43	10.20	
Cu (mg kg^{-1})	17.80	19.51	16.38	1500
Zn (mg kg^{-1})	141.67	128.31	127.81	2800
Ni (mg kg^{-1})	21.92	28.61	23.26	420
Cd (mg kg^{-1})	2.87	3.93	1.93	39
Cr (mg kg^{-1})	2.255	2.581	1.943	1000
Pb (mg kg^{-1})	42.67	38.54	40.31	300

Source: [a](CONAMA, 2006).

and organic matter content, after composting, and the effects of these characteristics on soil properties, mainly after successive amendments. Several studies have already been conducted using tannery sludge aiming to verify the effect of the amendments on soil pH, EC and organic matter content in tropical soils (Ferreira et al., 2003; Teixeira et al., 2006; Araujo et al., 2008; Araujo, 2011; Vieira et al., 2011). Araujo et al. (2008) observed that the amendment of tannery sludge with high calcium and sodium content increased soil pH and EC. However, there are no studies about the effect of composted tannery sludge on soil pH, EC and OM mainly after consecutives amendment in soil.

The objective of this work was to evaluate the effect of CTS, after three years of consecutive amendments on a Typic Quartzipsamment soil, on soil pH, electric conductivity and soil organic matter content.

MATERIALS AND METHODS

The experiment was carried out under field conditions at the "Long-Term Experimental Field" from Agricultural Science Center, Teresina, Piauí state (05° 05' S; 42° 48' W, 75 m). The regional climate is dry tropical (Köppen and Geiger, 1928) and it is characterized by two distinct seasons: rainy summer and dry winter, with annual average temperatures of 30°C and rainfall of 1.200 mm. The rainy season extends from January to April when 90% of total annual rainfall occurs. The soil is classified as Typic Quartzipsamment (Haplic Arenosol, FAO classification) (clay, 10%; silt, 28%; sand, 62% at a depth of 0 to 0.2 m). The experimental area was uncultivated since 2001 and during this period (2001 to 2009) it was covered with native vegetation. In 2009, the experimental area was prepared to start a long-term experiment with composted tannery sludge.

The experiments were conducted in 2009, 2010 and 2011 with five treatments: 0 (without CTS amendment), 2.5, 5, 10 and 20 t ha^{-1} CTS (dry basis). The experiment was arranged in a completely randomized design with four replications. The plots were marked out (20 m^2 each and 12 m^2 of useful area for soil and plant sampling) including rows spaced 1.0 m apart.

Tannery sludge was collected from the wastewater treatment plant of a tannery located at Teresina, Piauí State, Brazil. The compost was produced with tannery sludge, sugarcane straw and cattle manure in the ratio 1:1:3 (v:v:v) for 85-day composting. The size of pile was 2 m long by 1 m wide and 1.5 m high. The pile was turned twice during the first and second weeks and once a week during the rest period. At the end of the composting 20 subsamples were randomly collected from the pile to produce a composite sample of the compost. The chemical characteristics of composted tannery sludge (CTS) were determined by the EPA 3051 method (USEPA, 1986) and are shown in Table 1. Bio-available heavy metals content were determined by EPA method 3050B (USEPA, 1986).

In each year, CTS was applied ten days before cowpea (*Vigna unguiculata*) sowing. It was spread on the soil surface with incorporation into the 0.20 m layer with a harrow. Cowpea was grown at a density of 5 plants m^{-1} (about 62,000 plants ha^{-1}). Soil samples (0 to 0.2 m depth) were collected 60 days after CTS amendment using five subsamples per plot.

The soil samples were air dried, sieved (2.0 mm) and stored at room temperature before analyses. Soil pH was estimated in water (1:2.5 v:v) and measured by potenciometry (Tedesco et al., 1995). Soil electric conductivity was evaluated in water (1:2 v:v) according to Richards (1954) and measured by condutivimetry. Soil organic matter (SOM) was determined by wet combustion using a mixture of 5 mL potassium dichromate 0.167 mol L^{-1} and 7.5 mL of concentrated sulfuric acid under heating (170°C for 30 min) (Yeomans and Bremner, 1988).

The data were submitted to the analysis of variance (ANOVA) followed by Student Newman-Keuls test (5% level) and regression analyses using the ASSISTAT program (version 7.4 beta).

RESULTS AND DISCUSSION

The chemical properties of CTS used annually indicate the stability in your composition (Table 1). The high

Table 2. Soil pH, electric conductivity (EC) and organic matter (SOM) at the first and after three years of CTS amendment.

Rates (ton ha^{-1})	CTS amendment		Difference (%)
	First year	After three years	
	Soil pH		
0	6.7 ± 0.5[a b]	6.6 ± 0.4[c]	- 1.5
2.5	6.9 ± 0.4[b]	7.2 ± 0.5[b]	+ 4.3
5	7.1 ± 0.4[ab]	7.2 ± 0.5[b]	+ 1.3
10	7.2 ± 0.5[ab]	7.5 ± 0.4[ab]	+ 4.1
20	7.4 ± 0.3[a]	7.8 ± 0.6[a]	+ 5.4
	EC (dS m^{-1})		
0	0.25 ± 0.15[c]	0.28 ± 0.09[d]	+ 12
2.5	0.31 ± 0.12[b]	0.40 ± 0.11[c]	+ 29
5	0.44 ± 0.17[b]	0.61 ± 0.18[b]	+ 38
10	0.80 ± 0.23[a]	1.17 ± 0.20[a]	+ 46
20	0.92 ± 0.27[a]	1.32 ± 0.34[a]	+ 43
	SOM (g kg^{-1})		
0	8.2 ± 1.7[c]	7.8 ± 1.5[c]	- 4.0
2.5	10.2 ± 1.9[bc]	14.2 ± 2.3[b]	+ 39
5	11.8 ± 1.5[b]	18.4 ± 2.6[a]	+ 56
10	13.4 ± 2.0[a]	19.3 ± 2.0[a]	+ 44
20	14.3 ± 2.3[a]	20.1 v 2.7[a]	+ 41

Values are represented as Mean ± standard error. Values followed by the same letter within each column are not significantly different at 5% level by Student's test. Mean are calculated by the difference between the values of each parameter at first and after three years of CTS amendment.

values of CTS, pH and Ca content are related to hydroxides and carbonates used during tanning process (Santos et al., 2011). The organic matter from animal leather contributes for high organic matter content in the CTS. The CTS showed high Cd, Pb, Cr and Ni content (Table 1) and, except for Cr, these metals contents are below the maximum limits established by Brazilian regulation (CONAMA, 2006). The high contents of Cd, Cr, Ni and Pb are found because these metals are present in the chemical products used during the tanning process. In the case of Cr content, it was two times higher than the upper limits for Cr by CONAMA (2006) (Table 1) and it occurs due to the original tannery sludge which is not pre-treated in the industry. However, as observed in Table 2, soil pH increased above 7.0 as CTS rates increased and, when soil pH is alkaline, heavy metals stay inert in the soil under forms with low mobility (Gonçalves et al., 2013). Specially for Cr, in pH values above 5.0, Cr is in the insoluble form of $Cr(OH)_3$ (Teixeira et al., 2006) reducing the toxic potential. In addition, as the CTS pH is alkaline, Cr is found in the trivalent form (Cr^{3+}), which is more stable and has a low solubility and mobility (Alcântara and Camargo, 2001).

The soil pH did not increase expressively after three years of CTS amendment, while that of the soil EC and SOM showed an expressive increase after CTS amendment (Table 2). Compared with the values from the first year of amendment, three years later, the soil pH

increased about 5%, with amendment of 20 ton ha^{-1} CTS. The increase in soil EC ranged from 29 to 46%, with amendment of 2.5 and 10 ton ha^{-1} CTS, respectively. The expressive increases in soil EC was due to the high sodium content found in the CTS that contributes to promote soil salinity. On the other hand, the pH of CTS ranged from 7.2 to 7.8 (Table 1) and it contributed to lower increase in soil pH as compared with soil EC.

For SOM content, comparing the first year of amendment and three years later, the values increased from 39 to 69%, with amendment of CTS rates of 2.5 to 20 ton ha^{-1} (Table 2). These results can be attributed to the high OM content in this waste which can contributes to improve the soil quality since SOM is recognized as the main soil conditioner and a source of plant nutrients. Also, CTS presents high C content and it contributes for soil biological activity that decomposes soil organic matter. During three years of CTS amendment, the soil pH did not increase significantly, while that of the soil EC and SOM increased linearly and quadraticaly, respectively (Figure 1). Although no significant, the soil pH increased 0.95, 1.45, 1.75 units per rates of CTS amendment in 2009, 2010 and 2011, respectively, as compared with control. The regression equations for soil pH in 2009, 2010 and 2011 showed constant increases of 0.03, 0.05 and 0.04 units per ton of CTS applied, respectively. It means that CTS may present a potential to elevate soil pH which is attributed to the high carbonate

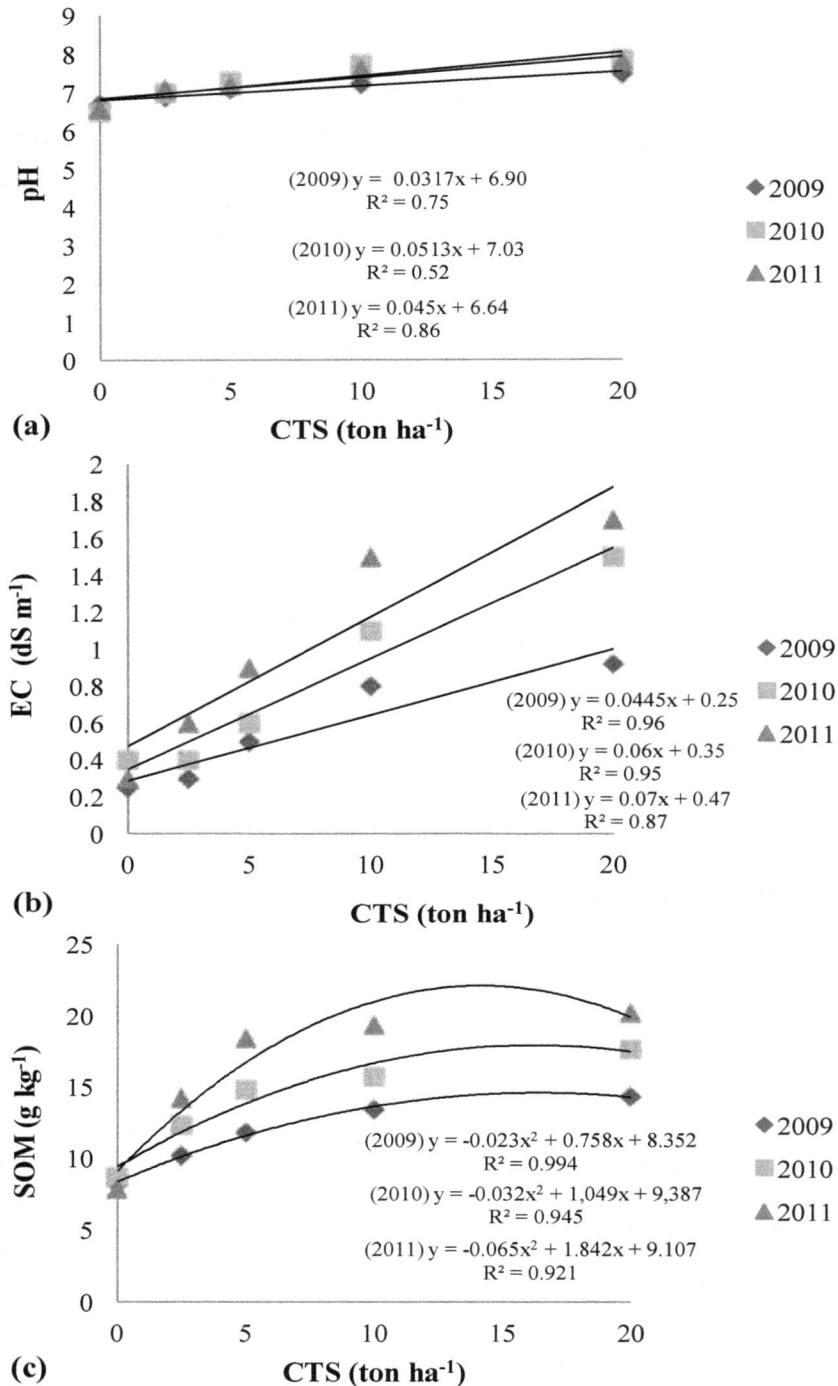

Figure 1. Soil pH (a), electric conductivity (EC) (b) and soil organic matter (SOM) (c) in a Typic Quartzipsamment after three years of CTS amendment.

content and to the hydroxides content of CTS (Teixeira et al., 2006). These values were higher than 0.01 units per ton and lower than 0.06 units per ton found by Ferreira et al. (2003) and Martinez (2012), respectively, with tannery sludge. However, the tannery sludge used by these authors presented lower and higher pH values and Ca

contents than CTS used in this study.

Our results indicated that CTS may be used as soil corrective and it contribute to neutralize Al^{3+} as also reported in other studies (Konrad and Castilhos, 2002; Ferreira et al., 2003; Teixeira et al., 2006). On the other hand, for soil pH above 7.0 may have unbalance in soil chemical properties, mainly decrease micronutrients availability (Novais et al., 2007).

The soil EC increased CTS amendment in 2009, 2010 and 2011, respectively, as compared with the control. These results indicate that CTS may increase soil salinity after its amendment in long-term as also reported in other studies (Ferreira et al., 2003; Teixeira et al., 2006). The salinity effect of CTS is due to the presence of salts, such as sodium chloride and others. The regression equations for soil EC with the data obtained in 2009, 2010 and 2011 showed a constant of increases of 0.04, 0.06 and 0.07 dS m^{-1} per ton of CTS applied, respectively, and these constants were significantly lower than those found by Martinez (2012), after tannery sludge amendment in sand soil, that varied between 0.10 and 0.18 dS m^{-1} per ton of tannery sludge. However, the tannery sludge used by that author contained about three times more sodium than the CTS used in this study.

The measurements of EC indicate the concentration of soluble salts in soil (Oliveira et al., 2002) and according to Richards (1954) the values of EC higher than 2.0 dS m^{-1} in saturation extract, indicate saline soils. In this case, our results showed that, after three years of CTS amendment, the values of EC were below this limit. The permanent monitoring of soil EC during soil use of CTS is important to avoid soil salinization.

The SOM increased more than 100% after CTS amendment in the years 2009, 2010 and 2011 as compared with control. The regression equations for SOM, using the data obtained in the three years, showed a maximum values with the amendment of 10 ton ha^{-1} CTS. These results show that CTS presents potential to increase SOM content because of its high organic content, improving soil physical, chemical, biochemical and biological properties. In addition, successive amendments of CTS would cause increases in SOM as obtained with sewage sludge (Oliveira et al., 2002), fibber and resins sludge (Trannin et al., 2008) and composted urban wastes (Mantovani et al., 2005) and tannery sludge (Santos et al., 2011).

Conclusions

The amendment of 5 ton ha^{-1} composted tannery sludge, during three years, does not increase soil pH and electric conductivity. Also, in this rate the soil organic matter increases after three years of CTS amendment. On the other hand, the amendment of 10 and 20 ton ha^{-1} composted tannery sludge increases soil alkalinity and salinity.

REFERENCES

Alcântara MAK, Camargo AO (2001). Movement of trivalent chromium as influenced by pH, soil horizon and sources of chromium. Rev. Bras. Eng. Agric. Amb. 5:497-501.

Araújo ASF, Monteiro RTR, Carvalho MES (2007). Effect of composted textile sludge on growth, nodulation and nitrogen fixation of soybean and cowpea. Biores. Technol. 97:1028-1032.

Araujo FF (2011). Disponibilização de fósforo, correção do solo, teores foliares e rendimento de milho após a incorporação de fosfatos e lodo de curtume natural e compostado. Acta Scient. Agron. 33:355-360.

Araujo FF, Tiritan CS, Pereira HM, Caetano Júnior O (2008). Maize growth and soil fertility after tannery sludge and phosphate amendment. Rev. Bras. Eng. Agric. Amb. 12:507-511.

Conselho Nacional do Meio Ambiente, CONAMA (2006). Define critérios e procedimentos para o uso de lodos de esgoto gerados em estações de tratamento de esgoto sanitário e seu productos derivados. Resolução nº 375 Diário Oficial da União: DF, No 167, pp 141-146. Acessed in Nov, 02 2012.

Ferreira AS, Camargo FAO, Tedesco MJ, Bissani CA (2003). Effects of tannery and coal mining residues on chemical and biological soil properties and on corn soybean yields. Braz. J. Soil Sci. 27:755-763.

Gonçalves ICR, Araújo ASF, Nunes LAPL, Melo WJ (2013). Soil microbial biomass after two years of consecutive application. Acta Sci. Agron. (In Press).

Konrad EE, Castilhos DD (2002). Soil chemical changes and corn growth as affected by the addition of tannery sludges. Braz. J. Soil Sci. 26:257-265.

Köppen W, Geiger R (1928). Klimate der Erde. Gotha: Verlag Justus Perthes. Wall-map 150cmx200cm.

Mantovani JR, Ferreira ME, Cruz MCP, Barbosa JC (2005). Alterações nos atributos de fertilidade em solo adubado com composto de lixo urbano. Rev. Bras. Ci. Solo 29:817-824.

Martinez AM (2012). Impacto do lodo de curtume nos atributos biológicos e químicos do solo. 2005. 74 f. Dissertação (Mestrado em Solos e Nutrição de Plantas) – Curso de Pós-graduação em Agronomia, USP-ESALQ. Acesso em 27 de maio de 2012. <http://www.teses.usp.br/teses/disponiveis/11/11140/tde-02082005-132525>

Novais RF, Alvarez VH, Barros NF, Fontes RLF, Cantarutti RB, Neves JCL (2007). Fertilidade do Solo. Viçosa: Sociedade Brasileira de Ciência do Solo. p. 1017.

Oliveira FC, Mattiazzo ME, Marciano CR, Rossetto R (2002). Organic carbon, electric conductivity, pH and CEC changes in a typic hapludox after repeated sludge amendment. Braz. J. Soil Sci. 26:505-519.

Richards LA (1954). Diagnosis improvements of saline and alkaline soils. Washington, Departament of Agriculture, 1954. p. 160.

Santos JA, Nunes LAPL, Melo WJ, Araújo ASF (2011). Tannery sludge compost amendment rates on soil microbial biomass in two different soils. Eur. J. Soil Biol. 47:146-151.

Silva JDC, Leal TTB, Araújo ASF, Araujo RM, Gomes RLF, Melo WJ, Singh RP (2010). Effect of different tannery sludge compost amendment rates on growth, biomass accumulation and yield responses of Capsicum plants. Waste Manag. 30:1976-1980.

Tedesco MJ, Gianello C, Bissani CA, Bohen H, Volkweiss SJ (1995). Análise de solo, plantas e outros materiais. 2.ed. Porto Alegre, Universidade Federal do Rio Grande do Sul. p.147.

Teixeira KRG, Gonçalves Filho LAR, Carvalho SEM, Araújo ASF, Santos VB (2006). Efeito da adição de lodo de curtume na fertilidade do solo, nodulação e rendimento de matéria seca do caupi. Ci. Agrotec. 30:1071-1076.

Trannin ICB, Siqueira JO, Moreira FMS (2008). Atributos químicos e físicos de um solo tratado com biossólido industrial e cultivado com milho. Rev. Bras. Eng. Agríc. Amb. 12:223-230

UNITED STATES ENVIRONMENTAL PROTECTION AGENCY, USEPA (1986). Test method for evaluating solid wastes. Washington, p.255. (Report Number, SW- 846)

Vieira GD, Castilhos DD, Castilhos RMV (2011). Atributos do solo e crescimento do milho decorrentes da adição de lodo anaeróbio da estação de tratamento de efluentes da parboilização do arroz. R. Bras. Ci. Solo 35:535-542.

Yeomans JC, Bremner JM (1998). A rapid and precise method for routine determimation of organic carbon in soil. Comm. Soil Sci. Pl. Anal. 19:1467-1476.

13

Effect of different levels of farmyard manure and nitrogen on the yield and nitrogen uptake by stevia (*Stevia rebaudiana* Bertoni)

Zahida Rashid, Mudasir Rashid, Suhail Inamullah, Souliha Rasool and Fayaz Ah. Bahar

Department of Agronomy, Punjab Agricultural University, Ludhiana, 141 004, Punjab, India.

Two independent field experiments were conducted at Punjab Agricultural University, Ludhiana, Punjab, India in 2006 and 2007 in a loamy sand soil normal in soil reaction and electrical conductivity, low in organic carbon and available nitrogen, medium in phosphorus and potassium status in a complete randomized design with four replications to study the effect of different levels of farmyard manure (FYM) and nitrogen on the yield and nitrogen uptake of *Stevia rebaudiana* Bertoni. Four levels of farmyard manure (0, 15, 30 and 45 t/ha) and four levels of nitrogen (0, 20, 40 and 60 kg/ha) were tested. Biomass yield (kg leaves/ha) and several other yield parameters (dry leaf yield, number of leaves per plant, leaf area index and dry matter accumulation per plant) were highest in plants grown at the highest level of farmyard manure (45 t FYM ha^{-1}). Maximum nitrogen uptake was also recorded under 45 t FYM ha^{-1} in both experiments. Plants grown at 40 and 60 kg N ha^{-1} produced significantly higher number of branches, number of leaves per plant, and showed higher leaf area index and nitrogen uptake compared to lower nitrogen levels. Dry leaf yield and dry matter accumulation per plant was highest at 60 kg N ha^{-1}.

Key words: Stevia, farmyard manure, nitrogen, yield, nitrogen uptake.

INTRODUCTION

Sugar forms an indispensable ingredient in the food habits of human being. The main source of sugar has for long been cane sugar with beet sugar contributing a small percentage. In India, the production of cane sugar is of the order of 240 million tonnes whereas that of beet sugar is 19500 tones. Though these sugars have sweetening qualities, they have been found to contribute to calories and are not advised for the consumption by diabetic patients. For these people, the world of sweetness has seen a sweeter change in the recent past with the introduction of stevia sugar obtained from leaves of stevia containing compounds about 250 to 300 times sweeter than the table sugar (Kumar, 2002). The chemicals

of interest are stevioside, rebaudioside - A and at least six other compounds that have glucoside groups attached to a three carbon ring central structure. Stevioside concentration usually ranges from 3 to 10% of the leaf dry weight, whereas rebaudioside - A is less concentrated, ranging from 1 to 3%. Stevioside could be equivalent to the sweetening power of 28 tonnes per acre of sucrose sugar (Shock, 1982).

Stevia rebaudiana Bertoni, also known as the "sweetest plant of the world", belongs to the family Asteracea and is native to South American centre of diversity. Many countries have shown interest in cultivation of stevia and research activities have been initiated. Incorporation of this species into agricultural production systems however, depends upon a thorough knowledge of the plant and its agronomic potential (Ramesh et al., 2007). Nutritional requirements of this crop are low to moderate (Goenadi, 1985) since this crop is adapted to poor quality soils in its natural habitat at Paraguay. When placed under commercial culture, for economic purpose, manuring is necessary (Donalisio et al., 1982). Since plant leaves are the profitable part of this crop it is expected that a higher nutrient supply will result in higher foliage yield. However, the nutritional physiology of stevia is still poorly investigated, with only a few published studies. Here we investigate the effects of farmyard manure (FYM) and nitrogen level on the yield and nitrogen uptake of stevia. The results of this study provide additional information on the species nutritional requirements in order to obtain a better foliage yield.

MATERIALS AND METHODS

Two independent field experiments were conducted at the Student's Research farm of the Department of Agronomy, Punjab Agricultural University, Ludhiana, India one in 2006 and the other in 2007. The research farm is located 30°50'N, 75°52'E at an altitude of 247 m above mean sea level. The farm soil was loamy sand in texture, with organic carbon content and pH of 0.25/0.31% and 7.89/7.30, respectively in the years 2006/2007. The available nitrogen in the soil was 187 and 261 kg/ha, phosphorus was 15.3 and 18.6 kg/ha and potassium was 154 and 167 kg/ha in 2006 and 2007, respectively. The texture of the soil was analysed by International Pipette method (Piper 1966), organic carbon was analysed by Walkley and Black method (Piper, 1966), Available phosphorus was analysed by 0.5 M $NaHCO_3$ extractable (Olsen et al., 1954) and available potassium was analysed by ammonium acetate extractable K (Jackson, 1967). New experimental fields were designated for the 2007 experiment to avoid any residual effect of the farmyard manure applied in the previous year. The experiment was laid out in a completely randomized block design with four replications, in a total of 16 treatments consisting of four levels of farmyard manure (0, 15, 30 and 45 kg/ha) and four levels of nitrogen (0, 20, 40 and 60 kg/ha) and their interaction. The nutrient content of farmyard manure in the 2006 experiment was 0.49% N, 0.20% P_2O_5 and 0.38% K_2O, whereas in the 2007 experiment it was 0.51% N, 0.23% P_2O_5 and 0.40% K_2O. The required amount of farmyard manure as per treatment was weighed, added and well mixed in the soil a few days prior to the transplant of the seedlings. A basal dose of Phosphorus and potassium (40 kg/ha each) were added through single super phosphate and muriate of potash at the time of transplanting while nitrogen was applied as per treatment through urea. Nitrogen was applied in two equal splits, that is, half at the time of transplanting and remaining one-half was top dressed at 45 days after transplanting. On January 2006 and 2007, seeds of *S. rebaudiana* were sown on a seed bed by rubbing seeds with vermi-compost and covering them with farmyard manure and soil as the seeds have a very light weight. In April 2006 and May 2007, three-month old seedlings were then transplanted to the experimental plots, where they were grown under the experimental levels of farmyard manure and nitrogen. After transplanting, a light irrigation was given until the seedlings were well established, after which full irrigation was given to fulfill the water requirement of the crop. Any gaps between the plant and the soil were filled in order to maintain a straight plant position. During the experiment, the plants were protected from insects and checked regularly for any potential diseases. Choloropyriphos 20 EC at 5 L/ha was supplied by

irrigation to control the termites and indofil at 0.3% (30 g in 10 L of water) was sprayed to control the leaf spot disease caused by *S. steviae*. Hand-weeding was done at 45, 75 and 95 days after transplanting. The crop was harvested manually in September 2006 and 2007 with the help of sickles. Plants were cut 15 cm above the ground level. Growth and yield parameters (number of branches per plant, number of leaves per plant, leaf area index) were recorded every month until harvest, whereas data on crop yield (biomass of dry leaves) and dry matter accumulation per plant were recorded at harvest, when the experiment ended. These parameters were chosen as important yield parameters of stevia because the leaves are the economic part of the plant. The plants harvested from each treatment were dried at 60°C till constant weight, ground, powdered and analyzed for nitrogen content using the alkaline permanganate method, according to Subiah and Asija (1956). Nitrogen uptake was estimated by multiplying yield and nitrogen content. The results are presented for each individual experiment (2006 and 2007). The data were statistically analyzed with the procedure described by Cochran and Cox (1967) and adapted by Cheema and Singh (1991) in statistical package CPCS-1 for significant differences between treatments.

RESULTS AND DISCUSSION

Number of branches per plant

Crop performance to a great extent is governed by the number of branches per plant. It is, therefore imperative that if the number of branches per plant is higher, the numbers of leaves are expected to be higher; ultimately the leaf yield will be higher. The number of branches per plant was significantly influenced by the different levels of farmyard manure and nitrogen. In the 2006 experiment, the application of 15 t FYM ha^{-1} resulted in a significantly higher number of branches plant^{-1} (18.6) compared to control (16.7), whereas in 2007 significant increase in the number of branches plant^{-1} were noticed at 30 t FYM ha^{-1}. Nitrogen fertilization influenced the number of branches per plant significantly during both the years. There was an increase in the number of branches per plant with each level increase in nitrogen. The highest number of branches per plant (18.9 and 19.0 during 2006 and 2007, respectively) was recorded under 60 Kg N ha^{-1} which was statistically at par with 40 and 20 Kg N ha^{-1} (18.8, 18.7 and 18.3, 17.9, respectively) and all the three levels of nitrogen produced significantly higher number of branches than control (16.8 and 16.3, respectively) during both years.

Number of leaves per plant

Green leaves are the site of photosynthetic activity taking place in the plants. The number of leaves per plant would also substantiate the fact that increased number of leaves per plant would contribute to the final yield of the plant particularly the crops like stevia in which only leaves are used for commercial product. The number of leaves per plant was significantly influenced by the different levels of farmyard manure and nitrogen. During 2006, the

Table 1. Influence of farmyard manure and nitrogen levels on yield attributes of *Stevia rebaudiana*.

Treatment	No. of branches/plant		Leaf no./plant		leaf area index (cm²/cm²)		Dry matter/plant(g/plant)	
	2006	2007	2006	2007	2006	2007	2006	2007
FYM (t ha⁻¹)								
0	16.7	15.9	286.0	249.3	7.4	6.7	63.2	59.2
15	18.6	16.9	339.7	360.2	7.6	6.8	69.7	66.0
30	18.5	18.5	451.2	439.1	7.7	8.1	78.2	74.9
45	18.7	18.9	588.3	494.1	8.2	8.4	83.6	80.2
CD (P=0.05)	1.5	0.9	27.5	28.3	NS	0.4	9.8	9.7
Nitrogen (kg ha⁻¹)								
0	16.8	16.3	354.1	359.1	7.7	6.9	47.2	43.2
20	18.3	17.9	394.1	367.8	7.7	7.3	65.4	61.3
40	18.8	18.7	432.9	386.7	7.8	7.8	81.9	78.4
60	18.9	19.0	484.0	429.0	7.8	8.0	97.3	100.3
CD (P=0.05)	1.2	1.4	27.5	28.3	NS	0.4	9.8	9.7
FYM × N	NS	NS	NS	NS	NS	NS	NS	NS

*CD, Critical difference; NS, non-significant.

maximum number of leaves per plant (588.0) was recorded with 45 t FYM/ha which was significantly higher than all other lower levels of farmyard manure, whereas, during 2007 the highest number of leaves count (494.1) was registered under 45 t FYM/ha which was statistically at par with 30 t FYM/ha but significantly higher than other lower levels (Table 1). The trend of increasing the number of leaves per plant with the application of farmyard manure was also recorded by Goenadi (1985). The application of increasing levels of farmyard manure increased the yield, which might have been due to the balanced availability of nutrients to the plants that resulted in a favorable soil environment. These favorable conditions increased the nutrient availability and the water holding capacity of the soil resulting in enhanced growth and yield.

The number of leaves per plant was significantly higher in plants grown at 40 and 60 kg N ha⁻¹, than at 0 and 20 kg N ha⁻¹. The maximum number of leaves per plant (484.0 and 429.0) was recorded with 60 kg N/ha which was significantly higher than other lower levels of nitrogen during both the years. The leaves count obtained under 60 kg N/ha were 302.3, 248.6, 137.1 and 244.8, 133.9 and 55 leaves per plant less than control, 20 and 40 kg N/ha during 2006 and 2007, respectively. Buana and Goenadi (1985) also reported increased number of leaves per plant with increased levels of nitrogen.

Leaf area index (cm²/cm²)

Leaf area index (LAI) is an important growth indices determining the capacity of plant to trap solar energy for photosynthesis and has marked influence on growth and yield of plant. The effect on leaf area index remained non-significant under different farmyard manure levels at all the growth stages during 2006. However, during 2007 the highest leaf area index (8.4) was obtained under 45 t farmyard manure/ha which was statistically at par with 30 t farmyard manure/ha (8.1) but significantly higher than 15 t farmyard manure/ha (6.8) and control (6.7) In case of nitrogen the maximum leaf area index (7.8) was recorded at 60 and 40 kg N/ha during 2006. In 2007, the maximum leaf area index (8.0) was obtained at 60 kg N/ha which was statistically at par (7.8) with 40 kg N/ha but significantly higher than 20 kg N/ha (7.3) and control (6.9).

Dry matter accumulation per plant (g/plant)

Dry matter accumulation by the crop is one of the important growth parameter of the crop to be considered for determination of the economic yield while assessing the effect of different treatments. Dry matter accumulation per plant significantly increased with the increment of farmyard manure in the two experiments. No significant differences were detected in dry matter accumulation between 30 and 45 t FYM ha⁻¹, the levels at which values were maximal for both years. These results corroborate the findings of Goenadi (1985). Similarly different levels of nitrogen significantly affected dry matter accumulation per plant during both the years. Maximum dry matter per plant (100.3 and 97.3 g/plant) was recorded at 60 kg N/ha which was significantly better than 40 and 20 kg N/ha and the lowest (47.2 and 43.1 g/plant) was obtained under control. These results are in close conformity with

Table 2. Influence of farmyard manure and nitrogen levels on yield and N uptake of *S. rebaudiana*.

Treatment	Dry leaf yield (kg/ha)		N uptake (kg/ha)	
	2006	2007	2006	2007
FYM t/ha				
0	802	596	16.98	14.58
15	846	680	21.33	17.00
30	909	793	25.71	22.15
45	1155	927	32.89	29.57
CD(P=0.05)	159.0	118.1	4.40	3.64
N kg/ha				
0	797	606	18.33	15.18
20	914	677	24.05	18.45
40	919	828	25.46	24.09
60	1080	885	29.06	25.59
CD(P=0.05)	159.0	118.1	4.40	3.6

the results obtained by Rakesh et al. (2012). Higher levels of nitrogen in the soil contributed to higher dry matter accumulation, which is an essential pre-requisite for photosynthetic ability in a given canopy and in turn might have helped other synthetic process during developmental sequence.

Dry leaf yield

No significant differences were detected in the dry leaf yield of plants grown at 30 and 45 t FYM ha^{-1}, when values were highest (respectively 909 and 1155 kg ha^{-1} in 2006 and 793 and 927 kg ha^{-1} in 2007) (Table 2). Plants grown at 60 kg N ha^{-1} produced the highest dry leaf yield (1080 kg ha^{-1}), which was significantly higher compared to all other levels of nitrogen. The increase in the dry leaf yield at 60 kg N ha^{-1} was 35.5, 18.1 and 17.5% higher over control, 20 and 40 kg N ha^{-1}, respectively, in 2006. The highest dry leaf yield, which was recorded with 60 kg N ha^{-1} (885 kg ha^{-1}) was not significantly different from that recorded at 40 kg N ha^{-1} (828 kg ha^{-1}) during 2007. Yet, the dry leaf yield obtained under these treatments was significantly higher than at 20 kg N ha^{-1} and in the control. Similar increase in dry leaf yield of stevia with nitrogen was also reported by Chalapathi et al. (1997).

The application of increasing levels of farmyard manure increased the yield, which might have been due to the balanced availability of nutrients to the plants that resulted in a favorable soil environment. These favorable conditions increased the nutrient availability and the water holding capacity of the soil resulting in enhanced growth and yield.

Maheshwar (2005) reported that application of 105:30:45 kg NPK/ha recorded significantly higher dry leaf yield due to maximum number of leaves per plant and branches per plant as compared to lower doses of

nitrogen under loamy soil in Karnataka, India. Similarly, Lima Filho et al. (1997) observed that shortly before or at flowering, production of 1 t of dry leaves of *S. rebaudiana* required 64.6 kg N/ha, 7.6 kg P/ha and 56.1 kg K/ha. Lee et al. (1980) reported increase in leaf yield with moderate application of N, P and K fertilizers in Korea. There are, however, reports that stevia crop show yield reduction at high rates of fertilizer.

Nitrogen uptake

The uptake of nitrogen by stevia increased with increasing supply of farmyard manure. Maximum uptake was recorded under 45 t farmyard manure ha^{-1}, whereas uptake was lowest in plots where no farmyard manure was supplied. The higher uptake of nitrogen at 45 t farmyard manure ha^{-1} was attributed to a slow and prolonged availability of nutrients to the crop. In the nitrogen fertilization treatments, the highest N uptake was recorded when 60 kg N ha^{-1} was added, and was not statistically different from that obtained at 40 kg N ha^{-1}. However, the N uptake under these two treatments was significantly higher than that obtained at 20 kg N ha^{-1} and in the control for both years. The higher N uptake obtained at 40 and 60 kg N ha^{-1} may be related to the higher biomass yield obtained at the same N levels. The uptake of nitrogen increased probably because it was being used for plant growth. Angkapradipta et al. (1986) also reported that increased supply of nitrogen resulted in increased plant N content and nitrogen uptake by stevia.

Conclusion

From the results of present investigation, on the basis of dry leaf yield and total glycoside yield it may be concluded that the stevia crop may be supplied with 45 t farmyard manure/ha and 60 kg N/ha.

REFERENCES

Angkapradipta P, Waristo T, Faturachin P (1986). The N, P and K fertilizer requirements of *Stevia rebaudiana* Bert. On latosolic soil. Menera perkebunan 54:1-6.

Buana L, Goenadi DH (1985). A study of growth patterns of stevia cutting. Menera perkebunan 53:124-133.

Cheema HS, Singh B (1991) Software Statistical package CPCS-1. Department of Statistics, Punjab Agric Univ, Ludhiana, India.

Chalapathi MV, Shivraj B, Parama,VRR, Ramakrishna VR (1997). Nutrient uptake and yield of stevia (*Stevia rebaudiana* Bertoni) as influenced by method of planting and fertilizer level. Crop Res. 14(2):205-08.

Cochran WG, Cox GM (1967).Experimental Designs. Asia Publishing House New Delhi.

Donalisio MG, Duarte FR, Souza CJ (1982). Estevia (*Stevia rebaudiana*). Agronomico, Campinus (Brazil) 34:65-68.

Goenadi DH (1985). Effect of FYM, NPK and liquid fertilizers on *Stevia rebaudiana*. Menera perkebunan 53:23-30.

Jackson ML (1967). Soil chemical analysis. Prentice Hall of India Pvt. Ltd. New Delhi.

Lee JI, Kang KH, Park HW, Ham YS, Park CH (1980). Studies on the new sweetening source plant, *Stevia rebaudiana* in Korea.II. Effects of fertilizer rates and planting density on dry leaf yields and various agronomic characteristics of Stevia rebaudiana. Research Reports of the office of Rural Development (Crop Suwon) 22:138-44.

Lima Filho OF, Malavilta E, De Sena, JOA, Carneiro, JWP (1997). Uptake and accumulation of nutrients in stevia (*Stevia rebaudiana*). II. Micronutrients. Scientia Agricola 54:23-30.

Maheshwar HM (2005). Effect of different levels of nitrogen and dates of planting on growth and yield of stevia (*Stevia rebaudiana*). M.Sc Thesis. Department of horticulture, University of agricultural sciences, Dharwad, Karnataka, India. P. 100.

Piper CS (1966). Soil and Plant analysis. Hans Publishers, Bombay India

Rakesh K, Saurabh S, Kulasekaran R, Ramdeen P, Vijay Lata P, Bikram S, Rakesh DS (2012). Effect of agro-techniques on the performance of natural sweetener plant-stevia (*Stevia rebaudiana*) under western Himalayan conditions. Indian J. Agron. 57 1):74-81.

Ramesh K, Singh .V, Ahuja PS (2007). Production potential of *Stevia rebaudiana* (Bert.) Bertoni under intercropping systems. Arch. Agron. soil Sci. 53(4):443-58.

Kumar VR (2002). Medicinal and Aromatic Plants. Green Cross Media International, Bangalore. pp. 90-91.

Shock CC (1982). Rebaudi's natural non-caloric sweeteners, Cal Agric 36;4-5.

Subiah BV, Asija HL (1956). A rapid procedure for the estimation of the available nitrogen in soils. Curr Sci. 25:259-260.

14

The effect of integrated organic and inorganic fertilizer rates on performances of soybean and maize component crops of a soybean/maize mixture at Bako, Western Ethiopia

Abebe Zerihun[1], J. J. Sharma[2] and Dechasa Nigussie[2] and Kanampiu Fred[3]

[1]Bako Agricultural Research Center, P. O. Box 03, Bako West Shoa, Ethiopia.
[2]Haramaya University, P. O. Box, 138, Dire Dawa, Ethiopia.
[3]International Maize and Wheat Improvement Centre (CIMMYT), P.O. Box 1041-00621, Nairobi, Kenya.

The experiment was conducted to determine the best compatible soybean varieties in intercropping systems and the most economically optimum integrated fertilizer rate. The factorial experiment consisted of two soybean varieties (Didessa and Boshe) treated with eight levels of combined organic and inorganic fertilizer applications in three replications. Both sole soybeans and maize under recommended fertilizer recommendation were also included for comparison purposes. The result indicated that there were significant differences in leaf area index, plant height and grain yield of maize due to integrated fertilizer application, but not in harvest index. However, statistically significant variations were observed on nodule number per plant, leaf area per plant and yield of intercropped soybeans as a result of soybean varieties and the interaction of varieties with fertilizer application. Higher nodules and leaf areas per plant were recorded in Didesa variety than Boshe. This could be due to varietal difference, integrated fertilizer application and cropping systems as well. Yield advantage obtained due to various combinations of fertilizer rates ranged from 6 to 28% over the yield of sole maize. Monetary advantage (MA) obtained due to intercropping systems ranged from the lowest Birr 1927 ha[-1] to Birr 8446 ha[-1] under various proportions of fertilizer applications. Application of both recommended NP and farmyard manure (FYM) resulted in the highest (Birr 8446 ha[-1]) MA followed by recommended NP (Birr 4583 ha[-1]). However, an integrated use of 12 t ha[-1] FYM with $28/12N/P_2O_5$ saved up to 75% cost of commercial fertilizer for both years and cost for application in the next year.

Key words: Varieties, organic and inorganic fertilizers, intercropping.

INTRODUCTION

Sustainable agriculture is successful management of resources to satisfy changing human needs while conserving natural resources. However, area of cultivable land per unit household is dwindling from time to time due to population pressure. This leads to intensive crop production per unit area of land. Intercropping is one of the intensive cropping systems which ensure sustainable utilization of limited land resources (Tesfa et al., 2001). The extent and importance of intercropping increases as farm size decreases and the smaller the farm size the more complex the combinations. In tropics, cereal/legume intercropping is commonly practiced because of yield advantages, greater yield stability and lower risks of crop failure, which are often associated with monoculture

(Nielsen et al., 2001; Tusbio et al., 2005). In intercropping systems, legumes can provide N for intercropped cereals through N transfer (Rochester et al., 2001). The same author indicated that soybean crop is capable of supplying nitrogen for its growth and intercropped cereals through symbiotic nitrogen fixation, and hence reduces the need for expensive and environment polluting nitrogen fertilizer. Maize is a staple food crop for smallholder farmers in western Ethiopia which is suitable for intercropping with legume crops. (Aschalew et al., 1999). The same author indicated that maize is believed to be the most dependable crop to bring about food self-reliance and self-sufficiency, being the highest yielding compared to all cereal crops grown in the country.

Declining soil fertility is fundamental impediment to agricultural growth and a major reason for slow growth in food production in sub-Saharan Africa. Low soil fertility due to monoculture cereal production systems is recognized as one of the major causes for declining per capita food production. Therefore, soil fertility replenishment is increasingly viewed as one of the critical to the process of poverty alleviation (Asfaw et al., 1998). This is generally true for Ethiopian agro-ecologies, particularly for a dominant maize based mono cropping system of western Oromiya. Sustainable crop production, therefore, requires a careful management of all nutrient sources available in a farm, particularly in maize based cropping systems. These include inorganic fertilizers, organic manures and integration of legume crops in cereal based mono cropping (Wakene et al., 2007). The objective of this study was therefore; to investigate the effect of intercropping maize and soybean on nodulation and yield traits of the companion crops under combined application of FYM and NP fertilizer.

MATERIALS AND METHODS

The experiment was conducted for two consecutive years (2010-2011) at Bako Agricultural Research center (BARC). The centre is located in the Western part of Ethiopia which lies at a latitude of 9° 6′ N; longitude of 37° 9 E and at an altitude of 1650 m above sea level. It has a warm humid climate with annual mean minimum and maximum air temperatures of 13.5 and 29.7°C, respectively. The area receives average annual rainfall of 1237 mm with maximum precipitation being received in the months of May to August. The soil of the experimental site was reddish-brown, Nitosol, which is acidic with a pH of 5.2-5.6. The experimental site was low in available nitrogen and phosphorus contents which could be because of mono-cropping history of the experimental site. Soybean varieties and integrated organic and inorganic fertilizers were the two main factors. Two soybean varieties (Didessa, and Boshe) were used. Didessa is a medium maturity type (135-145 days to maturity) whereas Boshe is early maturity type. Both crops are highly adaptable to areas of mid and low altitudes. Two soybean varieties (Didessa and Boshe) were combined with eight levels of combined inorganic and organic fertilizers (110/46+0, 0+16, 110/46+16, 110/46+ 4, 83/35+4, 55/23 + 8, 28/12+12 and control, in kg ha^{-1} N/P_2O_5 + t ha^{-1} FYM, respectively) in intercropping. In addition, sole maize with recommended N/P_2O_5 and FYM and sole soybean varieties with recommended N and

P_2O_5 were included in the experiment. The levels of manure and inorganic fertilizers (N and P_2O_5) were based on recommendation for sole maize in the area, which are 110/46 N/ P_2O_5 kg ha^{-1} and 16 t ha^{-1} FYM for hybrid maize varieties. The experiment was a randomized complete block design (RCBD) and replicated three times.

Manure application

In year one (2010), decomposed and dried FYM (20% moisture content) was applied per treatments and incorporated in to the soil manually, three weeks before maize planting. The plot was retained permanently to repeat the experiment in 2011 to evaluate the residual nutrient availability of applied manure.

Inorganic fertilizer application

Both in 2010 and 2011 years, at planting of maize, half of the N and full dose of phosphorus was uniformly drilled into the maize rows and mixed with the soil to avoid contact of the seed with the fertilizer. The remaining half of N was applied per treatment at knee height growth stage of maize. For sole soybeans, 100 DAP kg ha^{-1} was drilled into the furrows at the time of sowing.

Planting

A maize variety "BH-543" was sown on May 26, 2010 and June 3, 2011, respectively. The hybrid variety was released by BARC which requires 1000-1200 mm annual rainfall having 148 days to physiological maturity. The potential yield of the variety is 8.5 to 11 t ha^{-1} at research station and 4.7 to 6 t ha^{-1} at on farm. The size of each plot was 3.75 × 3.00 m. intercropped soybean and maize that consisted of five maize rows of 75 cm inter row spacing 30 cm intra row spacing. Three weeks after sowing of maize (in 2010), the two soybean varieties were intercropped on 15 June, 2010 in between two rows of maize. However, in second year, soybean was planted on 28 June, 2011 after maize planting. Intercropped soybeans were spaced at 75 cm with intra row spacing of 10 cm. For sole soybean varieties, however, inter and intra row spacing was 40 × 10 cm, respectively.

Data collection: Maize

Leaf area (LA) was measured at 50% days to tasseling (90 days after planting). LA was taken from ten plants and three representative active leaves per plant. Leaf length and maximum width were measured. Area of each leaf was determined by multiplying length by maximum width and constant factor as described by Burren et al. (1974).

Biological and grain yield (t ha^{-1})

All maize stocks from each harvestable plot were cut just at the ground level and the aboveground biomass including the cobs was measured. Grain yield from each net plot was also measured and finally standard moisture contents, 12.5%. Similarly, yield of soybean from each plot was measured and the moisture content of the grain was determined using a moisture tester and adjusted to standard moisture content (10%).

Harvest index (%)

It was determined as a ratio of economic yield to biological yield.

Table 1. The effect of varieties and integrated fertilizer rate in leaf area index and plant height of maize component crop in crop mixture.

Treatment	LAI at tasseling		Pooled means	Plant height (cm)		Pooled mean
	2010	2011		2010	2011	
Soybean varieties						
Didessa	3.69	3.4	3.6	238.3	221	230
Boshe	4.07	3.4	3.7	245.1	222	233
LSD (P<0.05)	0.22	NS	NS	5.2	NS	NS
NP$_2$O$_5$ kg ha^{-1} + FYM t ha^{-1}						
110/46 +0	4.43	3.59	4.01	245.8	223	234.5
0 +16	3.49	3.16	3.32	234.4	218	226.2
110/46 +16	4.03	4.0	4.39	247.2	219	232.9
110/46 + 4	4.27	3.64	4.04	248.9	224	236.2
83/35 + 4	3.95	3.59	3.77	241.7	217	229.4
55/23 + 8	3.99	3.30	3.65	247.8	229	238.4
28/12 + 12	3.73	3.47	3.60	241.9	223	232.4
control	3.11	2.46	2.78	225.7	218	222.3
LSD(P<0.05)	0.43	0.69	0.79	10.3	NS	13.1
CV (%)	9.5	17	13	3.6	5.9	6.9
Intercropped vs. sole crop						
Inter crop maize	3.87	3.4	3.64	241.7	221	231.6
Sole maize (NP)	3.47	3.1	3.26	232.2	221	226.7
Sole maize(FYM)	3.35	2.7	3.32	235.6	218	227
LSD(P < 0.05)	NS	NS	NS	NS	NS	NS
CV (%)	14	20	16.7	4.8	5.6	5.1

LSD = Least significant difference (P< 0.05); CV = coefficient of variation; NS =not significant; sole M with NP= sole maize sown by recommended NP fertilizer; sole M with FYM= sole maize sown by recommended farm yard manure.

Soybean

Leaf area index and nodule number

Leaf area (LA) of soybeans was measured at the time of flower initiation by using leaf area meter. Nodules were collected at the time of 50% flower initiation by digging from five plants in each plot. Effective nodules from sampled plants were counted based on their colour (pink colour) and the mean value of five plants was recorded.

RESULTS AND DISCUSSION

Maize

Leaf area index and plant height

There was significant (P<0.01) differences across the years in LAI due to the effect of integrated fertilizer applications of the maize intercrop at time of tasseling, but there was no significant difference due to interaction effect of varieties by fertilizer application (Table 1). The result of pooled means indicated that the highest LAI was recorded when consecutive recommended 110/46 kg ha^{-1} and 16 t ha^{-1} FYM (in 2010) was applied to the system.

This result with the support of other findings indicated that there was positive effect of residual nutrients on growth parameters of the following crops (Ayoola and Makinde, 2008). However, there was a decreasing trend from year one to year two, which might be due to decreasing nitrogen availability from applied FYM. The lowest LAI was recorded at zero fertilizer application in each year, which is in agreement with findings of Faisalabad et al. (2010).

A significant variation was observed in year one due to varieties and integrated fertilizer application while it was significantly unaffected in year two. But the pooled mean result indicated that the highest plant height was recorded when successive application of 83/35 N/P$_2$O$_5$ kg ha^{-1} with 4 t ha^{-1} in year 2010 were applied to the permanent plot while the lowest was obtained from unfertilized plot (Table 1). Associated soybean varieties did not significantly affect leaf area index and plant height of the maize in contrast with its sole crops both under recommended organic and inorganic fertilizer rates. However, higher leaf area index was attained in intercropped maize than the soles, indicating that the soybean varieties might have contributed available nitrogen through biological nitrogen fixation (Tamado and

Table 2. The effect of varieties and integrated fertilizer application on biomass, grain and harvest index of associated maize in intercropping system.

Treatment	Biomass (t ha^{-1})		Pooled mean	Grain yield (t ha^{-1})		Pooled mean	Harvest index (%)		Pooled mean
	2010	2011		2010	2011		2010	2011	
Varieties									
Didessa	22.41	20	21.2	9.11	9.1	9.1	41	46	43
Boshe	22.80	19.4	21.1	9.57	9.0	9.3	42	47	44
LSD (P<0.05)	NS	NS	NS	NS	NS	NS	NS	NS	NS
NP$_2$O$_5$ kg ha^{-1} + FYM t ha^{-1}									
110/46 + 0	22.52	20.2	21.3	9.52	9.6	9.6	43	47	45
0 +16	24.20	19.5	21.9	10.0	8.7	9.4	42	44	43
110/46 + 16	25.80	20.9	23.4	10.5	9.9	10.2	42	48	45
110/46 + 4	22.45	18.7	20.6	9.47	9.3	9.4	42	50	46
83/35 + 4	22.21	21.4	21.8	8.97	9.3	9.1	40	44	42
55/23 + 8	23.13	19.4	21.3	9.74	9.3	9.5	42	49	45
28/12 + 12	23.85	20.5	22.2	9.86	9.3	9.6	42	45	43
Control	16.69	17.2	16.9	6.88	7.2	7.0	40	43	41
LSD(P<0.05)	1.26	NS	1.92	1.14	NS	2.1	NS	NS	NS
CV (%)	9.6	12.5	11.1	10.4	17.7	14.1	11.1	16.5	13
Intercropped vs. Sole crop									
Intercrop M	22.6	19.74	21.2	9.5	9.1	9.21	41.5	44	44
SM (NP)	21	19.4	20.2	9.4	9.1	9.23	44.7	46	46
SM(FYM)	23.7	18.6	21.2	10.4	8.33	9.36	44.3	47	44
LSD(P < 0.05)	NS	NS	NS	NS	NS	NS	2.7	NS	NS
CV (%)	9.1	13	14	13.4	18	16.1	5.4	17	12

SM = Sole maize; IM = intercrop maize; LSD = least significant difference (P< 0.05); CV = coefficient of Variation; NS = Not significant; sole M with NP = sole maize sown by recommended NP fertilizer; sole M with FYM= sole maize sown by recommended farm yard manure.

Eshetu, 2000). Similar to leaf area index, plant height of intercropped maize was not significantly affected by cropping system.

Biological and economic yield

The result also revealed that both biomass and grain yield of maize were significantly varied in 2010 as the result of integrated fertilizer application. In year 2011, however, it showed no significant different was observed across fertilizer treatments (Table 2). This similarly confirmed with other authors that the residual nutrient availability from the preceding FYM application which ranged from 4-16 t ha^{-1} of the recommended manure might significantly increase both biomass and grain yield (Getachew, 2009). The residual nutrient definitely save up to 100% cost of inorganic fertilizer in both years and cost of manure application in second year (2011), as observed from sole application of FYM. The pooled mean also indicated that both biological and economic yield obtained from permanent plots, which were applied by 4-16 t ha^{-1} FYM in 2010 with consecutive application of

28/12-110/46 kg ha^{-1} in each year, were not statistically different except in unfertilized plot (Table 2). This response indicated that repeated application rates of organic manure in every year did not significantly increase yield and yield components of the main crops though there is a gradual increase in nutrient availability from organic manure that would ensure supply of the crop requirements (Achieng et al., 2010). The lowest grain yield, however, was recorded from untreated plot. Moreover associated soybeans did not significantly increase yield of maize.

Harvest index

The result also indicated that harvest indices (HI) did not considerably vary across the treatments in each year (Table 2). However, there was an increasing trend across fertilizer treatment from year one to year two. This indicated that there is lower translocation of the nutrient to economic yield under sufficient nutrient availability than biological yield. In other words, higher nutrient availability might enhance the vegetative growth of the crops that

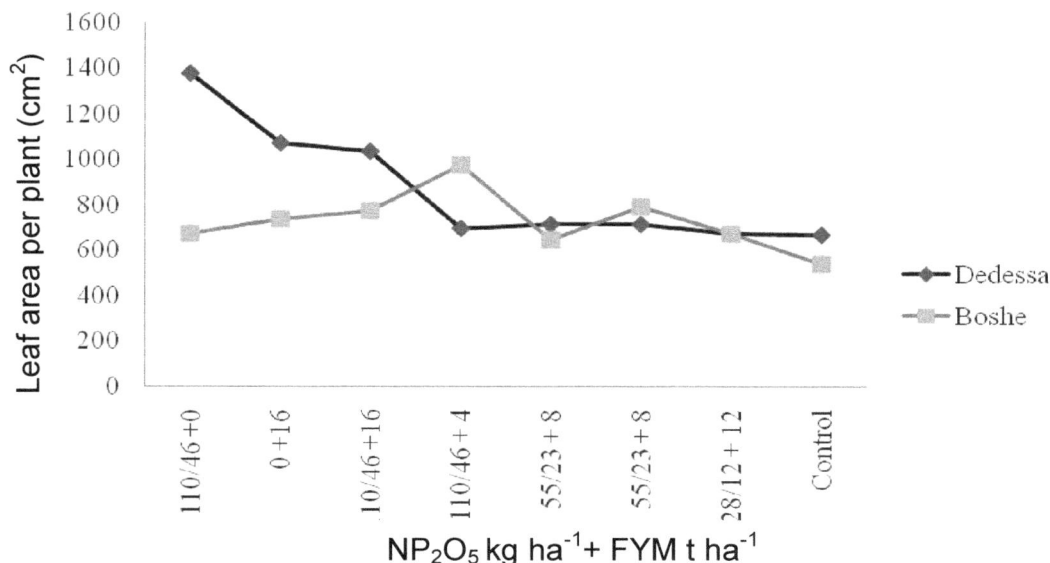

Figure 1. The effect of varieties and integrated fertilizer application on leaf area per plant (pooled mean) of associated soybean in intercropping (LSD at P<0.05=226).

may reduce the economic yield. Associated soybean varieties did not considerably affect the harvest index of the main crop.

Means of two year data revealed that the effect of cropping systems due to associated soybean varieties did not significantly vary on yield and yield traits (biomass yield, grain yield and harvest index) of the intercropped maize as compared with soles, which was treated under recommended organic and inorganic fertilizer application (Table 2). This result may indicate that maize is the main dominant crops that significantly compete with the associated soybean varieties. However, significant reduction in harvest index for intercropped maize was recorded as compared to the sole crops.

Soybeans

Leaf area per plant

The result of analysis of variance revealed that significant variations were observed across the years due to the effect of varieties and integrated fertilizer application (Figure 1). The two ways interaction of varieties by fertilizer application was also significantly affected in leaf area per plant of the companion crops. A significant higher leaf area per plant was recorded in 2010 year as compared to the second year. This variation was probably caused by variation imposed by time of planting of the companion crops after maize planting. Didessa variety generally produced higher leaf area than Boshe one across various fertilizer rates. This result is in agreement with other finding (Maheshbabu et al., 2008) that the higher leaf area was observed when higher

proportions of inorganic fertilizer rates were applied to the treatment, which might be because of higher availability of nitrogen. Untreated plots showed the lowest leaf area for both Didessa and Boshe varieties. The effect of cropping systems due to associated main crops was significantly reduced in leaf area per plant of the companion crops (Figure 2). The significant variations across the years were also observed, which might be the result of variation in cropping season. The mean of two years indicated that about 83 and 68% reduction in leaf area per plant of Didessa and Boshe varieties were recorded when compared with their respective sole crops (Figure 2). This reduction might be due to the maize shading effect which adversely affected light interception that result in reduced growth and expansion of associated soybean in intercropping and the same result was also reported by Demisew (2002). A part from this, sole Didessa variety had significantly higher leaf area per plant as compared to the sole Boshe (Figure 2).

Effective nodule number per plant

Nodule number per plant was significantly affected by both varieties and integrated fertilizer application across the years (Figure 3). Higher number of nodules was recorded in 2011 than in the first year. This higher number of nodules might be because of improved soil chemical properties caused by the residual effects of applied organic manure (Marschner, 1995). Didessa variety had significantly higher number of nodules when compared with Boshe. The difference in nodulation might be probably due to differential compatibility with effective indigenous *rhizobium* in the soil of the experimental field.

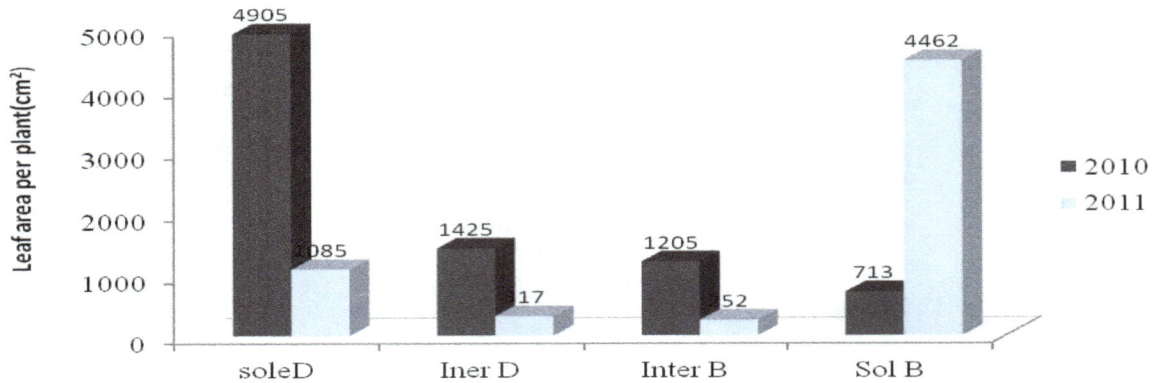

Figure 2. The effect of cropping systems on leaf area per plant of associated soybean varieties. Sole D= sole Didessa;Inter D= intercropped Didessa;Sol B= sole Boshe;inter B= intercropped Boshe.

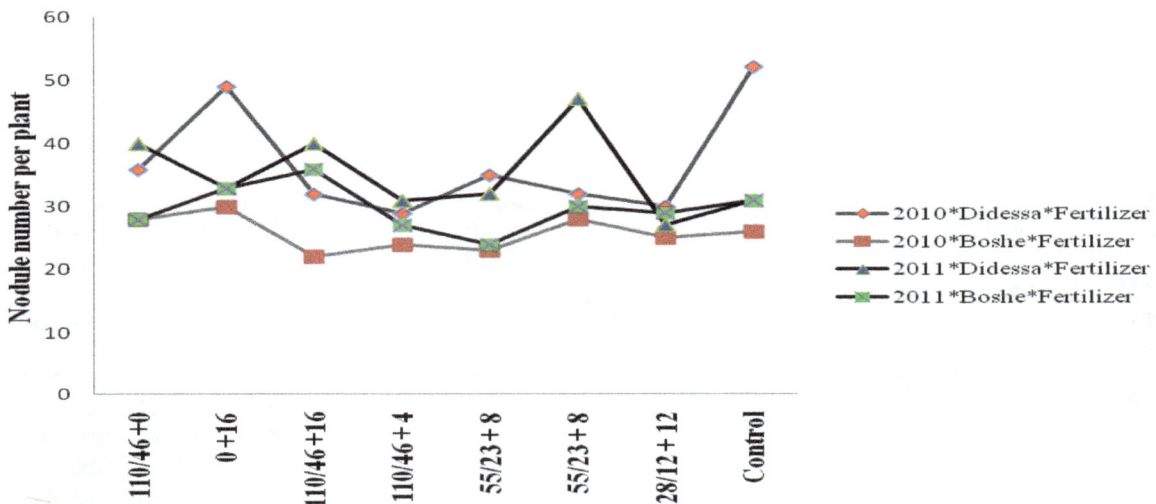

Figure 3. The effect of varieties and integrated fertilizer application on nodule number per plant soybean intercrops (LSD at P<0.05=12.4).

The highest nodule number 52 and 49 were recorded in 2010 from unfertilized plot and recommended application of FYM (2010), respectively (Figure 3).

Cropping systems caused a significant (P<0.05) reduction of nodule number when compared with the respective sole crops (Figure 4). A reduction in 39% for intercropped Didessa vareity and 22% for Boshe vareity were attained as compared with their respective sole crops (Figure 4). This result in agreement with other finding might possibly be the shading effects of maize that significantly reduced light interception potential of the associated soybeans and reduced the photosynthetic assimilate (Ghosh et al., 2006). Reduced assimilate might be resulted in limited food supply for associated *rhizobium* bacteria, and consequently their atmospheric fixation capacity were diminished (Tisdale et al., 1999). Moreover, Sole Didessa variety produced significantly

higher number of nodules per plant as compared to sole Boshe variety (Figure 4).

Grain yield

The result of analysis revealed that significant higher grain yield per hectare was obtained from each variety in year one as compared to year two for both associated companion crops (Figure 5). This might be caused by the variation in time of planting after maize planting, 20 days and 25 days after maize planting in 2010 and 2011, respectively (Addo-Quaye et al., 2011). Associated Didessa variety significantly produced higher grain yield than Boshe variety (Figure 5). This result with other finding might indicate that medium or late maturity types of soybeans considerably compete on common resources

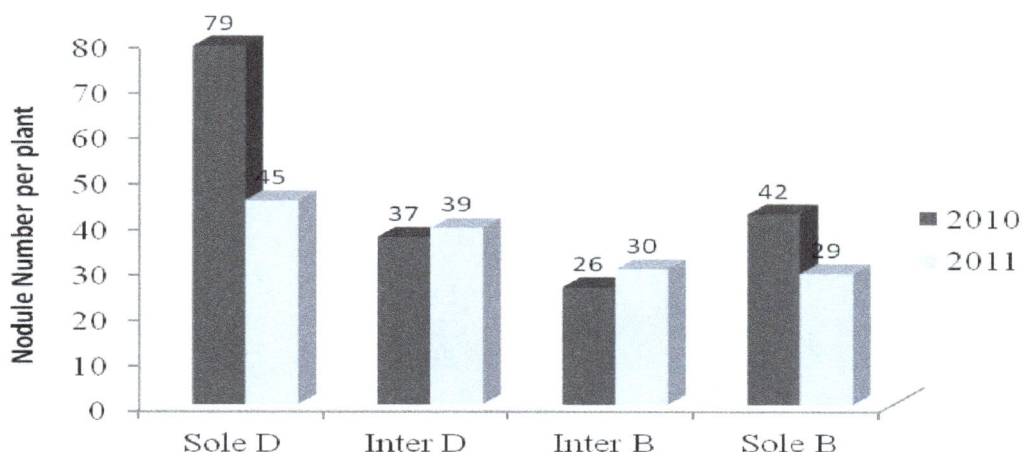

Figure 4. The effect of cropping systems on nodule number per plant of intercropped soybean varieties. Sole D= sole Didessa; Inter D= intercropped Didessa; Sole B= Sole Boshe;Inter= Intercropped Boshe.

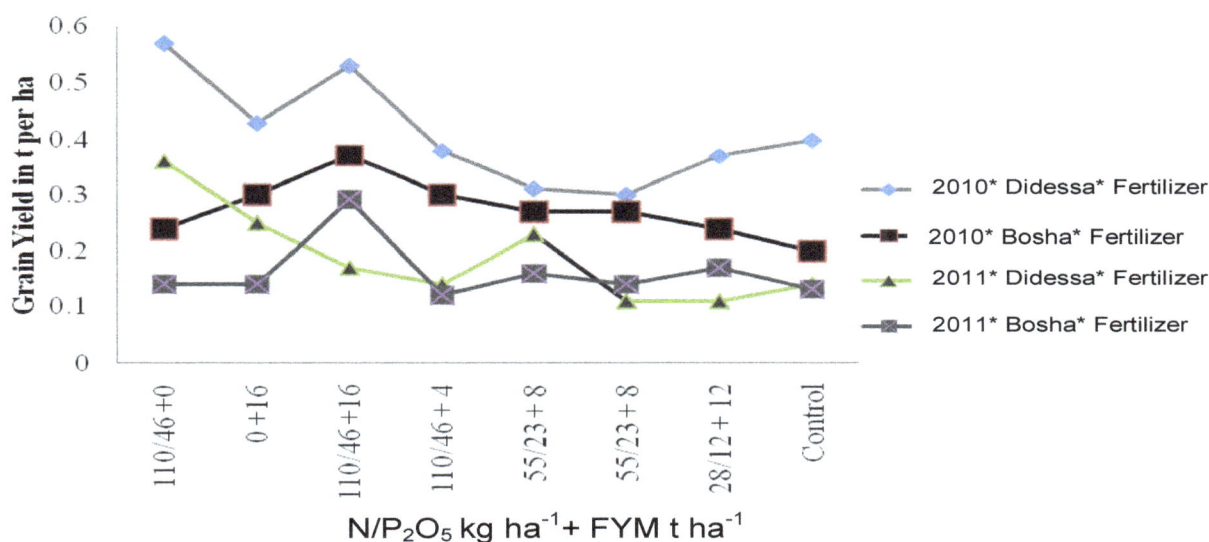

Figure 5. The effect of varieties and integrated fertilizer application on grain year of soybean intercrops (LSD at P<0.05=0.043).

or it might have escaped the period of more competition exerted, as the maize crop was approaching to maturity than the early types (Otieno et al., 2009). The highest grain yield was recorded for both associated crops when recommended NP_2O_5 (110/46 kg ha^{-1}) and FYM (16 t ha^{-1}) were applied while the lowest was obtained from untreated plot (Figure 5). However, a significant more grain yield of Didessa variety was obtained from unfertilized plot as compared to Boshe. Similar results with this finding indicated that the fixation capacity of the associated soybean under limited nutrient might be enhanced and utilized by the legumes so that the competitive ability of intercropped Didessa variety is very

high than Boshe even though maize is also competing for the same common resources (Muoneke et al., 2007).

The effect of cropping systems considerably influenced grain yield of intercropped soybean varieties when compared with their respective sole crops (Figure 6). A significant reduction in grain yield was observed as compared with the sole ones although population variations are another factor. The reduction of LAI, nodule number and other parameters caused by shading effect might contribute to reduction in grain yield (Figures 2 and 4). The study indicated that shading effect of the maize drastically reduced the light transmission that may significantly reduce photosynthetic assimilates (Ghosh et

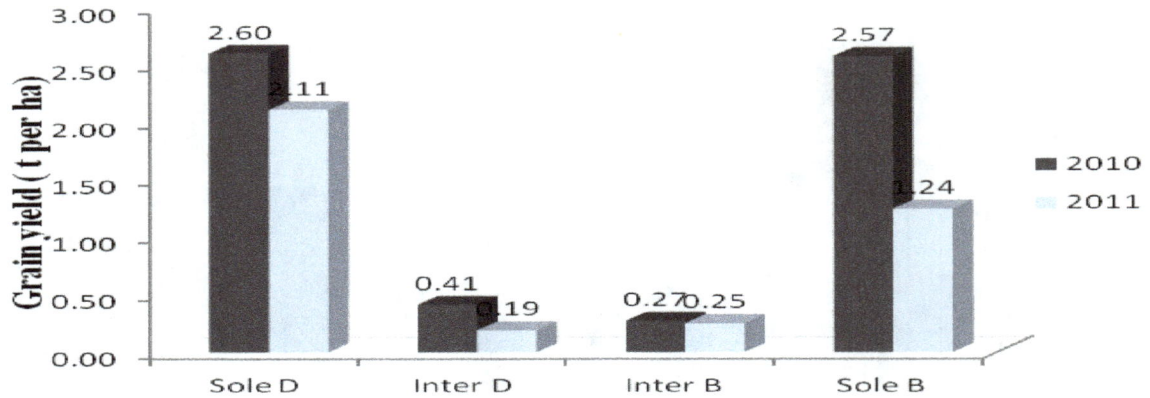

Figure 6. The effect of cropping systems on grain yield of intercropped soybean varieties. Sole D= sole Didessa; Inter D= Intercropped Didessa variety; Sole B= sole Boshe; Inter B= intercropped Boshe variety.

Table 3. Yield and Monetary advantage over sole maize monoculture as affected by integrated fertilizer application.

Treatments	Total land equivalent ratio			Monetary advantage (Eth birr ha^{-1})		
	2010	**2011**	**Mean**	**2010**	**2011**	**Mean**
Fertilizers						
110/46 +0	1.32	1.18	1.15	4419	4746	4583
0 + 16	1.17	1.04	1.10	5435	1148	3292
110/46 +16	1.25	1.32	1.28	7579	9314	8446
110/46 + 4	1.13	1.08	1.10	4467	2418	3442
83/35 + 4	1.02	1.10	1.06	687	3188	1937
55/23 + 8	1.12	1.08	1.10	3559	2418	2988
28/12 + 12	1.12	1.08	1.10	3906	2529	3218
Control	0.87	0.85	0.86	-3785	-4760	-4272

al., 2004).

Land equivalent ratio and monetary advantage

The result of pooled means evidently signified that yield advantage obtained due to various combinations of fertilizer rates ranged from 6 to 28% over the yield of sole maize obtained under recommended inorganic fertilizer. Mean of two years indicated that the highest LER (1.28) followed by 1.15 were recorded when 16 t ha^{-1} FYM in 2010 and consecutive application of 110/46 N/P$_2$O$_5$ kg ha^{-1} in both years were applied to the systems, respectively meaning that 28 and 15% greater area would be required under sole maize to produce the same yield as that of combined yield under intercropping system. The result with support of other finding (Demisew, 2002; Tolera et al., 2005), however, revealed the lowest LER (0.86) was without fertilizer uses to intercrops, validating that limited soil fertility significantly reduces the productivity of intercropping systems (Table 3). This result indicates that about 14% more area of land was required to produce the same amount of yield obtained from maize monoculture sown with the recommended NP rate.

The result of pooled means also indicated MA of intercropping ranged from the lowest Birr 1927 ha^{-1} to Birr 8446 ha^{-1} under various proportions of fertilizer applications. Application of both recommended 110/46 N/P$_2$O$_5$ kg ha^{-1} and 16 t ha^{-1} FYM resulted in the highest (Birr 8446 ha^{-1}) MA followed by recommended NP (Birr 4583 ha^{-1}). This attribute indicates that integrated fertilizer application with various proportions of NP with FYM significantly increased MA over the control. However, there was no gain in monetary advantage without fertilizer application. It was noted that due to total LER was less than one, negative value of MA was obtained in case of unfertilized intercrops. The negative value indicates a loss of Birr 4272 ha^{-1} from unfertilized intercrops as compared to the gross benefit obtained from maize monoculture sown with recommended NP.

Conclusion

Higher plant height and leaf area index was recorded under different rate of integrated fertilizer application when compared with the control. The result also revealed that higher leaf area index was attained in intercropped maize than the soles though not significant from each other. Similar to leaf area index, plant height of intercropped maize was not significantly affected by cropping system. Significant variation was also observed on yield of maize due to fertilizer rates. Application of recommended manure resulted in maximum number of effective nodules which did not significantly vary with the control. The effect of cropping system was significantly reduced in leaf area per plant of the companion crops (Figure 2). Significant variations across the years were also observed. The mean of two years indicated that about 83 and 68% reduction in leaf area per plant of Didessa and Boshe varieties were recorded when compared with their respective sole crops. From economic point of view, intercropping of maize with Didessa variety under application of 16 tha^{-1} organic manure resulted in the highest monetary advantage. However, an integrated use of 12 $t\ ha^{-1}$ FYM with 28/12 NP_2O_5 kg ha^{-1} saved up to 75% cost of commercial fertilizer and even cost of manure application for following years. Alternatively, integrated use of 55/23 N/P_2O_5 kg ha^{-1} with 8 $t\ ha^{-1}$ FYM also revealed better economic advantage with some additional yield from soybean (Didessa variety) without affecting yield of the maize.

ACKNOWLEDGMENTS

The author is grateful to Oromiya Agricultural Research Institute, Ethiopia, for sponsoring the study. His particular appreciation goes to Mr Terefe Daba and other field assistants who thoroughly collect all necessary data.

REFERENCES

Achieng JO, Ouma G, Odhiambo G, Muyekho F (2010). Effect of farmyard manure and inorganic fertilizers on maize production on Alfisols and Ultisols in Kakamega, western Kenya. Agric. Biol. J. North Am. 1(5):740-744.

Addo-Quaye AA, Darkwa AA, Ocloo GK (.2011). Yield and productivity of component crops in a maize-soybean intercropping system as affected by time of planting and spatial arrangement. ARPN J. Agric. Biol. Sci. 6(9):50-57.

Asfaw B, Heluf G, Yohannes U (1998). Effect of tied ridges on grain yield response of Maize (Zea mays L.) to application of crop residue and residual NP on two soil types at Alemaya, Ethiopia. South Afr. J. Plant Soil. 15:123-129.

Aschalew G, Legesse W, Gemechu K, Wende A, Benti T, Pexley K, Mosisa W (1999). Strategy for quality maize breeding and dissemination in Ethiopia. Maize Production Technology. pp. 42-46.

Burren L, Mock JJ, Anderson IC (1974). Morphological and physiological traits in maize associated with tolerance to high plant density. Crop Sci. 14:426-429.

Ayoola OT, Makinde EA (2008). Performance of green maize and soil nutrient changes with fortified cow dung. African Journal of Plant Science, 2 (3):19-22. Available online at http://www.academicjournals.org/AJPS

Demisew M (2002). Effect of planting density of haricot bean and nitrogen fertilization on productivity of maize and haricot bean additive intercrop systems. M.Sc. Thesis. Alemaya Uinversity, Alemaya, P. 105

Faisalabad S, Alireza V, Hossein AF (2010). Effects of planting density and pattern on physiological growth indices in maize (Zea mays L.) under nitrogenous fertilizer application. J. Agric. Exten. Rural Develop. 2(3):40-47

Getachew A (2009). Ameliorating effects of organic and inorganic fertilizers on crop productivity and soil properties on reddish- brown soils.Proceeding of the 10th Conference of the Ethiopian Society of Soil Science. 25-27 March 2009, held at EIAR, Addis Ababa, Ethiopia. pp. 127-150.

Ghosh PK, Ramesh P, Bandyopadhyay KK, Tripathi AK, Hati KM, Misra AK, Acharya CL (2004). Comparative effectiveness of cattle manure, poultry manure, phosphocompost and fertilizer-NPK on three cropping systems in vertisols of semi-arid tropics. I. Crop yields and system performance. Bioresour. Technol. 95:77–83

Ghosh PK, Manna MC, Bandyopadhyay, KK, Ajay AK, Tripathi RH. Wanjari KM, Hati AK, Misra CL, Acharya A, Subba R (2006). Inter specific Interaction and Nutrient Use in Soybean/Sorghum Intercropping System. Bioresour. Technol. 98:25–30.

Maheshbabu HM, Ravi H, Biradar P, NK, Babalad HB (2008). Effect of organic manures on plant growth, seed yield and quality of soybean. Karnataka J. Agric. Sci. 21(2):219-22.

Marschner H (1995). Mineral nutrition of higher plants. London, Academic Press, London. P. 889.

Muoneke CO, Ogwuche MAO, Kalu BA (2007). Effect of maize planting density on the performance of maize/soybean intercropping system in a guinea savannah agro ecosystem. Afr. J. Agric. Res. 2(12):667-677

Nielsen H, Ambus B, Jensen ES (2001). Evaluating pea and Barley cultivars for complementarities in intercropping at different levels of soil N availability. Exp. Agric.14:112-118

Otieno PE, Muthomi JW, Cheminingwa GN, Nderitu JH (2009). Effect of rhizobia inoculation, farm yard manure and nitrogen fertilizer on nodulation and yield of food grain legumes. J. Biol. Sci. 9:326-332.

Rochester JJ, Peaples MB, Hullugare RN, Gualt, RR, Constable GA (2001). Using legumes to enhance N fertility and improve soil conditions in cotton cropping systems. Field Crop Res. 70(1):27-41.

Tamado T, Eshetu M (2000). Evaluation of Sorghum, Maize and Common Bean cropping Systems in East Hararge, Ethiopia. Ethiopian J. Agric. Sci. 17:33-45.

Tesfa B, Tolessa D, Setegn G, Tamado T, Negash G, Tenaw W (2001). Development of Appropriate Cropping Systems for Various Maize Producing Regions of Ethiopia. Second National Maize Workshop of Ethiopia. 12-16 November, 2001. Addis Ababa.

Tolera A, Tamado T, Pant LM (2005). Grain yield and LER of maize-climbing bean intercropping as affected by inorganic, organic fertilizer and population density in Western Oromia, Ethiopia. Asian J. Plant Sci. 4(5):458-465.

Tisdale SL, Havlin JL, Beaton DB, Werner LN (1999). Soil fertility and fertilizers. An introduction to nutrient management. 6th ed. Prentice Hall New Jersey.

Tusbio M, Walker S, Ogindio, HO (2005). A simulation model a cereal-legume intercropping for semi- arid regions. Model applications. Field Crop Res. 93:23-33.

Wakene N, Fite G, Abdena D, Birhanu D (2007). Integrated use of organic and inorganic fertilizers for maize production. In utilization diversity in land use systems: Sustainable and organic approaches to meet human needs. Conference Tropentag, 2007, October 9-12, 2007, Witzenhousen, Kassel, Germany.

15

Micropropagation and field evaluation of seven strawberry genotypes suitable for agro-climatic condition of Bangladesh

Tanziman Ara, M. Rezaul Karim, M. Abdul Aziz, Rezaul Karim, Rafiul Islam and Monzur Hossain

Plant Breeding and Gene Engineering Laboratory, Department of Botany, University of Rajshahi, Rajshahi - 6205, Bangladesh.

Strawberry (*Fragaria x ananassa* Duch.) is one of the important and popular fruits in the temperate countries of the world due to its fragrance, taste and nutritional properties. Due to its popularity and increasing demand in Bangladesh, an experiment was conducted to establish a rapid *in vitro* clonal propagation of seven strawberry genotypes and their field evaluation under Bangladesh condition. Runner tips of seven strawberry genotypes viz. AOG, JP-2, JP-3, Camarosa, Sweet charley, Giant Mountain and Festival were cultured *in vitro* for multiple shoot proliferation and root induction. Proliferation of runner tips was obtained on Murashige and Skoog (MS) basal medium containing three different concentrations (1.0, 1.5, 2.0 mg/l) of BA with 0.5 mg/l 6-ferfuryl amino purine (KIN) or gibberellic acid (GA$_3$). The best shoot proliferation was obtained from cultures grown on medium supplemented with 1.5 mg/l BA with 0.5 mg/l KIN. Microcuttings were rooted on half strength MS medium with 0.5 to 1.5 mg/l indole butyric acid (IBA) or indole acetic acid (IAA). Maximum rooting (90 to 98%) with 9 to 12 roots/cultures was achieved at 1.0 mg/l IBA. The plantlets, thus developed were hardened and successfully stablished in soil. AOG was found to be the most responsive genotype followed by JP-2 and JP-3.

Key words: Strawberry, shoot multiplication, adventitious rooting, field establishment.

INTRODUCTION

Strawberry (*Fragaria x ananassa* Duch.) belongs to the Rosaceae family. It is a perennial, stoloniferous herb. Strawberries have traditionally been a popular delicious fruit for its flavour, taste, fresh use, freezing and processing. It contains relatively high quantities of ellagic acid having a range of biological activity and especially the fruit contains higher vitamin C concentration than orange or lemon. It is produced in 73 countries worldwide on 200,000 hectors and produced 31 lac metric tons strawberry (FAO, 2008). It has been commercially cultivated in Canada, USA, Japan, Spain, Germany, Korea, Italy, Poland, Thailand and so many countries in the world (Biswas et al., 2007).

Strawberry is traditionally propagated vegetatively by rooted runners but this method is not proved suitable due to incidence of many diseases infection and environmental hazards and resulting in the gradual degeneration of cultivers performance. Karhu and Hakala (2002) observed that micropropagated strawberry plants were comparatively better in different characters (crown size, number of runners, flowering time and yield of berries) than conventionally propagated runner plants.

Conventionally, strawberry is propagated by runners (Sakila et al., 2007), which is very labour intensive; time consuming and results in the transmission of viral diseases (Gautam et al., 2001).

In contrast of these, mass multiplication *in vitro* through tissue culture results high yield in disease free plant material (Mohan et al., 2005) and proved to be the best alternative approach to conventional propagation method (Mahajan et al., 2001). The standardization of protocol and procedure of micropropagation of strawberry was successfully attempted by many (Kaur et al., 2005; Sakila et al., 2007; Gantait et al., 2010). But complete field performance of micropropagated plants was not studied enough where extensive field evaluation is necessary for commercial utilization of tissue culture (Smith and Hamill, 1996). Moreover, the conventional way of production is not adequate to meet the commercial demand. Micropropagated strawberry plant has been introduced to prevent most of the plant and soil transmissible diseases (Biswas et al., 2008).

For better strawberry production photoperiod 10 to 20 h, day temperature 12 to 30°C and number of short days 12 to 24 are essential (Michel et al., 2006). Bangladesh is a sub tropical country and here in winter average day temperature is 15 to 25°C, photoperiod 12 to 16 h and short days about 30 to 50 days (Biswas et al., 2008). Therefore in winter season, strawberry can be grown and nowaday it becomes very popular due to attractiveness of fruits, fragrance and nutritional quality. Since last few years, strawberries are cultivated but the main constrain of its cultivation is to maintain plant materials due to hot summer in Bangladesh. Therefore, in the present investigation an attempt was made to develop an efficient method of *in vitro* plant regeneration of strawberry for mass production of planting materials for commercial cultivation in Bangladesh.

MATERIALS AND METHODS

Runnes with tips of seven strawberry varieties viz. AOG, JP-2, JP-3, Camarosa, Sweet charly, Giant Mountain and Festival were collected from strawberry germplasm stocks maintained in Akafuji Agrotechnologies, Rajshahi, Bangladesh. AOG, JP-2 and JP-3 are Japanese varieties and Camarosa, Sweet charly, Giant Mountain and Festival are American varieties. Fresh runner tips from 2 months old mature strawberry plants (Figure 1A) were collected during the first week of November 2007. Runner tips of seven strawberry varieties were washed first under running tap water for 30 min and treated with 1% Tween 80 for ten minutes followed by repeated rinsing with sterile distilled water. Further sterilization was done under aseptic condition in laminar air flow cabinet. Explants were surface sterilized with 50% (v/v) ethyl alcohol (1 min) followed by 0.1% (w/v) $HgCl_2$ (4 min).

Finally, the explants were washed thoroughly (five times) with sterilized distilled water and cut into appropriate size (1.5 cm) and cultured on MS basal medium supplemented with specific concentration of growth regulators viz. 6-benzyladenine (BA), 6-ferfuryl amino purine (KIN) and gibberellic acid (GA_3) adding 30 g/l sugar (market sugar) and 0.8% agar (British Drug House, England). The pH of the medium was adjusted to 5.7 before autoclaving at 1.06 kg/cm^2 and 121°C for 20 min. The cultures were incubated in growth chamber 16/8 light/dark cycle at 25±2°C. Proliferated multiple shoots after elongation were cut and individual shoots were placed in half strength MS medium containing different concentration of indole butyric acid (IBA) or indole acetic acid (IAA) for root induction. All chemical compounds including macro and micro nutrients, organic acids and inorganic acids, sugar, agar, KOH, $HgCl_2$, ethanol etc. used in the present study were the reagent grade products of either BDH, England or MERCK, India. The vitamins, amino acids (Glycin), growth regulators were mostly products of Sigma Chemical Company; USA and Phytotec (USA) and a small portion of thiamine was a product of British Drug House (BDH), England. Data on shooting and rooting efficiency were recorded after 5 weeks of culture initiation. For each treatment 10 to 12 explants were used and the experiments were repeated three times.

Three-week-old rooted shoots were taken out from the culture tubes, thoroughly washed in water to remove agar gel and then transferred to plastic pot containing garden soil and compost (3:1 v/v) and were kept under transparent plastic shed to control the moisture condition. After one week plants were taken out from the shed and successfully survived plants were transferred to the field. Data were recorded from randomly selected ten plants/variety on plant height, no. of leaves/plant, no. of stolon/plant, no. of nodes/stolon, canopy size (cm^2), no. of flowers/plant, no. of fruits/plant and fruit weight/plant (g) to compare the varietal performance of the seven strawberry genotypes. No. of fruits/plant was recorded after 80 days of plantation and other characters were recorded after 60 days of plantation.

RESULTS AND DISCUSSION

Runner tips of seven strawberry varieties were inoculated on MS medium fortified with different concentration of BA (0.1, 0.5 and 2.0 mg/l) with KIN (0.1 and 0.5 mg/l) or gibberellic acid (GA_3) (0.1 and 0.5 mg/l). Within 8 to 14 days of culture multiple shoots emerged directly from the explants. The explants cultured in medium with 1.5 mg/l BA either with KIN or GA_3 initiated shoots 2 to 4 days early than other treatments. The rate of shoot proliferation ranged from 45 to 80% and the highest rate of response for all genotypes was obtained at 1.5 mg/l BA + 0.5 mg/l KIN combination (Table 1, Figure 1G). AOG showed the maximum frequency (88%) of shoot formation followed by JP-2 (86%) and JP-3 (80%). The number of shoots per explant ranged from 8.00 to 9.00 in 1.0 mg/l BA + 0.5 mg/l KIN and 5.00 to 6.00 in 1.5 mg/l BA + 0.5 mg/l GA_3.

In other treatments, the number of roots/plant was low. When BA concentration was increased from 1.0 to 1.5 mg/l, the number of shoots/explant increased but further increase of BA number of shoots decreased. JP-3 produced maximum number of shoots/explant in 1.5 mg/l BA + 0.5 mg/l KIN followed by AOG and JP-2. Hu and Wang (1983) reported that high concentration of cytokinin reduced the number of micropropagated shoots. Similar results have already been reported in *Fragaria indica* Andr. (Bhatt and Dhar, 2000). Also, this result is in consistent with the findings in papaya (Conover and Litz, 1978) as well as in *Eucalyptus grandis* (Teixetra and Silva, 1990). The developing shoots were elongated by

Figore 1. Micropropagation of strawberry from runner tips. Source of explant (A); runner tips use as explants (B); shoot proliferation on MS + 1.5 mg/l + 0.5 mg/l KIN after 7 days (C), after 15 days (D-E), after 25 days (F) and after 35 days (G) of culture. Rooted shoots on 1.0 mg/l IBA after 3 weeks (H) and after 4 weeks (I) of culture. Regenerated plantlets in thump pots after 20 days of transplantation (J). Plants with flowers (K) and fruits (L) of strawberry.

subculturing on the same combinations of growth regulators. Later on elongated shoots were excised and used for root induction.

Bhatt and Dhar (2000) reported multiple shoot regeneration from Indian wild strawberry using MS supplemented with 4.0 mg/l BA and 0.1 mg/l napthalene acetic acid (NAA). Some workers also reported shoot regeneration in strawberry using MS medium containing BA in combination with KIN (Lee and de Fossard, 1977; Sobczykiewicz, 1980; Lis, 1990; Boxus, 1999; Neeru et al., 2000; Mereti et al., 2003). Our results indicated that, low concentration of BA alone or with KIN were found suitable for shoot initiation and further multiplication. This difference may be attributed by the difference of genotype and physiological condition of the explants. The daughter shoots (3 to 4 cm length) were excised and

transferred to root induction media. Both IBA and IAA were found to be effective for adventitious root induction and frequency of root induction ranged from 78 to 95%. Out of two concentrations of IBA or IAA 1.0 mg/l was proved to be superior where the shoots produced roots early and between the two auxins IBA showed better performance for root induction than IAA (Figure 1H). AOG produced highest number of roots per shoot with highest frequency (95%) of root induction in medium fortified with 1.0 mg/l IBA (Table 2). No difference was observed in root length in this experiment. Similar effects of IBA were also observed in *Calotropis gigentea* (Roy and De, 1986), *Capsicum annum* (Agarwal et al., 1989) and *Prunus* sp. (Mante et al., 1989).

Rooted plantlets ware taken out from culture tubes and washed thoroughly with tap water to remove the culture

Table 1. Effects of different concentrations of BA with KIN or GA$_3$ on multiple shoot induction from runner tips of seven genotypes of strawberry.

Growth regulators conc. (mg/l)	Genotypes	% of explants showing shoot proliteration	Days to shoot formation	No. of shoots/ explant	Shoot length (cm)
	AOG	50	10 to 12	3.33	3.3
	JP-2	45	10 to 12	2.67	3.0
	JP-3	40	10 to 12	3.00	3.0
1.0 mg/l BA+ 0.5 mg/l KIN	Camarosa	43	10 to 12	2.00	3.2
	Sweet charly	43	10 to 12	2.67	3.0
	Giant Mountain	45	10 to 12	3.00	3.1
	Festival	45	10 to 12	2.00	3.0
	AOG	88	8 to 10	8.33	2.2
	JP-2	86	8 to 10	9.00	2.0
	JP-3	80	8 to 10	8.33	2.0
1.5 mg/l BA+ 0.5 mg/l KIN	Camarosa	77	8 to 10	6.00	2.1
	Sweet charly	76	8 to 10	6.67	2.0
	Giant Mountain	75	8 to 10	6.33	2.1
	Festival	75	8 to 10	6.67	2.0
	AOG	55	10 to 12	2.67	3.3
	JP-2	52	10 to 12	2.33	3.2
	JP-3	50	10 to 12	2.00	3.2
2.0 mg/l BA+ 0.5 mg/l KIN	Camarosa	53	10 to 12	2.00	3.2
	Sweet charly	51	10 to 12	2.67	3.2
	Giant Mountain	50	10 to 12	2.00	3.2
	Festival	50	10 to 12	2.00	3.2
	AOG	56	10 to 12	2.67	4.2
	JP-2	51	12 to 14	2.00	4.0
	JP-3	51	12 to 14	1.67	4.0
1.0 mg/l BA+ 0.5 mg/l GA$_3$	Camarosa	55	10 to 12	2.33	4.1
	Sweet charly	55	12 to 14	2.00	4.0
	Giant Mountain	55	10 to 12	2.33	4.0
	Festival	50	12 to 14	1.67	3.9
	AOG	58	8 to 10	7.33	3.6
	JP-2	56	8 to 10	5.33	3.3
	JP-3	55	9 to 10	5.67	3.2
1.5 mg/l BA+ 0.5 mg/l GA$_3$	Camarosa	55	8 to 10	6.00	3.3
	Sweet charly	54	8 to 10	5.00	3.0
	Giant Mountain	52	9 to 10	5.33	3.4
	Festival	55	9 to 10	5.00	3.1
	AOG	50	10 to 12	4.00	2.0
	JP-2	45	12 to 14	3.00	3.9
	JP-3	45	12 to 14	3.00	3.8
2.0 mg/l BA+ 0.5 mg/l GA$_3$	Camarosa	50	10 to 12	3.00	2.0
	Sweet charly	45	12 to 14	3.33	3.9
	Giant Mountain	50	10 to 12	3.33	2.0
	Festival	45	12 to 14	3.00	3.9

Data were collected after 5 weeks of culture.

Table 2. Effects of different concentrations of IBA and IAA on in vitro rooting of microshoots of seven varieties of strawberry. Data were recorded after 5 weeks of culture.

Growth regulators conc. (mg/l)	Genotypes	% of microcuttings rooted	Days to root formation	No. of roots/ microcutting	Root length (cm)
	AOG	88	12 to 14	7.33	2.2
	JP-2	87	15 to 20	6.33	2.0
	JP-3	87	12 to 15	6.33	2.0
MS + 0.5 mg/l IBA	Camarosa	85	15 to 20	7.00	2.1
	Sweet charly	84	15 to 20	6.33	2.2
	Giant Mountain	85	15 to 20	7.00	2.1
	Festival	86	15 to 20	6.67	2.0
	AOG	95	7 – 10	12.00	2.2
	JP-2	90	8 to 10	10.67	2.1
	JP-3	90	8 to 12	10.00	2.5
MS + 1.0 mg/l IBA	Camarosa	92	8 to 11	9.67	2.6
	Sweet charly	90	8 to 10	9.33	2.2
	Giant Mountain	92	8 to 10	9.00	2.4
	Festival	90	8 to 12	9.33	2.6
	AOG	89	10 to 12	6.67	2.3
	JP-2	88	10 to 12	3.67	2.2
	JP-3	88	10 to 12	3.00	2.2
MS + 0.5 mg/l IAA	Camarosa	87	10 to 14	2.67	2.2
	Sweet charly	87	10 to 12	2.67	2.2
	Giant Mountain	88	10 to 12	3.00	2.2
	Festival	88	10 to 12	2.67	2.1
	AOG	85	12 to 14	6.67	2.0
	JP-2	82	12 to 14	6.33	2.0
	JP-3	80	12 to 14	6.00	2.0
MS + 1.0 mg/l IAA	Camarosa	82	12 to 14	6.33	2.1
	Sweet charly	80	12 to 14	6.00	2.0
	Giant Mountain	81	12 to 14	6.33	2.1
	Festival	78	12 to 14	5.67	2.0

medium from the roots. Washed plantlets were sprayed with fungicide and planted to normal and sterilized soil in thump pot (Figure 1J). After 7 days, the hardened planlets were planted in soil. After two months of planting, different agronomic characters of seven genotypes of strawberry were noted. It was exhibited from the result that plant growth showed wide range of variation for different agronomic characters except plant height (Table 3). JP-2 showed best performance in no. of leaves/plant (22.8 ± 2.27), no. of nodes/stolon (4.5± 0.17), no. of fruits/plant (6.0 ± 0.37), canopy size (cm^2) (382.72 ± 8.1) and fruit yield/plant (19.43 ± 0.28). JP-3 showed best performance in plant height (16.68 ± 0.32) and no. of stolons/plant (4.4 ± 0.73). AOG showed best performance in no. of flowers/plant (16.7 ± 0.62). Lowest plant height was found in Sweet charly (15.24 ± 0.27) and lowest no.

of nodes/stolon (2.2 ± 0.13) was found in Giant Mountain and Festival. Lowest no. of leaves/plant (12.8 ± 0.96), no. of stoloplant (2.8 ± 0.36), canopy size (cm^2) (299.43 ± 14.1), no. of flowers/plant (10.9 ± 0.31), no. of fruits/plant (2.0 ± 0.26) and fruit yield/plant (8.56 ± 0.22) were found in Festival. In this study, it is observed that Japanese varieties were more responsive under in vitro and ex vitro condition than American varieties. The protocol reported here is reproducible; it has a potential for allowing a large scale multiplication of this important and new fruit plant in Bangladesh.

ACKNOWLEDGEMENT

The authors thankfully acknowledge the Ministry of

Table 3. Field performance of micropropagated plantlets of seven strawberry genotypes. Morphological and yield data were recorded at 60-80 days after transplantation of plantlets in the field.

Genotypes	Plant height (Mean ± SE)	No. of leaves/plant (Mean ± SE)	No. of stolons/plant (Mean ± SE)	No. of nodes/stolon (Mean ± SE)	Canopy size (cm²) (Mean ± SE)	No. of flower/plant (Mean ± SE)	No. of fruits/plant (Mean ± SE)	Fruit yield/plant (g)
AOG	15.71 ± 0.17	20.5 ± 2.23	4.2 ± 1.18	3.0 ± 0.33	374.86 ± 20	16.7 ± 0.62	5.9 ± 0.35	17.61 ± 0.18
JP-2	15.81 ± 0.18	22.8 ± 2.27	4.3 ± 0.83	4.5 ± 0.17	382.72 ± 8.1	14.9 ± 0.87	6.0 ± 0.37	19.43 ± 0.28
JP-3	16.68 ± 0.32	19.1 ± 1.88	4.4 ± 0.73	3.3 ± 0.15	380.92 ± 7.5	14.8 ± 0.65	4.2 ± 0.29	15.54 ± 0.35
Camarosa	15.85 ± 0.15	18.9 ± 1.69	4.3 ± 0.56	2.5 ± 0.17	369.93 ± 3.9	12.3 ± 0.5	3.1 ± 0.31	12.87 ± 0.62
Sweet charly	15.24 ± 0.27	17.9 ± 1.57	4.6 ± 0.62	2.6 ± 0.16	334.28 ± 16.3	12.5 ± 0.7	2.7 ± 0.26	10.76 ± 0.22
Giant Mountain	16.29 ± 0.27	16.9 ± 1.43	3.4 ± 0.45	2.2 ± 0.13	356.54 ± 6.66	11.0 ± 0.37	3.1 ± 0.38	10.67 ± 0.37
Festival	15.25 ± 0.37	12.8 ± 0.96	2.8 ± 0.36	2.2 ± 0.13	299.43 ± 14.1	10.9 ± 0.31	2.0 ± 0.26	8.56 ± 0.22
F value	0.3463^{NS}	6.07**	8.29**	11.04**	19.77**	20.75**	15.12**	0.0881^{NS}

** = significant at 5% and 1% level of probability, NS = Non significant.

Science and Information and Communication Technology of the Peoples Republic of Bangladesh for financial grant.

REFERENCES

Agarwal S, Chandra N, Kothari (1989). Plant regeneration and tissue culture of pipper (*Capsicum annum l.* ev. Mathania). Plant Cell Tissue Org. Cult. 16:47-55.

Bhatt ID, Dhar U (2000). Micropropagation of Indian wild strawberry. Plant Cell Tissue Org. Cult. 60:83-88.

Biswas MK, Hossain M, Ahmed MB, Roy UK, Karim R, Razvy MA, Salahin M, Islam R (2007). Multiple shoots regeneration of strawberry under various colour illuminations. American-Eurasian J. Sci. Res. 2(2):133-135.

Biswas MK, Islam R, Hossain M (2008). Micropropagation and field evaluation of strawberry in Bangladesh. J. Agric. Technol. 4(1):167-182.

Boxus P (1999). Micropropagation of strawberry via axillary shoot proliferation. In: Plant Cell Culture Protocols. Methods in Molecular Biology. Part III. Plant Propagation *In Vitro*. Hall RD (ed.) Humana Press Inc. Totowa NJ 111:103-114.

Conover RA, Litz RE (1978). *In vitro* propagation of papaya. Hort. Sci. 13:241-242.

FAO (2008). FAOSTAT Agricultural Statistics Database. http://www.Fao.org.

Gantait S, Mandal N, Bhattacharyya S, Das PK (2010). Field performance and molecular evaluation of micropropagated strawberry. Rec. Res. Sci. Tech. 2(5):12-16.

Gautam H, Kaur R, Sharma DR, Thakur N (2001). A comparative study on *in vitro* and *ex vitro* rooting of micropropagated shoots of strawberry (*Fragaria × ananassa*). Plant Cell Biotechnol. Mol. Biol. 2:149-152.

Hu CY, Wang PJ (1983). Meristem shoot tip and bud culture, In Evans *et al.* (Eds) Hand book of Plant Tissue Culture. Macmillan. New York. pp. 177-227.

Karhu S, Hakala K (2002). Micropropagated strawberries in the field. Acta. Hort. (ISHS) 2:182.

Kaur R, Goutam H, Sharma DR (2005). A low cost strategy for micropropagation of strawberry (*Fragaria × ananassa* Duch.) cv. Chandler. Acta Hort. 696:129-133.

Lee ECM, Fossard RA (1977). Some factors affecting multiple bud formation of strawberry (*Fragaria X ananassa* Duch.) *in vitro.* Acta Hort. (ISHS) 78:187-196.

Lis EK (1990). *In vitro* clonal propagation of strawberry from immature achenes. Acta. Hort. (ISHS) 280:147-150.

Mahajan R, Kaur R, Sharma A, Sharma DR (2001). Micropropagation of strawberry cultivar Chandler and Fern. Crop Improv. 28:19-25.

Mante S, Scorza R, Cordts JM (1989). Plant regeneration from cotyledons of *Prunus persieg, P. domestica* and *P. cerasus.* Plant Cell Tissue Org. Cult. 19:1-11.

Mereti M, Grigoriadou K, Levantakis N, Nanos GD (2003). In vitro rooting of strawberry tree (*Arbutus unedo* l.) in medium solidified by peat- perlite mixure in combination with agar. Acta. Hort. (ISHS) 616:207-210.

Michel JV, Anita S, Svein OG (2006). Interactions of photoperiod, temperature, duration of short-day treatment and plant age on flowering of *Fragaria X ananassa* Duch. Cv. Korona. Scientia Hort. 107:164-170.

Mohan R, Chui EA, Biasi LA, Soccol CR (2005). Alternative in vitro propagation: use of sugarcane bagasse as a low cost support material during rooting stage of strawberry cv. Dover. Braz. Arch. Biol. Tech. 48:37-42.

Neeru S, Ranjan S, Singh OS, Gosal SS (2000). Enhancing micropropagation efficiency of strawberry using bandage in liquide media. J. Appl. Hort. 2(2):92-93.

Roy AT, De DN (1986). *In vitro* plantlets regeneration of the petrocrop *Calotropis gigantia.* Bioenergy Society of India, New Delhi. pp. 123-128.

Sakila S, Ahmed MB, Roy UK, Biswas MK, Karim R, Razvy MA, Hossain M, Islam R, Hoque A (2007). Micropropagation of strawberry (*Fragaria x ananassa* Duch.) a newly introduced crop in Bangladesh. American-Eurasian J. Sci. Res. 2:151-154.

Smith MK, Hamill SD (1996). Filed evaluation of micropropagated and conventionally propagated ginger in subtropical Queensland. Aust. J. Exp. Agric. 36:347-354.

Sobczykiewicz D (1980). Heat treatment and meristem culture for the production of virus-free strawberry plants. Acta Hort. (ISHS) 95:79-82.

Teixetra SL, Da Silva LL (1990). *In vitro* propagation of adult *Eucalyptus grandis* Hill Ex. Maiden from epicormic shoots. Abstracts VII th. Int. Cong. on Plant Tiss. and Cell Cult. (IAPTC), Amsterdam. p. 218.

16

Soil and nutrient losses along the chronosequential forest recovery gradient in Mabira Forest Reserve, Uganda

Kizza C. L.[1*], Majaliwa J. G. M.[1], Nakileza B.[1], Eilu G[1]., Bahat I[1], Kansiime F.[1] and Wilson J.[2]

[1]College of Agricultural and Environmental Sciences, Makerere University, Uganda.
[2]Centre for Ecology and Hydrology, Bush Estate, Penicuik, Midlothian, EH26 0QB, U.K.

Information on the effect of Mabira Forest Reserve degradation on water, soil and nutrient losses is scanty. This study was carried out to quantify runoff, soil and nutrient losses in six restoring forest regimes namely: 0 to 3, 10 to 20, 20 to 30, 30 to 40, 40 to 50, and >55 years of last disturbance. In each, a plot measuring 150 by 50 m was demarcated. Within this plot, three sub-plots each of 20 × 2 m and located 50 m apart were established for runoff, soil and nutrient losses measurement for period of two rain seasons. A runoff trap pre-calibrated to collect 1% of runoff was installed on each sub-plot. In addition, the forest regimes were characterized for some physio-chemical properties. Overall, annual runoff and soil loss across the forest regimes were low ranging between 20 to 160 m^3ha^{-1} and 10 to 380 kg ha^{-1}, respectively. Runoff and soil loss followed a parabolic trend along the forest age with the minimum in the 20 to 30 years forest regime. Annual nitrogen, phosphorus and potassium losses significantly varied from 0.11 to 4.26, 0.01 to 0.18 and 0.08 to 6.63 kg ha^{-1} respectively across the forest regimes. Bulk density was highest in the 30 to 40 year regime.

Key words: Soil erosion, bulk density, forest degradation.

INTRODUCTION

Whereas soil erosion and its control have been well documented worldwide, it still contributes significantly to soil functional degradation especially in Sub Saharan Africa (Smaling, 1993; Sanchez et al., 1997). In addition to affecting ecosystem services provided by the soil resource, erosion also reduces soil nutrient availability. Over the years, maintenance of vegetation cover - like forests, has been one of the common ways of controlling soil erosion (Coelho et al., 2001). However, in Uganda, rampant degradation of forests has persisted leading to grave environmental and ecological concerns (Banana and Sembajjwe, 2000). Regrettably, forest degradation has been attributed largely to anthropogenic activities other than climatic or natural causes. Notable of the

human-induced causes of forest loss are conversion of forests to agriculture, overgrazing and firewood and timber cutting (World Wildlife Fund and McGinley, 2007). Limited access to alternative sources for the numerous forest products by both the surrounding and distant communities is a leading driver to over exploitation and disturbance vulnerability of forests.

Kayanja and Byarugaba (2001) reported that uncontrolled degradation and forest conversion to other land use types were seriously threatening Uganda's forests. Forest cover in Uganda has declined from 3,090,000 ha (12.7%) of the total country area at the beginning of the 20th Century, to 730,000 ha (3.6%) of Uganda's land area (Forest Department, 1999). In Mabira Forest Reserve (MFR), encroachment that took place mainly in the 1970's for settlement and agriculture resulted into 25% forest cover loss (BirdLife International, 2009). In MFR, the government intervened to reverse the

*Corresponding author. E-mail: kizluswata@gmail.com.

trend of forest loss, and the encroachers were gradually evicted from the reserve in mid 1980s. This was achieved through either persuasion or forceful removal of all the encroachers. Because of the gradual removal of encroachment, the Reserve contains several blocks of different regeneration ages, which are the focus of this study.

Mabira Forest Reserve (MFR) is a tropical natural forest located between 32° 52' to 33° 07' E and 0° 24' to 0° 35' N, near the L. Victoria shoreline covering an area of 306 km^2 (Forest Department, 1999). Its altitude ranges between 1070 and 1340 m above sea level (asl) (Nature Uganda, 2001). Approximately 3.5 km^2 of the area comprises isolated hills lying above 1250 m asl. The slopes are generally undulating, and with numerous flat-topped hills and shallow valleys.

The soils are generally ferralitic with texture ranging from loamy to sandy clay loams (AES Nile Power, 2001), with isolated cases of waterlogged clays in the valley bottoms. The underlying rocks are composed of micaceous schists and shales of the Buganda-Toro system with ridges of quartzite and amphibolite. Its climate is tropical humid with bimodal rainfall from March to June and September to November for the long and short duration rains respectively with annual mean precipitation ranging between 1250 and 1400 mm. The annual mean minimum and maximum temperatures range between 16 to 17°C and 27 to 29°C, respectively (Lamto et al., 2010). The forest is a vital catchment for L. Kyoga because it is the origin of two rivers namely Musambya and Sezibwa (Environmental Alert, 2007) that drain into the lake.

The forest is rich in biota harbouring 47% of Uganda's total plant species (Baranga, 2007) many of which are on the list of globally endangered species (FAO, 1986). Its rich fauna includes 151 species of forest birds, 2 species of diurnal forest primates, 39 species of forest swallowtail and 218 species of butterflies including Charaxes (Baranga, 2007). Within MFR, there are 27 villages (Baranga, 2007) commonly known as enclaves, where subsistence farming is the primary activity for the 3,506 families within (BirdLife International, 2009). The most commonly grown crops are maize, cassava, bananas, sugarcane and beans. Additionally, a large section of the community is involved in some illegal activities like charcoal burning, pitsawing and collection of poles for construction as well as collecting medicinal and other plants to obtain further income.

Whereas several studies have been conducted in MFR and elsewhere in Uganda (Earth Trends, 2003; Pomeroy and Tushabe, 2004; Poulsen et al., 2005; Winterbottom and Eilu, 2006), these previous studies have focused largely on biodiversity, and there is limited information on the dynamics of soil physical and chemical properties, and nutrient losses in regenerating forests in Uganda. Therefore, the current existence of forest regimes at different recovery stages offers the opportunity for this study to assess the effect of forest regeneration age on runoff, soil and nutrient losses from Mabira Forest Reserve.

MATERIALS AND METHODS

Runoff, soil and nutrient losses measurements

Studies were conducted during the short season (September to December 2007) and long season (March to June 2008). Six sets of forest regimes representing 0 to 3, 10 to 20, 20 to 30, 30 to 40, 40 to 50, and over 55 years since the last disturbance were identified for this study. Because of difficult of getting several forest regimes with similar ages, replication was done within a forest regime. Within each forest regime, three runoff plots each measuring 2 × 20 m were installed 50 m apart to measure runoff using a pipe sampler (Ngigi et al., 2005). The erosion trap area was sealed off using several pieces of iron sheets measuring 3.3 m long and 0.25 m width. The plots were geo-referenced and a study area map drawn (Figure 1) in ArcView a GIS environment.

A clean container was tightly tied at the end of the collection pipe located at the posterior end of the erosion trap to collect the runoff via a small slit calibrated to take 1% the total runoff (Ngigi et al., 2005). However, each trap was calibrated using a known volume of water and determining the transmission coefficient. Total runoff (water and sediments) that collected in the container were quantified using a measuring cylinder every morning of the day following a rainfall event. The runoff samples were then put into clean plastic bottles and periodically delivered to the Department of Soil Science, Makerere University analytical laboratory for subsequent analyses. Event runoff was multiplied by the divider transmission coefficient to obtain the plot runoff, and then extrapolated to hectare basis, by multiplying the obtained number by 250. Runoff loss in a season was obtained by summing up runoffs registered for the different storms in that season. The two season runoff collections were eventually summed up to obtain the annual runoff per hectare.

In the laboratory, each runoff sample was thoroughly homogenized and immediately split into two known volumes. The first portion was filtered using a Whatman No. 1 filter paper and its contents oven-dried at 105°C for 24 h to determine the dry weight of the soil. The amount of soil loss in the plot was obtained by multiplying the sediment concentration by runoff (MacDonald et al., 2003) and transmission coefficient and then expressed on ha basis. The seasonal soil loss was computed by summing up the soil losses from all the season collections. Finally, annual soil loss was determined by combining the losses of the two seasons. The second portion of a known volume was also filtered as the first portion, air dried and used for macro nutrient loss determination. Three macronutrients namely: nitrogen, phosphorous and potassium were measured for assessment of nutrient loss. All the three nutrients were determined following standard procedures described by Okalebo et al. (2002). Seasonal nutrient losses were estimated by multiplying by seasonal soil loss and computed on a hectare basis. Finally, annual nutrient losses were computed by summing seasonal losses.

General physical and chemical site characterization

Soil samples were taken around the runoff plots in order to characterize the study area in terms of its physiochemical properties. Bulk density was determined using the core oven drying method (Blake and Hartage, 1986). Four cores each measuring 5 cm long by 5 cm internal diameter were used to take samples from

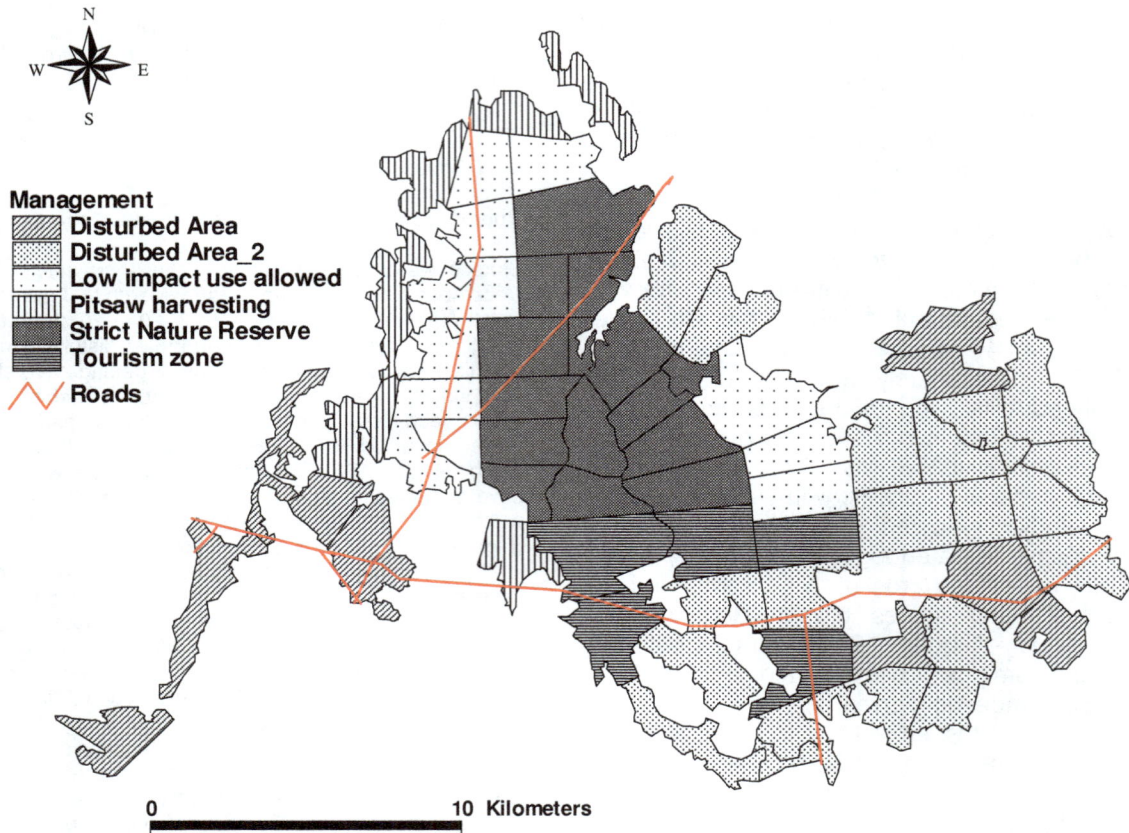

Figure 1. Mabira Forest Reserve showing different forest land uses and study area.

two soil depths of 0 to 15 cm and 15 to 30 cm for top and sub soils respectively. The core samples were taken about 3 m away from the mid-length of each sub-plot. Composite soil samples were taken around the plot using a soil auger. A total of four samples were collected at each sub-plot, two from 0 to 15 cm and the rest from 15 to 30 cm. These samples were used to probe some of the physiochemical base of the area. The soils were analyzed for pH, organic matter (OM), total nitrogen (N), available phosphorus (P), exchangeable bases (sodium, calcium, magnesium and potassium), micro nutrients (iron, zinc, manganese and copper) and texture (Okalebo et al., 2002).

Rainfall determination

Rainfall was measured for two seasons using an electronic rain gauge with DT1 - One Rain Channel Data Logger manufactured by Environmental Manufacturers (EM) Ltd of United Kingdom (www.waterrauk.com/pages/product/Raingauge_Loggers.asp). The gauge was installed in the forest regime of 20 to 30 years located near the Eco-Tourism Centre at Najjembe. Data was downloaded every fortnight throughout the study period. Adjacent to this rain gauge, a locally fabricated manual rain gauge was installed. Similar manual gauges were installed in all the other forest regimes. The fabricated manual gauges were calibrated against the electronic one. Water that collected in the manual gauge was measured immediately after a rain event. Seasonal rainfall was determined by summing up all the season rain events. The two seasons' rainfall was then summed up to obtain annual rainfall per forest regime.

Data analysis

Data were entered in MS Excel (ref) and then imported into Genstat (VSN International, 2000) for statistical analysis. Mean soil, runoff, and nutrient losses from the different treatments were separated by a one-way analysis of variance (ANOVA) in Genstat.

RESULTS AND DISCUSSION

Rainfall characteristics of the various forest restoring regimes

There were two sets of forest regimes without intra-rainfall variation but with significant inter-variation. The first set of all forest stand ages of at least 20 years had significantly higher amount of annual rainfall than the other set of young forests (Figure 2) averaging 2000 and 1000 mm respectively. The higher amount of rainfall in the relatively older forests compared to young ones was also reported by Turner et al. (2009). This is an indication of the crucial role of mature forests in hydrological control. It also signifies the hydrological disruption as a result of forest ecosystem interference (Wang et al., 2009).

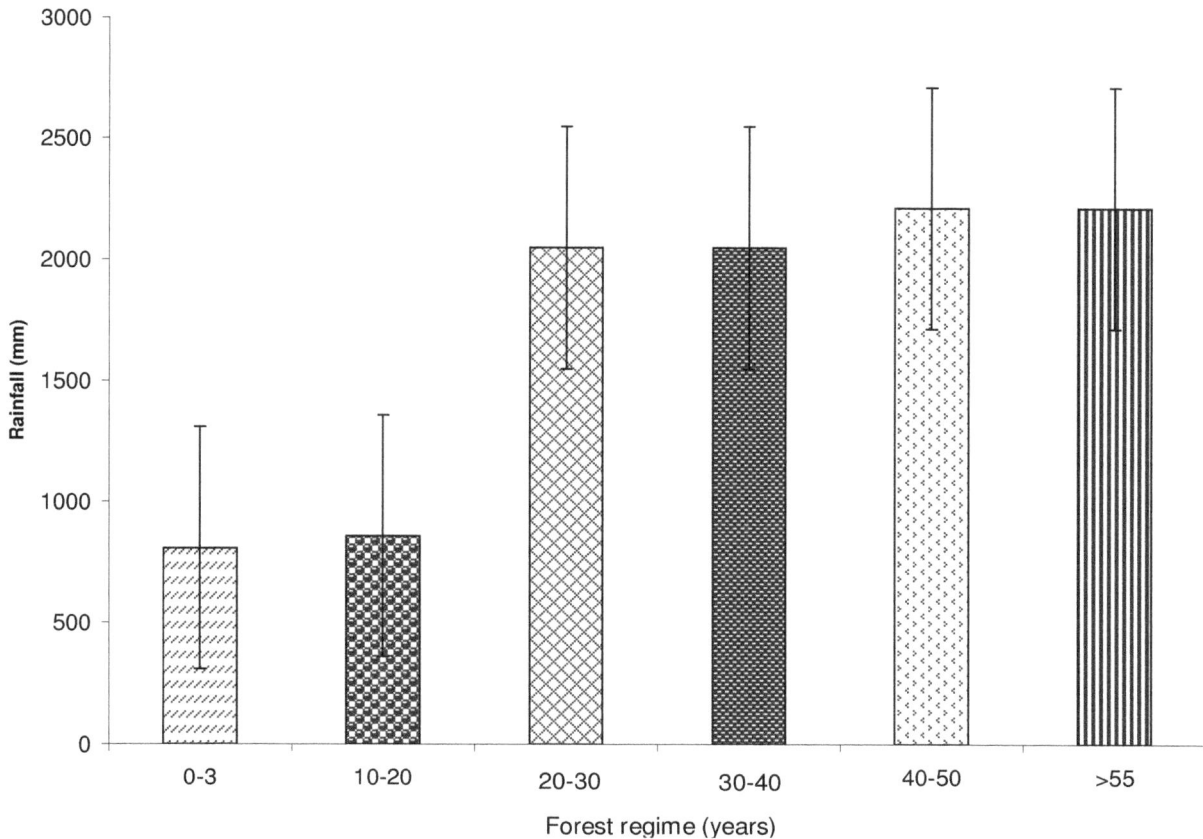

Figure 2. Annual rainfall in Mabira forest restoring regimes.

General physio-chemical characteristics of the study area

The soils in Mabira Forest Reserve were all rich in most of the nutrients required for normal or optimum plant growth (Table 1) though with a lot of variation among the forest regimes. For instance, soil pH ranged between 5.66 and 7.02 with highest observed in forest regime of 20 to 30 years while least in that of 40 to 50 years. Organic matter also differed with age of forest where it was significantly higher in those of over 40 years than in the younger ones. Phosphorus did not vary so much among the forest regimes except for the block of 20 to 30 years that had more but overall, the amount was below optimum values across all the forests. Potassium, magnesium and calcium were very high for all soils and differing significantly ($p < 0.05$) among the forest ages with the 20 to 30 year forest stand age having the highest values except for magnesium that generally decreased with forest age. Soils in all the regimes were mainly clay loam (CL) except for the regimes of at least 45 years ago. Though there were significant variations in the physio-chemical properties, there was no clear trend attributable to forest stand age apart from organic matter that increased with forest stand age. This differed from the

findings of Sharma et al. (1985) under *Alnus nepalensis* plantation where most of the physiochemical properties improved with forest age. Therefore, the observable differences except organic matter could not be explained by age differences alone but could be possibly due to the intrinsic soil characteristics.

The bulk densities of Mabira soils were very low, but varied significantly with soil depth and forest regime (Table 2). Consistently, sub soils had higher bulk densities over the top soils. Similarly, porosity varied with soil depth and forest regime but unlike the bulk density, the top soils were more porous than the sub soils thus in agreement with the findings of Dudley et al. (2002). The low bulky density and high porosity in the top soil is generally due to modification by the rich organic matter in the upper layers owing to the littering effect of the trees. This is in agreement with earlier findings of Haque and Karmakar (2009). The high porosity in the 0 to 3 year forest stand is attributed to macro pores created as a result of decaying effect of the roots of the trees formally in the forest before it was degraded as was also reported by Noguchi et al. (1997). The high bulk density and low porosity in the sub soils is characteristic of a relatively more developed soil structure of the mineral soil than the organic soil on the top (Murty et al., 2002). The generally

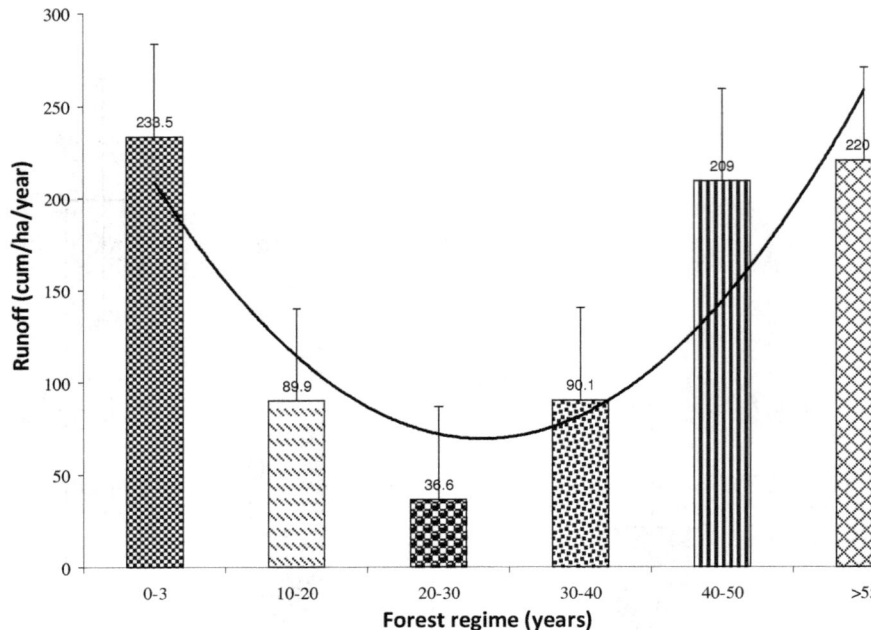

Figure 3. Annual runoff in Mabira chronosequential forest stands.

low bulk density and high porosity in Mabira Forest soils could also be attributed to the influence of the litter and roots in modifying soil structure and soil pores (Dudley et al., 2002).

Runoff from the various Mabira forest regimes

The annual runoff ranged between 36.6 and 233.5 $m^3ha^{-1}year^{-1}$ averaging 93 m^3/ha per annum (Figure 3). Two pair-wise sets of forest regimes without intra-variations but with significant inter-variations were observed. The first set with the highest amount of runoff was observed in three forest stand ages of 0 **to** 3, 40 **to** 50 and over 55 years and the low amount in forest stand ages of 10 **to** 20, 20 **to** 30 and 30 **to** 40 years with runoff exhibiting a parabolic trend across forest stand ages. The average runoff for the two forest stand age categories was 214 and 72.2 $m^3ha^{-1}year^{-1}$, respectively. The high runoff in the youngest forest of 0 **to** 3 year**s** was similarly reported by Tiwari (2009) for agricultural land. This can be attributed to limited vegetation cover that would have otherwise acted as runoff breakers through prevention of detachment and modification of infiltration (Young, 1989, Gyssels et al., 2005). However, the unusually high runoff in the old forest stands is difficult to explain but as Coelho et al. (2001) asserted, it could be due to the persistent water saturation most of the time that reduced the soil infiltration capacity thus favoring runoff. This was so despite the mature forests having sandy clay loams as opposed to the clay loams in the young forests. The latter are expected to be slightly more liable to logging than the

former.

Generally, runoff was very low in all the forest stands compared to that reported under agricultural land in the Lake Victoria Basin (Majaliwa, 1998; Johnny, 2004; Mulebeke, 2004; Majaliwa et al., 2003) and elsewhere (Floor, 2000; Gafur et al., 2003). However, similar runoff results were reported earlier by Aina (1993) under natural erosion from forested lands. The general low level of runoff implies that even under degraded conditions, the vegetation cover under forest can minimize runoff compared to the less vegetated areas.

The percentage rainfall that ended up as runoff ranged between 2 and 20% (Figure 4). The percentage rainfall amount converted into runoff also followed a parabolic trend with forest age. Whereas the youngest forest of 0 to 3 years had the least amount of rainfall, it exceedingly had a higher percentage of rainfall that resulted into runoff over all the other forests. The least run off to rainfall ratio was observed for the 20 to 30 years forest. The relatively low vegetation cover (canopy and ground cover) in the 0 to 3 years forest could have encouraged runoff over infiltration unlike in other forest stands (Coelho et al., 2001). The moderately high rate of rainfall resulting into runoff in the other forest stand was subjective to the consistently high moisture content in those forests almost throughout the year.

Soil loss from the various Mabira forest regimes

The annual soil loss across the forest ages ranged from10 to 513 kg ha^{-1} year^{-1}. Like runoff, the forest of 0 to

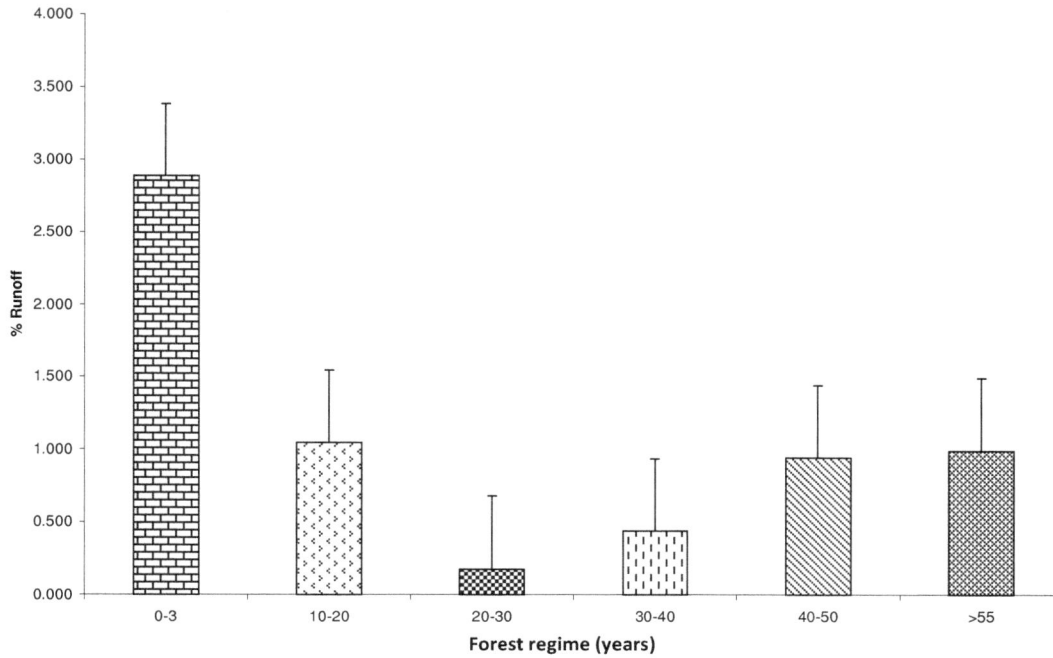

Figure 4. Percentage runoff to rainfall.

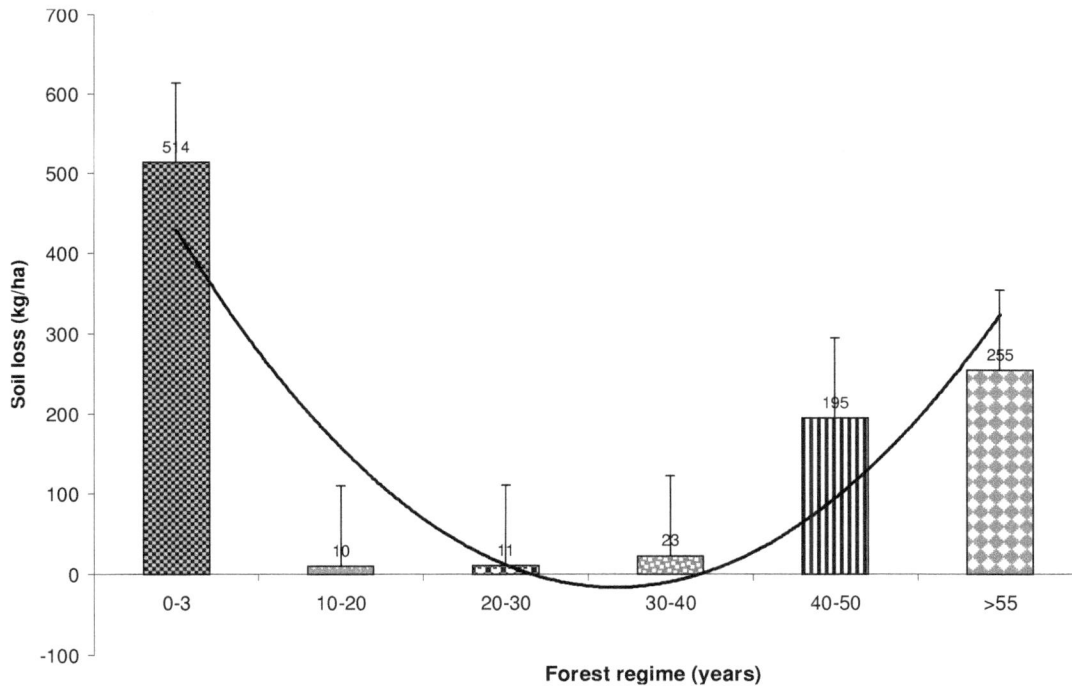

Figure 5. Annual soil loss from Mabira chronosequential forest stands.

3 years experienced the highest soil loss. However, if this young forest (0 to 3 years) is excluded, soil loss increased exponentially with forest age but if retained, a parabolic trend is exhibited with the least soil loss between 10 and 40 years forests (Figure 5). As was also reported by de Vente et al. (2005) and Gyssels et al. 2005), soil in the youngest forest of 0 to 3 years was more exposed to raindrop impact, detachment and thus

Table 1. Chemical and textural properties of MFR regenerating regimes.

Parameter	Units	Forest stand age					
		0-3	10-20	20-30	30-40	40-50	>55
Av. P	mg/kg	9.20[a,b]	5.80[b]	12.90[a]	7.20[b]	7.30[b]	6.80[b]
OM	%	4.08[e]	3.53[e]	4.79[e]	3.26[e]	7.12[d]	7.02[d]
pH		5.72[h]	6.22[g]	7.03[f]	6.21[g]	5.66[h]	5.79[h]
K		0.57[j]	0.30[k]	0.88[i]	0.34[k]	0.69[j]	0.76[i,j]
Mg	cmol/kg	4.71[l]	4.27[lm]	3.92[n]	3.60[n]	1.61[o]	2.23[o]
Ca		8.37[q]	11.46[q]	24.93[p]	9.49[q]	5.23[r]	4.79[r]
Na		0.50[t]	0.57[s]	0.51[t]	0.44[t]	0.52[t]	0.51[t]
Fe		59.7[u]	24.3[w]	18.7[w]	39.80[v]	62.5[u]	56.5[u]
Cu	mg/kg	2.77[z]	6.46[x]	7.38[x]	3.38[y]	1.88[z]	3.52[y]
Mn		138.8[1]	141.2[1]	146.2[1]	142.6[1]	118.6[2]	149.5[1]
Zn		4.03[ii]	8.24[i]	7.88[i]	3.67[ii]	2.00[iii]	4.80[ii]
Sand	%	46	37	38	48	58	57
Clay		26	41	43	25	27	24
Class		CL	CL	CL	CL	S CL	S CL

Figures with similar letters appended along the rows are insignificantly different whereas those with differing letters are significantly differing.

Table 2. Bulk density and porosity for top and sub soils of Mabira chronosequential forest stands.

Forest stand age (year)	Bulk density (g/cm³)		Porosity (%)	
	Top soil	Sub soil	Top soil	Sub soil
0-3	0.74	0.79	72.3	70.7
10-20	0.82	0.93	69.6	65.5
20-30	0.82	0.84	69.6	68.8
30-40	0.95	1.05	64.9	61.1
40-50	0.81	0.87	70.2	68.3
>55	0.8	0.93	70.4	65.5

$LSD_{0.05}$ (Forest stand age) = 0.07; $LSD_{0.05}$ (soil depth) = 0.11

(displacement along with runoff as a result of limited vegetation cover and litter. The low soil loss in older forests could be due to improved vegetation cover as well as the binding action by plant roots (Gyssels and Poesen, 2003) and improvement of soil physical properties (Gyssels et al., 2005). The relatively high soil loss in the intact forest with good vegetation cover invalidated the findings of Stolte (1997) but tallied with those much earlier reported by Patric (1976).

The conditions that favored runoff could have contributed to the trend of soil loss in the forest regimes. Low water infiltration as a result of the soil being kept almost saturated with moisture in the old forest could have favoured both runoff and soil loss. Generally, soil loss was too small measuring about 1/400th and 1/1000th

of annual values reported by Majaliwa (1998) and Majaliwa et al. (2003).

Nitrogen, phosphorus and potassium losses

Table 3 shows the annual nutrient losses from the various forest regimes in Mabira Forest Reserve. Noteworthy, is the highest annual nitrogen and phosphorus losses experienced in the 0 to 3 years forest stand. The loss for the two nutrients drastically reduced in the next 10 to 20 years forest stand before gradually increasing with forest stand age. Potassium loss did not follow any particular trend though like N and P it was significantly more lost in 0 to 3 years forest stand. The

Table 3. Annual nutrient losses in Mabira chronosequential forest stands.

Forest stand ages (year)	Nutrients lost (kg ha⁻¹)		
	N	P	K
0-3	5.36	0.30	39.81
10-20	0.11	0.01	0.99
20-30	0.16	0.01	7.49
30-40	0.23	0.02	0.33
40-50	3.06	0.25	5.94
> 55	4.38	0.20	9.93

average annual N, P and K losses for the study area were 4.38, 0.200 and 9.93 kg ha^{-1} year^{-1} respectively.

The losses are generally very small in all the forest stand ages but in harmony and contrast with those obtained by Bormann et al. (1983) and Miller and Newton (1983) respectively. Most likely the same factors described earlier for influencing runoff and soil loss across the forest ages could have positively contributed to the nutrient losses.

CONCLUSIONS AND RECOMMENDATIONS

Excluding 0 to 3 years forest stand, runoff in Mabira forest was very low compared to the figures reported on agricultural land but increased with increasing years of last forest disturbance. The forest stand age of 0 to 3 years recorded the highest soil loss. However, after ten years of regeneration, there was a gradual increase in soil loss with increase in forest age. Results indicate that even under disturbance, the forest ecosystem was still capable of providing the function of runoff, soil and nutrient loss reduction. The average annual soil loss as well as nutrient losses in Mabira Forest was very low compared to losses reported in agricultural land. Of the three macro nutrients, potassium was more lost followed by nitrogen in runoff. Fortunately, the nutrient losses are too small to affect forest regeneration. The bulk density and porosity are in permissible ranges for soil and water conservation under a forest ecosystem.

Since the forest ecosystem is still superior in controlling soil, water and nutrient losses, it is recommended that forests should be conserved. It is also recommended that broader studies to cover watershed nutrient and soil losses be undertaken since this study only considered point losses.

ACKNOWLEDGEMENTS

The authors acknowledge the European Union for the funding under the FOREAIM Project (INCO-CT-2005-510790); The National Forest Authority Staff in Mabira Forest Reserve; and the field assistants in Mabira for collecting runoff data.

REFERENCES

AES Nile Power (2001). Description of soils within the region, Appendix B1. Bujagali Project Transmission System, EIS.

Aina AO (1993). Rainfall runoff management techniques for erosion control and soil moisture conservation. Runoff control by basin tillage techniques. In: Soil tillage in Africa: Needs and challenges; FAO Soils Bull. 3:69.

Banana AY, Sembajjwe WG (2000). The Importance of Security of Tenure and Rule Enforcement in Urban Forests. In: Gibson C., M. A. McKean and E. Ostron (Eds). Forest Resources and Institutions, Chapter 4. FAO Corporate Document Repository.

Baranga D (2007). Observations on resource use in Mabira Forest Reserve. Afr. J. Ecol. 45(1):2-6.

BirdLife International (2009). Important Bird Area factsheet: Mabira Forest Reserve, Uganda. Downloaded from Data Zone at http://www.birdlife.org on 9/9/2010.

Blake GR, Hartage KH (1986). Bulk density. In: Klute, A. ed. Methods of soil analysis. Part 1. Physical and mineralogical methods—Agron. Monogr. 2:9.

Bormann FH, Likens GE, Fisher DW, Pierce RS (1983). Nutrient Loss accelerated by clear-cutting of Forest Ecosystem. Science 159:(3817):882-884.

Coelho AT, Galvao TCB, Pereira AR (2001). The effects of vegetative cover in the erosion prevention of a road slope. Environ. Manage. Health, MCB UP Ltd. 12(1):78-87.

de Vente J, Poesen J, Verstraeten G (2005). The application of semi-quantitative methods am resrevior sedimentation rates for prediction of basin sediment yield in Spain. Esevier B.V. J. Hydrol. 305:63-86.

Dudley MD, Tate KW, McDouglad NK, Melvin RG (2002). Factors influencing soil surface bulk density on oak savannah Rangeland in Southern Sierra Nevada Foothills. USDA Forest Service Gen. Tech. Rep. PSW-GTR 184:131-138.

Earth Trends (2003). Biodiversity and Conserved Areas in Uganda. http://earthtrends.wri.org/pdf_library/country_profiles/bio_cou_800.pdf

Environmental Alert, (2007). Facts about Mabira. www.envalert.org/docs/BriefOnMabira.pdf

FAO, Food and Agricultural Organization, (1986). Early Agrometeorological crop yield forecasting. FAO Plant Production and Protection, by M. Frère and G.F. Popov. FAO, Rome, Italy. p. 73

Floor JA (2000). Soil: Erosion and Conservation. Accessed indicate date from: www.seafriends.org.nz/enviro/soil/erosion.htm

Forest Department (FD) (1999). The status of the biomass in Uganda. Biomass study report, Ministry of Environment, Water, Lands and Natural Resources, 1999.

Gafur A, Jensen JR, Borggaard OK, Petersen L (2003). Runoff and losses of soil and nutrients from small watersheds under shifting cultivation (Jhum) in the Chittagong Hill Tracts of Bangladesh. J. Hydrol. 279(1-4):293-309.

Gyssels G, Poesen J (2003). Importance of plant root characteristics in controlling concentrated flow erosion rates. Earth Surface Processes and Land forms 28(4):371-384.

Gyssels G, Poesen J, Bochet E, Li Y (2005). Impact of plant roots on resistance of soil to erosion by water: A review. Progress Phys. Geogr. 29(2):189-217.

Ngigi SN, Savenije HHG, Thome JN, Rockström J, Penning de Vries FWT(2005). Agro-hydrological evaluation of on-farm rainwater storage systems for supplemental irrigation in Laikipia district, Kenya. Agricultural Water Management 73:21-41. Elsevier.

Haque SMS, Karmakar NC (2009). Organic matter accumulation in hill forests of Chittagong region, Bangladesh. J. For. Res. 20(3):249-253

Johnny GM (2004). Soil erosion following forest operations in the Southern Piedmont of central Alabama. J. Soil Water Conserv. 59(4):180-185.

Kayanja FIB, Byarugaba D (2001). Disappearing forests of Uganda: The way forward. Curr. Sci. 81(8):936-947.

Lamto G, Okwakol MJN, Isabirye BE, Isabirye M, Kalema J (2010). Characterization of Mabira Forest CSM-BGBD benchmark area, Uganda. In: Rwakaikara-Silver, M.C; B.E. Isabirye; A.M. Akol; C. Nkwiine; M.J.N. Okwakol, J. Huising; P. Okoth; W. Brooijimans and T.B. Etyang (eds). Ecology and Management of Soil Biodiversity in Mabira Forest, Uganda. An Inventory. Towards Conservation and Sustainable Management of Below-Ground Biodiversity in Uganda, pp. 3-18.

MacDonald MA, Lawrence A, Shrestha PK (2003). Soil erosion. In: Schroth G. and F. L. Sinclair (Eds): Trees, Crops and Soil Fertility: CABI Publishing, CABI International, Willingford, Oxon, OX10 8DE, UK. Concepts Res. Methods pp. 325-343.

Majaliwa JGM (1998). Effect of vegetation cover development on soil loss from maize based cropping systems. MSc. Thesis, Makerere University.

Majaliwa JGM, Magunda MK, Tenywa MM, Semalulu O (2003). Soil erosion and pollution loading from agricultural land in Bukoora sub-catchment. Uganda J. Agric. Sci. 9:305-312.

Miller JH, Newton M (1983). Nutrient loss from disturbed forest watersheds in Oregon's Coast Range. Agro-Ecosyst. 8:153-167.

Mulebeke R (2004). Validation of a GIS-Use Model in a banana-based microcatchment of the L. Victoria basin. MSc thesis, Makerere University, Uganda.

Murty D, Kirschhaum MUF, Mcmurtrie RE, Mcgilyray H (2002). Does conversion of forest to agricultural land change soil carbon and nitrogen? A review of the literature. Global Change Biol. 8(2):105-123.

Nature Uganda (2001). Mabira Forest Reserve. International Bird Areas.

Noguchi S, Tsuboyama Y, Sidle RC, Hosoda I (1997). Spatially distributed morphological characteristics of macropores in forest soils of Hitachi Ohta Experimental Watershed, Japan. Springer Japan. J. For. Res. 2(4):207-215.

Okalebo JR, Gathua KW, Woomer P (2002). Laboratory Methods of Soil and Plant Analysis. A working manual. Tropical Soil Biology and Fertility Programme. Marvel EPZ, Kenya Press Ltd, Nairobi, Kenya. pp. 22-80.

Patric JH (1976). Soil erosion in Eastern Forest, Society of American Foresters. J. For. 74(10):671-677.

Pomeroy D, Tushabe H (2004). The State of Uganda's Biodiversity 2004. Makerere University Institute of Environment and Natural Resources/National Biodiversity Data Bank.

Poulsen AD, Hasashim ND, Eilu G, Liengola IB, Ewango CEN, Hart TB (2005). Composition and species richness of forest plants along the Albertine Rift, Africa. in I. Friis and Balslev (eds.). Plant diversity and complexity patterns. Local, regional and global dimensions. Biogiske Skrifter 55:129-143,

Sanchez PA, Shepherd KD, Soule MJ, Place FM, Buresh RJ, Izac AN, Mokwunye AU, Kwesiga FR, Ndiritu CG, Woomer PL (1997). Soil fertility replenishment in Africa: an investment in natural resource capital. In: Replenishing Soil Fertility in Africa, Soil Science Society of America Special Publication, Soil Science Society of America, Madison, WI. p. 51.

Sharma E, Amasht RS, Singh MP (1985). Chemical soil properties under five age series of *Alnus nepalensis* in Eastern Himalayas. Plant Soil 82:105-113.

Smaling E (1993). An Agro-Ecological Framework for Integrated Nutrient Management with Special Reference to Kenya. Landbouwuniversiteit te Wageningen, Waginingen, Netherlands. p. 19.

Stolte KW (1997). Soil Health. In 1996 National Technical Report on Forest Health, USDA, Forest Service. Washington DC.

Tiwari KR (2009). Runoff and soil loss responses to rainfall, land use, terracing and management practices in the Middle Mountains of Nepal, Acta Agriculturae Scandinavica, Section B - Plant Soil Sci. 59(3):197-207.

Turner TR, Duke S, Fransen B, Reiter M, Kroll AJ, Ward J, Bach J, Justice T, Bilby B (2009). Landslide density and its association with rainfall, forest stand age, and topography, december 2007 storm, Willapa Hills, Southwest Washington. Geological Society of America Abstracts with Program 41(7):335.

VSN International (2000). Software for biosciences. VSN International Ltd, 5 The Waterhouse, Waterhouse Street, Hemel Hempstead, HP1 1ES, UK.

Wang B, Yang Q, Liu Z (2009). Effect of conversion of farmland to forest or grassland on soil erosion intensity changes in Yanhe River Basin, Loess Plateau of China. Front. For. China. 4(1):68-74.

Winterbottom R, Eilu G (2006). Uganda Biodiversity and Tropical Forest Assessment. Report for United States Agency for International Development under the International Resources Group.

World Wildlife Fund and McGinley (2007). Central China loess plateau mixed forests. In: Encyclopedia of Earth. Eds. Cutler J. Cleveland (Washington, D.C.: Environmental Information Coalition, National Council for Science and the Environment).

Young A (1989). Soil fertility and soil degradation. In: Agroforestry for Soil Conservation. CAB International, Wallingford, Oxon, Ox 10 8DE.

17

Soil fertility status in some soils of Muzaffarnagar District of Uttar Pradesh, India, along with Ganga canal command area

Pramod Kumar, Ashok Kumar, B. P. Dhyani, Pardeep Kumar, U. P. Shahi, S. P. Singh, Ravindra Kumar, Yogesh Kumar, Amit Kumar and Sumit Raizada

Department of Soil Science, College of Agriculture S.V.P.U.A.T., Meerut 250 110(U.P.), India.

Surface and subsurface soil samples of Muzaffarnagar district were collected to characterize their chemical properties and accordingly to develop optimum land use plan to realize maximum agricultural productivity. The pH value of study area varied from 6.02 to 8.39 and 6.35 to 8.50 for surface and sub surface, respectively, electrical conductivity from 0.069 to 0.390 and 0.073 to 1.10 dSm^{-1}, organic matter content 7.241 to 15.221 and 3.695 to 10.179 g kg^{-1}, available nitrogen (N) 131.53 to 348.97 and 99.32 to 217.44 kg ha^{-1}, Phosphorus (P) 15.67 to 52.61 and 11.17 to 45.40 P$_2$O$_5$ kg ha^{-1}, potassium (K) 79.16 to 436.8 and 47.04 to 399.84 K$_2$O kg ha^{-1} for surface (0-15 cm) and subsurface soil (15-30 cm). Cationic micronutrients Zn, Cu, Fe and Mn varied from 1.636 to 6.164, 1.024 to 4.282, 0.672 to 5.802 and 0.332 to 2.652, and 113.13 to 11.232, 10.33 to 79.326, 10.272 to 38.572 and 29.578 to 77.882 mg kg^{-1} in surface (0 - 15 cm) and subsurface soil (15-30 cm) respectively. As per soil nutrient index (SNI), the soils of study area were found in low fertility category for nitrogen and medium with respect to phosphorus and potassium. A positive and significant correlation of NPK and micronutrients was found with organic matter content while significant and negative correlations exist between micronutrients and soil pH.

Key word: Soil fertility, organic matter, NPK, micronutrients, surface soil, soil nutrient index.

INTRODUCTION

Soil is one of the most important natural resource of a country and knowledge about its characteristics is essential for developing optimum land use plan for maximizing agricultural production. Soils differ greatly in their morphological, physical, chemical, mineralogical and biological characteristics. Since these characters control the response of soil to management practices, it is essential to have information about these characters of each soil. The knowledge of different macro and micro nutrient and their distribution in the root zone is important.

Among the macro and micro nutrient usually applied through chemical fertilizers, nitrogen seems to have the quickest and most pronounced effect thus nitrogen is considered as a potent nutrient element that should not only be converted but carefully regulated. Most of the nitrogen in the soil is associated with organic matter. In this form it is protected from rapid microbial release. Only 2 to 3% nitrogen is mineralized annually under normal condition. Assessment of soil quality generally consist physico-chemical properties and their interaction with one another. Variation in nutrient supply is a natural phenomenon and some where it may be sufficient while some where deficient. Within a soil, variability may exist depending upon the hydrological properties of the soil. Since the study is undertaken to the adjoining area of the Ganga canal therefore variability in nutrients in the soils which are far away from Ganga canal is obvious. Therefore both locations will require different management practices to sustain crop productivity and for this full information about the nutrients status is important. Deficiency of essential plant nutrients to different extent in soils of Muzaffarnagar had been reported by Sharma et al. (2003). Deficiencies of many plant nutrients are emerging fastly in intensively cultivated sandy alkaline soils of Haryana (Narwal, 2006). Since the quality of produce is significantly influenced by the nutrient supplying capacity of soils coincidently with time crops and their economic product become significantly suboptimal. Therefore to have sound information about the nutrient status of these soils this study was undertaken.

MATERIALS AND METHODS

The study area falls in Muzaffarnagar district of Western Uttar Pradesh. Ganga canal was considered as base line and on the left hand side (LHS) of Ganga canal from Purkaji to Khatauli was taken as the study area. Each bridge on the canal between these two points (Purkaji to Khatauli) was selected for sampling location. Samples were taken from the distance of 1000, 2000, 3000, 4000, and 5000 m away from canal.

Soil samples were collected from eight locations of Muzaffarnagar district under different cropping pattern. Soil samples from two depth at every location were collected with the help of auger and stored in polythene bags. Collected soil samples were air dried in shade, crushed gently with a wooden roller and then pass through 2.0 mm sieve to obtain a uniform representative sample. Samples were properly labeled with the aluminum tag and stored in polythene bags for analysis. The processed soil samples were analyzed by standard methods for pH and electrical conductivity (1:2 soil water suspensions), organic matter (Walkley and Black, 1934), available nitrogen (Subbiah and Asija, 1956), available phosphorus (Olsen et al., 1954), available potassium (Jackson, 1973) and cationic micronutrients (Fe, Mn , Cu and Zn) in soil samples with extracted with a Diethylene triamine pentaacetate (DTPA) solution (0.005 M) DTPA + 0.01 M $CaCl_2$+ 0.1 M triethanolamine, pH 7.3 as outlined by Lindsay and Norvell (1978). The concentration of micronutrients was determined by atomic absorption spectrophotometer (GBC Avanta PM). All the analysis of soil samples was carried out in the laboratory of Department of Soil Science, SVPUA&Tech, Modipurm , Meerut (U.P), India.

RESULTS AND DISCUSSION

Characteristics of soil

Soil reaction (pH)

Soil samples collected from surface and subsurface of eight different locations from the left hand side of Ganga

canal in Muzaffarnagar district were usually found neutral to alkaline in reaction (Table 1). Ramesh et al. (2003) also reported the similar results. They observed that the soil reaction was slightly alkaline to moderately alkaline (black soil) and acidic to moderately acidic (red soils). The pH value for surface soil (0 to 15 cm) and subsurface soil (15 to 30 cm) of different locations ranged from 6.02 to 8.39 and 6.35 to 8.50, respectively. According to classification of soil reaction suggested by Brady (1985), 27 samples were neutral (6.35 to 7.50), 26 samples were mildly alkaline (7.4 to 7.8), 25 samples were moderately alkaline (7.81 to 8.4) and 2 samples were strongly alkaline (8.4 to 9.0). The minimum pH 6.02 was observed in Jouli location, while maximum pH 8.39 was observed in Khatauli at surface (0-15 cm), while minimum pH 6.35 in Jouli and maximum pH 8.50 in Kamheda (TP) at subsurface soil (15-30 cm). The relative high pH of the soils might be due to the presence of high degree of base saturation. Gabhane et al. (2006) observed that, most soils of Vidarbha region of Maharashtra were neutral to moderately alkaline in reaction (pH 7.15 to 8.03). In general pH of the soils increased with depth.

Electrical conductivity

The electrical conductivity of the soils varied from 0.069 to 0.390 and 0.073 to 1.10 (dSm^{-1}) at surface and subsurface of soil (Table 2). On the basis of the limits suggested by Muhar et al. (1963) for judging salt problem of soils, most of the samples (99%) were found normal (EC < 1.0 dSm^{-1}) and remaining 1% samples were found in the category of soluble salt content critical for germination (EC 1 to 2 dSm^{-1}). Sangwan and Singh (1993) also noticed the similar trend, they found that the electrical conductivity varied from 0.6 to 2.5 dSm^{-1} with average value of 1.16 dSm^{-1} and were thus categorised as non-saline in character. The salt content increase with soil depth, the high content of salt may be due to irrigation with saline water.

Organic matter content

The organic matter content of the soils varied from 7.241 to 15.221 and 3.695 to 10.179 g kg^{-1} soil at surface and subsurface (Table 3). The maximum organic matter content 15.221 g kg^{-1} at surface (0 to 15 cm) was found in Kamheda while minimum 7.241 g kg^{-1} in Tajpur. In the subsurface soil maximum organic matter content 10.179 g kg^{-1} was found in Balda and minimum 3.695 g kg^{-1} in Tajpur. Lower organic matter in the area may be due to prevailing high temperature and good aeration in the soil which increases the rate of oxidation of organic matter content. Aggarwal et al. (1990) reported that the organic carbon content of some Aridisols of western Rajasthan ranged from 0.14 to 0.40% in surface soil. Organic carbon was low and generally decreased with depth.

Table 1. pH variability in soil profile at different distance from Ganga canal.

S/N	Locations	Depth (cm)	Soil sampling distance (m) from Ganga canal				
			1000	2000	3000	4000	5000
1	Purkaji	0-15	7.45	7.62	7.25	7.35	8.15
		15-30	7.64	7.74	7.44	8.33	8.46
2	Kamheda (TP)	0-15	7.56	7.56	7.65	7.35	7.40
		15-30	7.98	7.48	7.84	7.45	8.50
3	Balda	0-15	7.35	6.90	7.36	7.09	7.33
		15-30	7.87	7.94	8.11	7.43	7.88
4	Bhopa	0-15	7.20	7.75	6.55	7.20	8.05
		15-30	7.68	8.25	6.93	8.17	8.23
5	Jouli	0-15	7.55	6.02	6.20	6.23	7.74
		15-30	7.92	6.85	7.07	6.35	7.70
6	Janshath	0-15	7.10	7.15	6.72	7.42	7.31
		15-30	7.23	7.74	7.22	7.79	7.62
7	Tajpur	0-15	7.47	7.52	7.93	7.45	7.50
		15-30	7.74	7.86	8.11	7.84	7.77
8	Khatauli	0-15	8.18	7.85	6.98	7.98	8.39
		15-30	8.32	8.02	7.36	8.08	8.17

Table 2. Electrical conductivity (dSm^{-1}) variability in soil profile at different distance from Ganga canal.

S/N	Locations	Depth (cm)	Soil sampling distance (m) from Ganga canal				
			1000	2000	3000	4000	5000
1	(Purkaji)	0-15	0.126	0.191	0.132	0.133	0.189
		15-30	0.101	0.095	0.573	0.230	0.190
2	Kamheda (TP)	0-15	0.284	0.128	0.193	0.135	0.190
		15-30	0.189	0.110	0.092	0.575	0.193
3	Balda	0-15	0.254	0.108	0.145	0.093	0.190
		15-30	0.132	0.187	0.203	0.086	0.083
4	Bhopa	0-15	0.191	0.138	0.187	0.108	0.67
		15-30	0.105	0.081	0.109	0.132	1.10
5	Jouli	0-15	0.282	0.118	0.200	0.190	0.135
		15-30	0.189	0.084	0.095	0.089	0.096
6	Janshath	0-15	0.210	0.124	0.135	0.214	0.196
		15-30	0.289	0.073	0.081	0.091	0.259
7	Tajpur	0-15	0.128	0.211	0.350	0.346	0.39
		15-30	0.125	0.157	0.313	0.226	0.267
8	Khatauli	0-15	0.284	0.209	0.069	0.384	0.376
		15-30	0.248	0.226	0.108	0.101	0.186

Available macronutrients

Nitrogen

Available nitrogen in soils of study area varied from 131.53 to 348.97 and 99.32 to 217.44 kg N ha^{-1} at surface (0 to 15 cm) and subsurface (15 to 30 cm) (Table 4). The maximum available nitrogen 348.94 kg ha^{-1} was found in Bhopa and minimum 131.53 kg ha^{-1} in Tajpur in surface soil (0 to 15 cm) while in subsurface soil (15 to 30 cm) the highest available nitrogen 217.44 kg ha^{-1} was found in Balda and minimum 99.32 kg ha^{-1} in Khatauli location. On the basis of the ratings suggested by Velayutham and Bhattacharyya (2000) 95% samples were rated low (<280 N kg ha^{-1}) while 5% samples were in the medium range (281 to 560 N kg ha^{-1}). Available nitrogen was correlated significantly and positively (r=0.6356) with organic matter (Table 13). Walia et al.

Table 3. Organic matter (g kg^{-1}) variability in soil profile at different distance from Ganga canal.

S/N	Locations	Depth (cm)	Soil sampling distance (m) from Ganga canal				
			1000	2000	3000	4000	5000
1	(Purkaji)	0-15	12.118	13.448	14.630	10.345	11.232
		15-30	8.128	8.128	8.424	6.650	7.684
2	Kamheda (TP)	0-15	13.004	13.891	11.231	12.118	15.221
		15-30	7.389	9.015	6.798	8.424	9.606
3	Balda	0-15	14.778	13.005	12.709	11.970	13.744
		15-30	9.754	8.276	10.197	6.946	9.015
4	Bhopa	0-15	11.970	10.197	12.562	9.310	9.0148
		15-30	7.980	8.424	7.241	6.059	5.468
5	Jouli	0-15	10.936	11.084	11.970	11.232	9.754
		15-30	8.424	8.719	9.0148	7.833	6.502
6	Janshath	0-15	11.231	9.310	10.493	12.414	11.084
		15-30	7.389	6.650	7.537	9.015	6.355
7	Tajpur	0-15	8.128	8.424	7.833	7.241	11.527
		15-30	6.059	7.241	3.695	4.581	5.764
8	Khatauli	0-15	9.606	8.424	12.562	14.926	13.300
		15-30	6.502	5.764	9.015	8.571	6.650

Table 4. Available nitrogen (Kg ha^{-1}) variability in soil profile at different distance from Ganga canal.

S/N	Locations	Depth (cm)	Soil sampling distance (m) from Ganga canal				
			1000	2000	3000	4000	5000
1	(Purkaji)	0-15	226.83	236.23	237.57	198.65	210.57
		15-30	186.57	190.59	204.01	155.69	182.54
2	Kamheda (TP)	0-15	190.59	195.96	177.17	187.91	205.36
		15-30	165.09	166.43	155.69	163.75	177.17
3	Balda	0-15	292.59	252.33	244.28	201.33	287.23
		15-30	217.44	158.38	153.01	153.01	182.54
4	Bhopa	0-15	153.01	158.38	220.12	289.92	348.97
		15-30	142.27	139.59	155.69	169.12	212.07
5	Jouli	0-15	163.75	182.54	189.25	166.43	142.27
		15-30	147.64	161.06	166.43	153.01	112.74
6	Janshath	0-15	189.25	150.33	171.8	224.15	174.49
		15-30	159.72	131.53	139.59	166.43	142.23
7	Tajpur	0-15	139.59	131.53	139.59	159.72	166.43
		15-30	126.17	102.00	120.79	139.59	142.27
8	Khatauli	0-15	144.95	163.75	174.49	293.94	217.44
		15-30	99.32	131.54	144.96	180.49	150.33

(1998) reported that available N in the soils of Bundelkhand region accounted for 12 to 42% of total N in the range of 95 to 159 mg N kg^{-1} in surface soil and 51 to 159 mg N kg^{-1} in subsurface horizon. The continuous mineralization of organic matter in surface soils was responsible for the higher values.

Phosphorus

The available phosphorus in soils of the study area varied from 15.67 to 52.61 and 11.17 to 45.40 P_2O_5 kg ha^{-1} at surface (0 to 15 cm) and subsurface (15 to 30 cm) (Table 5). The maximum available phosphorus 52.61 P_2O_5 kg ha^{-1} was found in Kamheda (TP) and minimum 15.67 P_2O_5 kg ha^{-1} in Purkaji at surface (0-15 cm) whereas at subsurface maximum 45.40 P_2O_5 kg ha^{-1} was found at Kamheda and minimum 11.17 P_2O_5 kg ha^{-1} at Janshath. On the basis of limits suggested by Muhar et al. (1963) 42.5% samples were rated low (<20 P_2O_5 kg ha^{-1}), 56% medium (20-50 P_2O_5 kg ha^{-1}) and 1.50% high (> 50 P_2O_5 kg ha^{-1}) at surface and subsurface soils.

Table 5. Phosphorus (kg ha^{-1}) variability in soil profile at different distance from Ganga canal.

S/N	Locations	Depth (cm)	Soil sampling distance (m) from Ganga canal				
			1000	2000	3000	4000	5000
1	(Purkaji)	0-15	30.539	33.692	39.998	15.675	20.179
		15-30	13.873	14.774	17.927	11.171	13.423
2	Kamheda (TP)	0-15	38.647	42.701	33.692	38.647	52.611
		15-30	31.440	32.791	17.477	25.585	45.404
3	Balda	0-15	39.998	21.981	21.08	17.927	35.945
		15-30	30.089	15.522	12.072	12.112	17.927
4	Bhopa	0-15	40.449	36.845	40.899	34.593	30.539
		15-30	30.539	27.386	32.098	26.936	17.927
5	Jouli	0-15	30.108	35.944	37.296	34.593	18.828
		15-30	13.423	29.639	30.395	15.225	8.018
6	Janshath	0-15	28.738	17.026	21.531	33.241	29.188
		15-30	13.423	11.170	12.072	17.477	13.873
7	Tajpur	0-15	29.638	29.638	27.837	15.675	37.296
		15-30	19.638	17.927	15.225	8.919	20.134
8	Khatauli	0-15	18.378	17.026	28.287	38.647	38.197
		15-30	15.225	12.522	15.675	32.845	18.378

Potassium

Available potassium (K$_2$O) in the soils ranged from 79.16 to 436.8 and 47.04 to 399.84 kg ha^{-1} at surface (0 to 15 cm) and subsurface (15 to 30 cm) (Table 6). The maximum available potassium 436.8 kg ha^{-1} was found at Tajpur and minimum 79.16 K$_2$O kg ha^{-1} in Khatauli at surface soil (0 to 15 cm). Similarly maximum available K$_2$O 399.84 kg ha^{-1} was found in Tajpur and minimum 47.04 kg ha^{-1} in Khatauli at subsurface. According to Muhar et al. (1963) 25% samples were categorized as low (< 125 kg ha^{-1}), 67.5% medium (125 to 300 kg ha^{-1}) and 2.5 high (> 300 K$_2$O kg ha^{-1}). A significant and positive correlation (r=+0.3505) was observed between organic matter and available K (Table 13). Meena et al. (2006) observed similar significant and positive correlation between organic carbon and available K (r=0.420). This might be due to creation of favorable soil environment with presence of high organic matter.

Soil nutrient index

Considering the concept of "Soil Nutrient Index" the soils of study area were found in category of low for nitrogen and Medium for phosphorus and potassium (Table 7).

Available cationic micronutrients

Copper

DTPA-extractable available copper in the surface and subsurface soil of eight different locations was found to be sufficient. The DTPA-extractable Cu (mg kg^{-1} soil) in surface (0 to 15) and sub surface (15 to 30 cm) soils varied from 1.636 to 6.164 and 1.024 to 4.370 mg/kg soil, respectively (Table 8). The maximum available Cu 6.164 mg kg^{-1} soil for surface soil (0 to 15 cm) was found in Belda location and minimum 1.024 mg kg^{-1} soil in Tajpur location. Maximum available Cu 4.370 mg kg^{-1} in soil was found in subsurface soil of Belda location and minimum 1.024 mg kg^{-1} soil in Bhopa (Table 12). Considering the critical limit 0.20 mg/kg soil as suggested by Lindsay and Norvell (1978) all the soil samples of eight different locations were found to be sufficient in available Cu. A decreasing trend in available Cu content with increasing depth was noticed in all eight different locations.

Zn

The available Zn estimated by DTPA in the surface and subsurface soil of eight different locations was found to be sufficient range as per criteria given by Lindsay and Norvell (1978) (Table 12). The minimum and maximum value of available Zn in surface soil and sub surface soil ranged from 0. 458 to 5.80 and 0.332 to 2.65 mg kg^{-1} soil respectively (Table 9). The maximum DTPA extractable available Zn 5.80 mg/kg soil was found in Janshath location and minimum 0.310 mg kg^{-1} soil in Purkaji location for surface soil. In case of sub surface soil maximum DTPA extractable zinc 2.65 mg kg^{-1} and minimum 0.33 mg kg^{-1} was found in Janshath and Khatauli location respectively. As per critical limits suggested by ISSS 24.44% samples were deficient < (0.6 mg kg^{-1}), 35.55% marginal (0.6 to 1.2 mg kg^{-1}) and 40% were sufficient (> 1.2 mg kg^{-1}).

Table 6. Potassium (kg ha^{-1}) variability in soil profile at different distance from Ganga canal.

S/N	Locations	Depth (cm)	Soil sampling distance (m) from Ganga canal				
			1000	2000	3000	4000	5000
1	(Purkaji)	0-15	217.28	221.76	219.52	110.88	133.28
		15-30	172.48	196.00	184.80	97.44	117.60
2	Kamheda (TP)	0-15	181.44	273.28	165.76	169.12	281.12
		15-30	150.08	178.08	133.28	148.96	217.28
3	Balda	0-15	213.92	159.04	153.44	125.44	197.12
		15-30	174.72	125.44	123.2	120.96	131.04
4	Bhopa	0-15	218.4	163.52	229.6	161.28	134.4
		15-30	188.16	107.52	192.64	109.76	104.16
5	Jouli	0-15	208.32	221.76	244.16	240.8	202.72
		15-30	170.24	181.44	194.88	193.76	162.40
6	Janshath	0-15	140	76.16	95.2	148.96	129.92
		15-30	114.24	67.20	73.92	125.44	103.04
7	Tajpur	0-15	197.12	248.64	159.04	133.28	436.8
		15-30	155.68	226.24	140.00	119.84	399.84
8	Khatauli	0-15	98.56	89.6	105.28	165.76	112
		15-30	61.60	47.04	87.36	135.60	96.32

Table 7. Nutrient index and fertility status class of study area soils.

S/N	Name of nutrient	Percent sample			Nutrient Index		Fertility status class
		Low	Medium	High	Observed	Proposed	
1	Nitrogen	95.00	5.00	Nil	1.05	<1.5	Low
2	Phosphorus	42.50	56.00	1.50	1.56	>2.5	Medium
3	Potassium	28.75	68.75	2.50	1.75	1.5-2.5	Medium

Table 8. DTPA-extractable copper (mg kg^{-1}) variability in soil profile at different distance from Ganga canal.

S/N	Locations	Depth (cm)	Soil samples distance (m) in Ganga canal				
			1000	2000	3000	4000	5000
1	(Purkaji)	0-15	3.46	3.676	3.932	3.128	3.224
		15-30	3.034	3.064	3.662	2.832	2.182
2	Kamheda (TP)	0-15	3.644	3.822	3.464	3.562	3.844
		15-30	3.004	3.206	3.042	1.732	3.262
3	Balda	0-15	6.164	4.224	3.468	2.872	6.024
		15-30	4.230	2.262	2.542	2.152	4.370
4	Bhopa	0-15	3.682	3.436	3.962	2.468	2.278
		15-30	2.028	1.928	2.038	1.540	1.024
5	Jouli	0-15	3.072	3.748	5.104	4.644	2.184
		15-30	2.256	2.794	4.282	3.544	1.904
6	Janshath	0-15	4.008	1.984	3.108	4.278	3.312
		15-30	2.428	1.564	2.188	3.560	2.414
7	Tajpur	0-15	1.838	2.366	1.8	1.636	2.428
		15-30	1.802	1.832	1.360	1.316	1.854
8	Khatauli	0-15	2.508	2.648	2.826	4.954	3.178
		15-30	1.300	1.746	2.016	3.332	2.040

Table 9. DTPA-extractable zinc (mg kg^{-1}) variability in soil profile at different distance from Ganga canal.

S/N	Locations	Depth (cm)	Soil samples distance (m) in Ganga canal				
			1000	2000	3000	4000	5000
1	(Purkaji)	0-15	1.538	1.644	1.906	0.738	1.31
		15-30	0.516	0.974	1.250	0.384	0.476
2	Kamheda (TP)	0-15	1.344	1.36	0.784	0.702	1.74
		15-30	0.822	0.842	0.378	0.368	1.124
3	Balda	0-15	1.902	1.598	0.956	1.36	1.634
		15-30	1.422	0.822	0.63	1.022	1.202
4	Bhopa	0-15	4.268	2.016	5.202	1.772	1.646
		15-30	1.728	1.036	2.172	1.204	0.974
5	Jouli	0-15	0.768	1.648	2.248	1.876	0.73
		15-30	0.526	0.732	1.652	0.972	0.454
6	Janshath	0-15	4.99	0.458	1.016	5.802	1.784
		15-30	2.268	0.612	0.36	2.65	0.762
7	Tajpur	0-15	1.258	1.02	0.76	0.748	1.338
		15-30	0.712	0.646	0.572	0.484	0.842
8	Khatauli	0-15	0.672	0.68	1.15	1.636	1.242
		15-30	0.332	0.444	0.720	0.804	0.954

Table 10. DTPA-extractable iron (mg kg^{-1}) variability in soil profile at different distance from Ganga canal.

S/N	Locations	Depth (cm)	Soil samples distance (m) in Ganga canal				
			1000	2000	3000	4000	5000
1	Purkaji	0-15	34.268	38.536	57.8	11.232	25.41
		15-30	24.480	29.550	33.564	10.334	20.044
2	Kamheda (TP)	0-15	41.64	70.782	26.26	41.506	77.6
		15-30	22.326	30.060	16.598	22.028	36.084
3	Balda	0-15	70.42	58.212	56.724	51.034	66.416
		15-30	56.136	39.430	39.816	26.286	44.426
4	Bhopa	0-15	72.026	52.034	82.66	26.582	28.45
		15-30	36.402	29.480	48.242	17.538	19.558
5	Jouli	0-15	39.44	91.9	113.13	72.824	16.074
		15-30	24.750	66.430	79.326	49.082	11.350
6	Janshath	0-15	50.308	28.798	45.9	68.95	48.036
		15-30	34.962	20.052	20.624	47.870	30.190
7	Tajpur	0-15	24.636	26.358	24.51	11.768	27.03
		15-30	18.566	13.156	18.640	12.362	8.858
8	Khatauli	0-15	35.7	28.362	37.79	88.99	47.38
		15-30	24.502	19.038	24.984	41.780	25.956

Fe

The DTPA- extractable available Fe in the surface and subsurface soil of eight different locations was found to be high. The DTPA extractable iron in surface (0 to 15 cm) as well as subsurface (15 to 30 cm) varied from 11.232 to 113.30, 10.334 to 79.33 mg kg^{-1} soil respectively (Table 10). The highest as well as lowest values of available Fe for surface as well as subsurface soils were found at Jouli and Purkaji locations for surface soil. According to the critical limit 4.5 mg/kg soil as purposed by Lindsay and Norvell (1978) all the surface soil sample and sub surface samples were sufficient in available Fe (Table 12). The amount of available Fe decreased with an increasing soil depth. On the basis of critical limit suggested by Indian Society of Soil Science 3.33% samples were marginal (4.5 to 9.0 mg kg^{-1}) and 96.66% were sufficient (> 9.0 mg kg^{-1}).

Table 11. DTPA-extractable manganese (mg kg^{-1}) variability in soil profile at different distance from Ganga canal.

S/N	Locations	Depth (cm)	Soil samples distance (m) in Ganga canal				
			1000	2000	3000	4000	5000
1	Purkaji	0-15	21.64	21.696	34.2	10.998	11.89
		15-30	15.224	15.990	16.620	7.882	8.468
2	Kamheda (TP)	0-15	33.024	33.578	30.214	22.642	38.03
		15-30	20.422	22.046	21.640	15.042	26.116
3	Balda	0-15	38.572	34.252	23.92	19.472	29.378
		15-30	23.780	23.448	19.798	17.180	21.018
4	Bhopa	0-15	27.138	20.864	29.08	19.564	10.272
		15-30	18.920	16.656	24.298	16.218	9.442
5	Jouli	0-15	22.530	35.812	24.380	26.432	24.492
		15-30	13.052	29.578	14.560	18.512	17.244
6	Janshath	0-15	32.57	16.9	18.852	45.046	22.11
		15-30	20.680	10.480	14.944	30.248	20.308
7	Tajpur	0-15	27.042	19.286	28.624	17.486	18.932
		15-30	20.422	13.442	20.588	10.250	12.630
8	Khatauli	0-15	25.256	15.040	51.442	21.398	33.614
		15-30	16.854	13.256	37.210	14.420	23.042

Table 12. Class of cationic from study area soils.

SN	Name of nutrient	Percent sample			Critical Limit(mg kg^{-1})		
		Deficient	Marginal	Sufficient (%)	Deficient	Marginal	Sufficient
1	Copper	-	-	100	< 0.2	0.2 - 0.4	> 0.4
2	Zinc				< 0.6	0.6 -1.2	> 1.2
3	Fe	-		100	< 4.5	4.5 - 9.0	> 9.0
4	Mn	-		100	< 3.5	3.5 - 7.0	> 7.0

Table 13. Correlation studies between OM to available N, total N, available P, K, and Cu, Fe, Mn, Zn under different locations (all values of different locations).

Locations	OM to macronutrient and micronutrient							
	Av. N	Total N	Av. P	Av. K	Zn	Cu	Fe	Mn
OM	0.6356	0.6773	0.7014	0.3505	0.4473	0.7686	0.6941	0.6265

Mn

The DTPA extractable available Mn in the surface and subsurface soil in eight different locations is sufficient to high since are well above the critical limit (1.0 mg/kg) as proposed by Lindsay and Norvell (1978).

The maximum and minimum DTPA extractable Mn in surface and sub surface varied from 10.272 to 51.442 and 7.882 to 37.230 mg kg^{-1} soil (Table 11). The maximum available Mn content 51.442 mg kg^{-1} soil was found in Khatauli location and minimum 10.272 mg kg^{-1} soil in Bhopa for surface soil (0 to 15 cm). Maximum extractable Mn content 37.23 mg kg^{-1} in subsurface was found in soil of Khatauli location and minimum 7.882 mg kg^{-1} in Purkaji.

According to critical limit 1.0 mg/kg purposed by Lindsay and Novell (1978) and (3.5 to 7.0 mg kg^{-1}) suggested by Indian Society of Soil Science all the soils were sufficient in available Mn (Table 12).

Relationship among available NPK, micronutrient with organic carbon and pH in soil at different locations

Correlation among the soil properties of eight different location soils of Muzaffarnagar district were work out (Tables 13 and 14). The organic matter was positively and significantly related with N, P, K and cationic micronutrients. A significantly and positive correlation of

Table 14. Correlation study between pH to DTPA extractable micronutrients under different locations (all value of different locations).

Locations	pH to DTPA extractable micronutrients			
	Zn	Cu	Fe	Mn
pH	-0.3779	-0.41601	-0.57899	-0.37915

organic matter with available nitrogen (r=+0.6356) and total nitrogen (r=+0.6773) was found. Similarly a positive and significant correlation was also observed between organic matter and available phosphorus (r=+0.7.14) and potassium (r=+0.3505).

Among the cationic micronutrients copper was related much strongly with organic matter (r=+(0.7686) followed by iron (r=+0.6941), manganese (r=+0.6265) and Zinc (r=+0.4447359) .These association showed that available copper, iron, zinc and manganese in these soil are largely influenced by organic matter.

The relationship between available micronutrients and soil pH were also worked out. The soil pH was significantly and negatively correlated with all the cationic micronutrients .Among the micronutrients studied Fe was found related much strongly with soil pH (r=-0.579) followed by Cu (r=-0.416), Mn (r=-0.379) and Zn (r=-0.3779).

Conclusion

Soil chemical composition of the study area did not followed a particular pattern with increasing distance from Ganga Canal which may be due to variation in management practices, cropping sequence and their yield potential. Study area was dominated by sugarcane, wheat, rice and fodder crops. About sixty seven percent area is under sugar cane. Physico-chemical characteristics and nutrient status of soil in Muzaffarnagar district of Uttar Pradesh as discussed previously indicates that soil of study area were neutral to strongly alkaline in reaction and non saline in nature. Nutrient status regarding to the available macro and micro nutrient in surface (0 to 15 cm) and subsurface (15 to 30 cm) depth of soil indicate that soils are low in available N and medium in available P and K in surface and subsurface soil and in general sufficient in available Zn, Cu, Fe and Mn in the surface and subsurface layer of the profiles.

REFERENCES

Aggarwal RK, Kumar P, Sharma BK (1990). Distribution of nitrogen in some Aridisols. J. Indian Soc. Soil Sci. *38:430-433*.

Brady NC (1985). The nature and properties of soils, 8th Edition Macmillan Publishing Co. Inc., New York.

Gabhane VV, Jadhao VO, Nagdeve, MB (2006). Land evaluation for land use planning of a micro watershed in Vidarbha region of Maharashtra. J. Indian Soc. Soil Sci. 54:307-315.

Jackson ML (1973). Soil chemical analysis, Prentice Hall of Index Pvt .Ltd, *New Delhi*, India, p. 498.

Lindsay WL, Norvell WA (1978). Development of DTPA soil test for zinc, iron, manganese and copper. Soil Sci. Soc. Am. J. 42:421-428.

Meena HB, Sharma RP, Rawat US (2006). Status of macro and micro nutrient in some soils of Tonk district of Rajasthan. J. Indian Soc. Soil *Sci.* 54:508-512.

Muhar GR, Datta NP, Shankara SN, Dever, F, Lecy, VK , Donahue, RR (1963). Soil testing in India. *USDA Mission to India.*

Narwal RP (2006). Annual progress report of AICRP on micro and secondary nutrients and pollutants elements in soils and plants. *CCSHAU, Hissar.* p. 85.

Olsen SR, Cole CV, Watanabe FS, Dean LA (1954). *Estimation of Available Phosphorus in Soils by Extraction with Sodium Bicarbonate.* USDA Circ. 939. United States Department of Agriculture, Washington D.C.

Ramesh K, Prasad VB, Rao MS (2003). Physical nature and fertility status if soils of Singarayakonda Madal in Prakasam district of Andhra Pradesh, *Andhra.* Agric. J. 50 (122) 54-55.

Sangwan BS, Singh K (1993). Vertical distribution of Zn, Mn, Cu and Fe in the semi-arid soils of Haryana and their relationship with soil properties. *J.Indian Soc. Soil Sci.* 41: 463-467.

Sharma RP, Singh M, Sharma JP (2003). Correlation studies on micronutrient vis- a vis soil properties in some soils of Nagpur district in semi- arid region of Rajasthan. J. Indian Soc. Soil Sci. 51:522 -527.

Subbiah BV, Asija, GL (1956). A rapid procedure for the determination of available nitrogen in soil. Curr. Sci. 1956;25:259–260

Velayutham M, Bhattacharyya T (2000). Soil resources management from natural resource management for agriculture production in India. p. 103

Walia CS, Narayan A, Uppal KS, Rao YS (1998). Distribution of various forms of nitrogen and C: N ratio in some land form of Bundelkhand region of Uttar Pradesh. J. Indian Soc. Soil Sci. 46:193-198.

Walkley AJ, Black IA (1934). Estimation of soil organic carbon by the chromic acid titration method. Soil Sci. 37:29-38

18

Elevation and variability of acidic sandy soil pH: Amended with conditioner, activator, organic and inorganic fertilizers

Mohd Hadi-Akbar Basri[1], Nasima Junejo[2,3], Arifin Abdu[1,3], Hazandy Abdul Hamid[2,3] and Mohd Ashadie Kusno[2,3]

[1]Department of Forest Management, Faculty of Forestry, Universiti Putra Malaysia, 43400 Serdang, Selangor, Malaysia.
[2]Department of Forest Production, Faculty of Forestry, Universiti Putra Malaysia, 43400 Serdang, Selangor, Malaysia.
[3]Institute of Tropical Forestry and Forest Products, Universiti Putra Malaysia, 43400 Serdang, Selangor, Malaysia.

Availability of nutrients for plant uptake is directly related to soil pH and as an indicator of soil fertility status. Determination of effects of soil amendments on soil pH should be a necessary part of fertilizer and fertility research. A pot experiment was carried out to determine and compare the effects of biochar, chicken manure, urea and zeolite on soil pH variability and elevation in a sandy loam acidic soil. A modified method was used to determine the soil pH in the pots. Soil pH in pot was measured by a glass micro-electrode and spatial variability was interpolated and mapped by using geographic information system (GIS) + software. Kriged Maps clearly showed the presence of variability and elevation in pH within each treatment. Furthermore, the position of patches with maxima and minima values for pH changed between all treatments used in the experiment. The highest elevation was found in zeolite treated soil followed by urea and biochar. However, a significant decrease was measured in soil pH in chicken dung treated soil. These findings could be the first step towards temporal stability of the pattern of spatial distribution of soil pH affected by the soil amendments (biochar, urea, chicken manure and zeolite).

Key words: Biochar, soil pH, urea, zeolite, spatial variability.

INTRODUCTION

Soil pH is one of the fundamental soil properties, which influences nutrient availability and many soil chemical processes. A soil with neutral or higher pH, is poor in micronutrient and causes higher greenhouse-gas emissions such as high ammonia volatilization loss (Kissel, 1988). In contrast, the soil with acidic pH (3 - 5.5) has high concentration of heavy metals, Al and Fe toxicities and lower CEC (Zhao and Xing, 2009; Srinivasarao et al., 2011). Changes in soil pH with added

fertilizer may have an effect on the reaction of fertilizer in soil; resulting in improvement of nutrient efficiency and reduction in N losses (Ayanaba and Kang, 1976; Antil et al., 1992; Ahmed et al., 2008a).

At present, fertilizer optimization for the improvement of soil fertility is a worldwide key concern. Mixing of organic and inorganic fertilizer with soil activator or conditioner may become a common practice which affects soil properties. Zeolite, biochar and chicken manure are

Table 1. The properties of soil series used in a glass house study to measure the effects of selected agriculture in puts on soil pH.

Texture (%)			pH	Total C	Total N	Exchangeable cation (cmol$_c$ kg^{-1})			
Clay	Sand	Silt	H$_2$O	(%)	(%)	Ca	Mg	K	CEC
33	50	17	5.3	2	0.1	0.9	4.0	0.2	5.4

Table 2. pH of Soil amendments were used to determine their effects on soil pH at application site.

Soil amendments	pH	CEC cmol/kg	Rate of application g/pot
Urea	8.2	--	17.6
Chicken dung	4.5	--	17.6
Biochar	9.2	42.85	17.6
Zeolite	5.6	171.74	17.6

highly recommended soil amendments to improve soil properties and nutrient optimization in nutrient poor soils (Glaser et al., 2002; Duncan, 2005; Ahmed et al., 2008b). The value of chicken manure as fertilizer in agriculture is well documented. For example, chicken manure has been used as organic matter in poor soils to improve soil properties and fertility status (Perkins et al., 1964; Nichols and Daniel, 1994). Zeolite is a broad spectrum crystalline aluminosilicates. It has used in agricultural lands to increase the pH in acidic situations; a medium of free nutrients, trap of heavy metals and to improve soil cation exchange capacity (Ahmed et al., 2008b; Omar et al., 2010; Zhang et al., 2010; Ramesh and Reddy, 2011). Biochar is attractive approaches to reduce environmental pollution these days as it have an ability to reduce leaching of nutrient, improve crop yield and lead to a sustainable management of fertilizer (Glaser et al., 2001). It can be a source of organic carbon made by plant materials, highly considered as a carbon sequester, soil conditioner and retainer of ammonium to reduce nitrous oxide emission (Clough and Condron, 2011; Singh et al., 2011). These materials were evaluated as bio resources to improve soil fertility but their information on effects of soil pH are still lacking and not well documented. On the other hand, urea is an inorganic N fertilizer which is widely used but is volatilised in soil of high pH due to hydrolysis and urease enzyme activity (Cabrera et al., 1991; Freney et al., 1993; Chen et al., 2010).

Soil and fertilizers evaluation for its effects on soil properties especially soil pH should be a part of research into fertilizers and amendments to avoid nutrient losses and failure of experiment (Cabrera et al., 1991; Christianson et al., 1993). Otherwise, the soil amended with a material of unknown soil pH may affect soil nutrient availability. The work reported here was done to study the effects of zeolite, biochar, chicken manure and urea on spatial variability and elevation of soil pH in a pot experiment. The information obtained by the presented research could be used to optimize soil nutrient management.

MATERIALS AND METHODS

Soil sampling and analysis

The experiment was conducted on a sandy loam soil under glass house conditions at Agricultural University Park, University Putra Malaysia. The average daily maximum temperature in the glasshouse is 35°C and the minimum temperature is 25°C. The soil (Typic Paleudult) was sampled from experimental area (bare) at the depth of 0 to 15 cm. The sampled soil was air dried, ground, sieved and analysed for its physical and chemical properties such as; soil pH measured in a 1 : 2 ratio (soil : distilled water) using a glass electrode; organic carbon was determined by the potassium dichromate and H$_2$SO$_4$ digestion method (Walkley and Black, 1934); CEC by leaching with 1 N ammonium acetate buffer (adjusted to pH 7.0); mineral-N (NH$_4^+$ -N and NO$_3$-N) by steam distillation techniques (Bremner, 1965) and total N by the salicylic acid digestion - Kjeldhal procedure (Bremner and Mulvaney, 1982). Mechanical analysis of the soils was done using pipette method and the USDA Textural Triangle was used to determine soil texture class (Table 1). The soil was filled in a plastic pot with measuring 32 cm (height) × 30 cm (diameter). Each pot was filled with 20 kg of soil (air-dried, crushed and sieved to pass a 2 mm sieve).

Soil amendments

Five agriculture inputs; control, chicken manure (pellet processed), zeolite, biochar (prepared from empty fruit bunches of oil palm) and urea were selected to evaluate their effects on soil pH elevation and variability. Before application to the pot, each amendment was analysed for their pH in water by pH meter (Table 2). The rate of agriculture inputs was same for all treatments which was 17.6 g for each of the material (Table 2). The amendments were mixed uniformly in the pot using a scoop.

Soil pH determination

The soil pH was measured at 150 points for each pot by glass microelectrode of digital pH meter (flat surface 3 mm Ag - AgCl model) using the method of Fan and Mackenzie (1993). A grid (15 × 10 cm) was placed on the surface area of the each pot and pH was measured at 1 cm intervals in both of the X and Y directions. The method was used previously for the determination of urea in petri dishes with 50 g of soil. In the present study, the quantities of

Table 3. Descriptive statistics of spatial variability of soil pH in a pot area (30 cm^2) influenced by application of selected agriculture inputs.

Variable	Mean	Min	Max	CV (%)
Control	5.37^c	4.6	5.7	10.11
Urea	5.79^b	5.3	6.8	12.21
Chicken dung	4.64^d	4.0	5.8	13.22
Biochar	5.84^b	5.1	6.5	10.23
Zeolite	6.51^a	5.9	7.4	11.22

Means with different letters indicated the significant differences in soil pH among various soil amendments application at p < 0.05.

amendments were modified and calculated to measure the variability in pot area at larger surface area than petri dish. The applied amount of agar solutions was used in this experiment was just 2 L for 20 kg soil which was evaluated in preliminary lab experiments. Furthermore, Fan and Mackenzie (1993) used only one granule to observe the hydrolysis or diffusion of urea on a microsite. But here in this experiment, 17.6 g of each fertilizer source and soil amendments were applied on surface. The purpose of modification of this method was to analyse the changes in soil pH on a pot surface area without any disturbance. Geostatistical mapping of a pot for soil pH could be novel approach to interpret clearly the possible variation and change in soil pH of a small experimental area.

Statistical and geostatistical analysis

Statistical analysis included investigation of mean values, coefficients of variation, maximum and minimum values. Spatial variability of soil pH was assessed by means of semi-variogram analysis. From models of spatial dependence between neighbouring data, the Kriging approximation was used for interpolation by using software (GS+ v. 9, Gamma design, Plainwell, MI) (Webster and Oliver, 2007; Balasudram et al., 2008; Dharejo et al., 2011; Junejo et al., 2012). The experimental design used was Completely Randomized Design (CRD) with four replications for each treatment. The differences among treatments were estimated by ANOVA followed by Tukey's mean test by using statistical analysis software (SAS version 9.5).

RESULTS AND DISCUSSION

The determination of soil properties showed that soil is sandy loam, strongly acidic, low in total N, moderate in total C, very low in K and Ca, high in Mg and low in CEC according to international soil interpretation values (Hazelton and Murphy, 2007). The soil pH is a considerable guide to understand nutrient deficiencies and toxicities such as; availability of phosphorus, nitrogen, potassium, calcium, sulphur reduced at < 5.0 (McKenzie et al., 2004).

Table 2 shows the pH of urea, chicken manure, biochar and Zeolite, which showed the basic pH of biochar, urea and the chicken manure were acidic; however, zeolite also has acidic pH with high CEC. The previous literature indicated that urea, biochar and chicken manure have higher pH which ranged from 7.5 to 9 (Ahmed et al., 2006; Steiner et al., 2010).

Soil pH observed were significantly different (P = 0.001) to each other in all the treatments. The soil pH of each treatment was ranged from 4.64 to 6.5. Mean pH was significantly higher in soil treated with zeolite and significantly lower in soil treated with chicken manure. Co-efficient of variation of soil pH treated with all the treatments had a narrow range from 10.11 to 13.22% (Table 3). Soil pH in all the treatment was modelled by exponential model. The range expressed as distance, and can be interpreted as the diameter of the zone of influence, which represents the average maximum distance over which a soil property of the two samples is related. Minimum range of 5.4 cm for control and maximum 6.5 cm was observed in the soil treated with zeolite. Co-efficient of determination indicating a good fit for models ranged from 0.46 to 0.56 for the soil pH treated with different treatments (Table 4).

Kriged maps showed variation in soil pH within 30 cm diameter pots. Soil pH in control ranged from 4.6 to 5.4, Urea ranged from 5.6 to 6.50, chicken manure ranged from 4.0 to 5.4, Biochar ranged from 5.6 to 6.5 and Zeolite ranged from 6.1 to 7.0 (Figure 1). The data in control (without any amendments) ranged from 5.3 to 5.7 showed a narrow range of variability in soil pH at soil surface of pot (Figure 1a). The urea treated pot show a soil pH which was ranged from 5.3 to 6.8 after 24 h of fertilizer applications (Figure 1b). These occur due to fast hydrolyses process of urea in soil which was resulted in an increase of soil pH. It is reported previously that the applied urea to the soil will rapidly hydrolyse to $(NH_4)_2CO_3$ facilitated by urease enzymes and subsequently to NH_4OH and CO_2 which results in pH increase at urea micro site and favours liberation of NH_3 (Ayanaba and Kang, 1976; Antil et al., 1992; Junejo et al., 2012). The addition of any type of organic matter always benefited buffer capacity of soils against the impact of acidity. However the addition of chicken manure reduced the soil pH which was ranged from 4.5 to 5.1 in this experiment (Figure 1c). Usually, the pH for the soil chicken manure mixtures were found to be neutral to slightly alkaline (Nichols and Daniel, 1994).

Table 4. Geostatistical statistics of spatial variability of soil pH in a pot area (30 cm^2) influenced by application of selected agriculture inputs.

Variable	Model	Nugget	Sill	Effective range	r^2	Nug sill
Control	Exponential	0.03	0.12	5.40	0.48	0.26
Urea	Exponential	0.06	0.18	6.00	0.52	0.33
Chicken dung	Exponential	0.06	0.17	4.90	0.48	0.34
Biochar	Exponential	0.08	0.23	5.70	0.46	0.33
Zeolite	Exponential	0.05	0.14	6.50	0.56	0.38

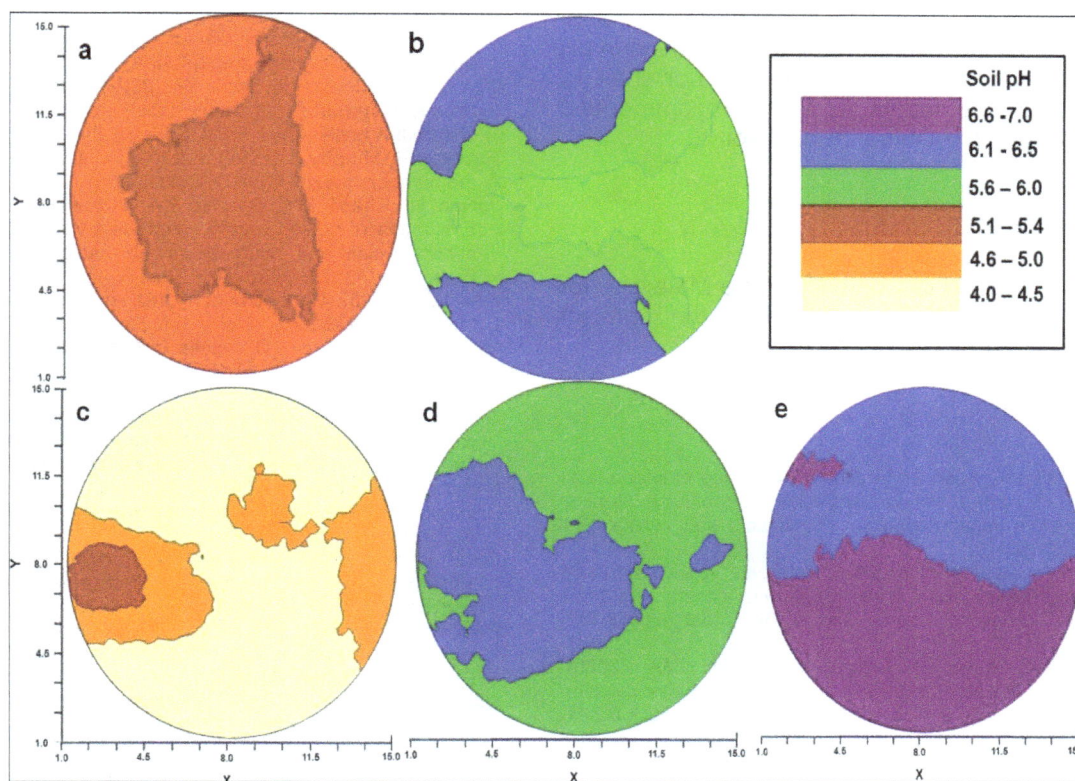

Figure 1. Spatial variability map of soil pH of a sandy loam soil affected by the application of a) control (b) urea (c) chicken dung (d) biochar (e) zeolite after 24 hours of application (the shades of same colour comes in the similar pH range).

Although, it was also documented that the addition of eight to ten years old chicken manure to soil can cause acidity or lower the soil pH (De Datta, 1995). The Figure 1d indicates the effect of biochar on micro-site soil pH variability on biochar treated soil surface. The EFB biochar pH is 8 as analysed and claimed by manufacturer. When EFB biochar added to the soil it will obviously expected to elevate the soil pH (Singh et al., 2011). The soil pH of biochar treated pot was increased from 5.2 to 6.2 which was less than the measured pH range of zeolite. Biochar have neutral to basic pH and many research studies revealed an elevation in soil pH after biochar application when the initial pH was low.

On alkaline soils, this may be an adverse effect but for acidic soils it is a positive property. Increase in pH of an acidic soil reduced aluminium toxicities and enhance nutrient availability (Ayanaba and Kang, 1976; Bremner and Mulvaney, 1978; Ahmed et al., 2008a; Dawar et al., 2011). The highest elevation of soil pH were found in zeolite treated soil with a pH variability range from 6.3 to 6.7 (Figure 1e). Addition of zeolite increases the soil pH due to its catalytic ability and parental material structure. Zeolite has been used instead of lime to increase soil pH and to improve the nutrient availability and CEC of soil for a long time (Nibou et al., 2009; Ramesh and Reddy, 2011).

Conclusion

The presented research experiments revealed and compare the effects of soil activator (zeolite), soil conditioner (biochar), organic matter (chicken manure), and inorganic fertilizer (urea) on soil pH, which is a fundamental property of soil. The outcomes of study revealed that the addition of these soil amendments having liming effects because of significant increase in soil pH except chicken manure application. The liming capacity of above mention materials can give better advantage to acidic soils that require liming, by being applied more frequently at lower application rates, reducing labour cost and time. In contrast, elevation in pH can cause nutrient losses in alkaline soils. Prior analysis of soil and agriculture inputs is highly recommended before proceeding to any amendment of soil.

ACKNOWLEDGEMENT

The authors are thankful to Universiti Putra Malaysia, for financial and technical support for the project Fundamental Research Grant Scheme 5524142.

REFERENCES

Ahmed OH, Aminuddin H, Husni MHA (2006). Reducing ammonia loss from urea and improving soil-exchangeable ammonium retention through mixing triple superphosphate, humic acid and zeolite. Soil Use Manage. 22:315-319.

Ahmed OH, Husin A, Hanif AHM (2008a). Ammonia volatilization and ammonium accumulation from urea mixed with zeolite and triple superphosphate. Acta Agric. Scand. Sec. B: Soil Plant Sci. 58:182-186.

Ahmed OH, Hussin A, Ahmad HMH, Rahim AA, Majid NMA (2008b). Enhancing the urea-N use efficiency in maize (Zea mays) cultivation on acid soils amended with zeolite and TSP. The Sci. World J. 8:394-399.

Antil RS, Narwal RP, Gupta AP (1992). Urease activity and urea hydrolysis in soils treated with sewage. Ecol. Eng. 1:229-237.

Ayanaba A, Kang BT (1976). Urea transformation in some tropical soils. Soil Biol. Biochem. 8:313-316.

Balasudram SK, Husni MHA, Ahmed OA (2008). Application of Geostatistical tools to quantify spatial variability of selected soil chemical properties from a cultivated Tropical peat. J. Agron. 7:82-87.

Bremner JM (1965). Total Nitrogen. In: Black CA, E.D., Ensminger LE, White JL, Clark FE, Dinauer RC (eds.) (Ed.), In, Method of soil analysis, Part 2. American Society of Agronomy, Madison, Wisconsin. pp. 1149-1178.

Bremner JM, Mulvaney CS (1982). Nitrogen-total. American Society of Agronomy and Soil Science Society of America Inc. Madison.

Bremner JM, Mulvaney RL (1978). Urease activity in soils. In: Burns, R.G. (Ed.), In "Soil Enzymes" Academic Press, London. pp. 149-196.

Cabrera ML, Kissel DE, Bock BR (1991). Urea hydrolysis in soil: Effects of urea concentration and soil pH. Soil Bio. Biochem. 23:1121-1124.

Chen D, Suter HC, Islam A, Edis R (2010). Influence of nitrification inhibitors on nitrification and nitrous oxide (N_2O) emission from a clay loam soil fertilized with urea. Soil Biol. Biochem. 42:660-664.

Christianson CB, Baethgen WE, Carmona G, Howard RG (1993). Microsite reactions of urea-nbtpt fertilizer on the soil surface. Soil Biol. Biochem. 25:1107-1117.

Clough TJ, Condron LM (2011). Biochar and the Nitrogen Cycle:

Introduction. J. Envior. Qual. 39:1218-1223.

Dawar K, Zaman M, Rowarth JS, Blennerhassett J, Turnbull MH (2011). Urea hydrolysis and lateral and vertical movement in the soil: Effects of urease inhibitor and irrigation. Biol. Fert. Soils 47:139-146.

De Datta SK (1995). Nitrogen transformations in wetland rice ecosystems. Fertil. Res. 42:193-203.

Dharejo KA, Anuar AR, Khanif YM, Samsuri AW, Junejo N (2011). Spatial variability of Cu, Mn and Zn in marginal sandy beach ridges soil. Afr. J. Agric. Res. 6:3493-3498.

Fan MX, Mackenzie AF (1993). Urea and phosphate interactions in fertilizer microsite: Ammonia volatilization and pH changes. Soil Sci. Soc. Am. J. 57:839-845.

Freney JR, Keerthisinghe DG, Chaiwanakupt P, Phongpan S (1993). Use of urease inhibitors to reduce ammonia loss following application of urea to flooded rice fields. Plant Soil 155-156:371-373.

Glaser B, Haumier L, Guggnenberger G, Zech W (2001). The 'Terra Preta' Phenomenon; a model for sustainable agriculture in the humid tropics. Naturwissenschafen 88:37-41.

Glaser B, Lehmann J, Zech W (2002). Ameliorating physical and chemical properties of highly weathered soils in the tropics with charcoal-A review. Biol. Fertil. Soils 35:219-230.

Hazelton P, Murphy B (2007). Interpreting soil tesr results; what do all the numbers mean? CSIRO, Collingwood.

Junejo N, Khanif MY, Dharejo KA, Hazandy AH, Abdu A (2012). Evaluation of coated urea for ammonia volatilization loss, nitrogen mineralization and microsite pH in selected soil series. Afr. J. Biotechnol. 11:366-378.

Kissel DE (1988). Management of urea fertilizers. North Central regional publication NCR-326.

McKenzie NJ, Jacquire D, Isbella R, Brown K (2004). Australian Soils and Landscapes- an illustrated compendium. CSIRO publishing: Melbourne.

Nibou D, Mekatel H, Amokrane S, Barkat M, Trari M (2009). Adsorption of Zn2+ ions onto NaA and NaX zeolites: Kinetic, equilibrium and thermodynamic studies. J. Hazard. Mater. 173:637-646.

Nichols DJ, Daniel TC (1994). Nutrient Runoff from Pasture after Incorporation of Poultry Litter or Inorganic Fertilizer. Soil Sci. Soc. Am. J. 58:1224-1228.

Omar L, Ahmed OH, Majid NMA (2010). Enhancing nutrient use efficiency of maize (Zea mays L.) from mixing urea with zeolite and peat soil water. Int. J. Phys. Sci. 6:3330-3335.

Perkins HF, Parker MB, Walker ML (1964). Chicken manure its production,composition and use as a fertilizer. Georgia Agricultural Experiment Station, Georgia.

Ramesh K, Reddy DD (2011). Zeolites and thier potential uses in agriculture. Adv. Agron. 113:219-240.

Singh BP, Hatton BJ, Singh B, Cowie AL, Kathuria A (2011). Influence of Biochars on Nitrous Oxide Emission and Nitrogen Leaching from Two Contrasting Soils. J. Envir. Qual. 39:1224-1235.

Srinivasarao C, Venkateswarlu B, Lal R, Singh AK, Vittal KPR, Kundu S, Singh SR, Singh SP (2011). Long-Term Effects of Soil Fertility Management on Carbon Sequestration in a Rice–Lentil Cropping System of the Indo-Gangetic Plains. Soil Sci. Soc. Am. J. 76:168-178.

Steiner C, Das KC, Melear N, Lakly D (2010). Reducing Nitrogen Loss during Poultry Litter Composting Using Biochar. J. Environ. Qual. 39:1236-1242.

Walkley A, Black IA (1934). An examination of Degtjaref method for determination of soil organic matter and a proposed modification of the chromic acid titration method. Soil Sci. Soc. Am. J. 37:29-38.

Webster R, Oliver MA (2007). Geostatistics for Environmental Scientists. John Wiley and Sons, press, UK.

Zhang M, Zhang H, Xu D, Han L, Niu D, Tian B, Zhang J, Zhang L, Wu W (2010). Removal of ammonium from aqueous solutions using zeolite synthesized from fly ash by a fusion method. Desalination 271:111-121.

Zhao X, Xing GX (2009). Variation in the relationship between nitrification and acidification of subtropical soils as affected by the addition of urea or ammonium sulfate. Soil Biol. Biochem. 41:2584-2587.

19

Effects of simulated acid rain on *Shorea macroptera* growth and selected soil chemical properties

I. M. Hilmi[1], K. Susilawati[1], O. H. Ahmed[1] and Nik M. Majid[2]

[1]Department of Crop Science, Faculty of Agriculture and Food Science, University Putra Malaysia Bintulu Sarawak Campus, 97008 Bintulu, Sarawak, Malaysia.
[2]Institute of Tropical Forestry and Forest Products (INTROP), Universiti Putra Malaysia 43400 Serdang, Selangor, Malaysia.

There is dearth of information on the effect of acid deposition on *Shorea macroptera* in Malaysia. Thus, this study was conducted to investigate the potential effect of simulated acid rain (SAR) on *S. macroptera* growth and selected soil chemical properties. Six treatments were evaluated in this study. Growth variables of *S. macroptera* were observed for 90 days. After 90 days, seedlings and soil were sampled and analyzed using standard methods. The seedlings height decreased with decrease in SAR pH. Chlorosis and necrosis were observed for low SAR pH (pH 3.5 and 4) treatments and this observation explains the reduction of dry matter production of the plants subjected to these treatments. Regardless of treatment, K, Ca, Mg and Na contents in the plants and soil were statistically similar. A similar observation was found for soil exchangeable Fe, Cu, Zn, acidity, Al and H. Thus, it can be concluded that SAR pH of 3.5, 4 and 4.5 affects *S. macroptera* height, biomass and selected nutrient contents in soil. *S. macroptera* is susceptible to acid deposition and it could be considered as one of the bio-indicators in Malaysia. A field study is recommended to validate the findings of this study.

Key words: *Shorea macroptera*, acid deposition, growth, soil nutrients.

INTRODUCTION

Rapid industrialization and unsustainable agricultural practices are some of the possible causes of acid deposition in Malaysia (Ayers et al., 2002). According to Wang et al. (2004), application of urea and animal manure causes ammonia (NH_3) accumulation in the atmosphere and long term accumulation may cause acid deposition. Acid deposition has adverse effect on soils and plants. It can reduce plant growth and yield due to foliar injury, low nutrient availability in soils, or exposure of plants to toxic substances that are released from the soil (Liu et al., 2011).

For deciduous trees, slow height increment and low biomass were shown at simulated acid rain (SAR) pH of 2.0 (John et al., 2012). Foliar damages were recorded for red spruce and Brassica napus at SAR pH of 2.5 and 3.0. Mountain birch showed reduced seeds germination when exposed to SAR at pH 4.0 (Ernst, 2012). The SO_4^{2-}, H^+, NO_3^- and NH_4^+ input from acid deposition may reduce soil pH. Then, it leads to soil acidification. Higher acidity in soil also increased the solubility of heavy metal (Heij et al., 1991). Due to the serious effect of acid deposition on plants, there is a need for a scientific study of this global problem in Malaysia.

Shorea macroptera is one of the commercial Dipterocarps species in Malaysia. This timber species is classified as hardwood by Malayan Grading Rules

(Symington, 2004) and commonly has high growth rate (ranging from 5.6 to 8.1 mm year^{-1}). This tree is in high demand in local or international markets for timber (Ang and Maruyama, 1995). The growth rate of *S. macroptera* is mostly affected by the surrounding environment such as sunlight, moisture and nutrients (Manokaran and Kochummen, 1993). Thus, this research was conducted to: (i) determine the effect of simulated acid rain on *S. macroptera* growth and (ii) determine the effect of simulated acid rain on selected soil chemical properties.

MATERIALS AND METHODS

Preparation of simulated acid rain

Original rainwater collected from our study area was used as a control (pH 6 ± 0.02) in this experiment. Preparation of simulated acid rain (SAR) was done using combination of nitric and sulphuric acid at a ratio of 3: 2 (v/v), with original rainwater to obtain pH of 3.5, 4.0, 4.5, 5.0, and 5.5. The ratio was taken from SO_2 and NO_3 composition in Malaysia (Ayers et al., 2002). The preparation of SAR was done using Liu et al. (2008) method. Application of SAR with different pH was carried out using a dripper at a velocity of 2.71 ml s^{-1}. About 618 ml of SAR was applied on the 90th day of the study. Treatments were applied every 6 days to mimic the dry condition weather of Malaysia (MMD, 2008). All treatments were applied in the morning (between 7 to 9 am) because high temperature and high irradiances reduce plants' ability to neutralize acid deposition. However, 150 mL rainwater was applied once every 2 days to avoid dryness of seedlings.

Pot study

A pot experiment was conducted in a greenhouse at Universiti Putra Malaysia Bintulu Sarawak Campus (UPMKB) (03°12.301'N, 113°04.032'E) Bintulu, Malaysia. *S. macroptera* plants were grown in a greenhouse to minimize infestation of pests and to also ensure uniformity of sunlight and water supply. *S. macroptera* seedlings were collected from a rehabilitated forest at UPMKB. The mean height of the seedlings ranged from 20 to 30 cm. The seedlings were grown for 7 months in a poly bag with 0.8 kg soils. Afterwards, the plants were selected based on height and number of leaves for subsequent experiment.

Nyalau series (*Typicpaleudults*) was sampled at 10 to 15 cm depth from an undisturbed area of UPMKB. The soil was air-dried and sieved to pass a 2.0 mm sieve. The soil was analysed for texture, water holding capacity, pH, cation exchange capacity (CEC), organic matter content, total organic C, total N, available P and exchangeable cation using standard procedures. A 6 kg air-dried soil was weighed into a 20" × 18" polybag, watered at 70% field capacity with tap water and afterwards, transplanting was carried out. The poly bags were arranged in a Completely Randomized Design (CRD) with 6 treatments and 6 replications.

Plant growth measurement and soil analysis

Plant height, number of leaves, chlorophyll content, chlorosis and necrosis were determined every 30 days. At day 90, plant and soil samples were taken and analyzed using standard procedures. The soil samples were analyzed for pH (Tan, 2005), exchangeable cation and available P (Mehlich, 1953) and available NO_3^- and NH_4^+ using distillation method (Keeney and Nelson, 1982). Exchangeable

acidity, Al and H were determined using titration method (Rowell, 1994). Meanwhile for SO_4^{2-}, soil was extracted using 0.5 M $NaHCO_3$ and was further analyzed using Ion Chromatograph-Mass Spectrometry (IC-MS) (PerkinElmer Inc., Model AI300).

Sampled *S. macroptera* plants were partitioned into leaves, stems and roots before being oven-dried at 60°C until constant weight was attained. Then, the dried samples were weighed for dry matter production using a digital balance. The plant samples were ground, ashed, after which K, Ca, Mg and Na contents were determined using Atomic Absorption Spectrophotometer (AAS) (PerkinElmer Inc., Model AAnalyst 800) (Cottenie, 1980). P content in the plant parts was determined using the Blue method (Murphy and Riley, 1962). Kjeldahl method was used to determine total N (Bremner, 1965) whilst nutrients uptake were calculated using a formula (Pomares-Gracia and Pratt, 1987).

Statistical analysis

Analysis of variance (ANOVA) at p ≤ 0.05 was used to detect effect of treatments while means of the treatments were compared using Tukey's test. All statistical analyses were conducted using Statistical Analysis System Version 9.2 (SAS, 2001).

RESULTS

Basic characteristics of soil and SAR

The soil used in this study was acidic (pH$_w$ 4.84 and pH$_{KCl}$ 3.58). The CEC of the soil was low (Table 1). The organic matter and ash content of the soil were 6 and 94%, respectively. The soil was a sandy clay loam with low available P and exchangeable cation. The chemical characteristics of the SAR used in this study are shown in Table 2.

Effect of SAR on *S. macroptera* growth

The height of *S. macroptera* was affected by SAR (Table 3). Seedling height treated with pH 3.5 (T1), 4 (T2) and 4.5 (T3) was lower compared to that of T5 (pH 5.5) which showed the highest increase compared to the other treatments, including the control (pH 6 ± 0.02). The number of leaves treated with T1 (pH 3.5) and T2 (pH 4.0) was lower than those of T3, T4 and the control (pH 6 ± 0.2). At day 30, significant number of leaves treated with low SAR pH was lower than those of T5 and the control. However, on day 90, there was no difference between these treatments (Table 3).

There was an increase in chlorophyll content due to application of T1, T2, T3 and T4 particularly in T2 (pH 4) which produced a sharp increase in chlorophyll content within 90 days and this might be due to reflex action of the SAR (Figure 1).

Total biomass was significantly affected by pH of SAR (Table 4). T4 (pH 5) produced higher total biomass compared to the other treatments. The foliar biomass of T1 (pH 3.5) and T2 (pH 4.0) was lower than those of T3, T4 and T5. The stem and roots biomass was the lowest

Table 1. Physico-chemical characteristics of Nyalau series.

Parameter	Value
pH in water	4.84
pH in KCl	3.58
Exchangeable K^+ (cmol kg^{-1})	0.54
Exchangeable Ca^{2+} (cmol kg^{-1})	2.25
Exchangeable Mg^{2+} (cmol kg^{-1})	0.02
Exchangeable Na^+ (cmol kg^{-1})	0.23
Exchangeable acidity (cmol kg^{-1})	3.30
Exchangeable Al (cmol kg^{-1})	1.35
Exchangeable H (cmol kg^{-1})	1.95
Cations exchange capacity (cmol kg^{-1})	10.2
Exchangeable NH_4^+ (mg kg^{-1})	49.04
Available NO_3^- (mg kg^{-1})	21.02
Available P (mg kg^{-1})	0.1
Available SO_4^{2-} (mg kg^{-1})	18.0
Total N (%)	0.11
Bulk density (g cm^{-3})	1.2
Total organic carbon (%)	3.48
Ash content (%)	94
Organic matter (%)	6.00
Soil texture (sandy clay loam)	
Sand (%)	65.57
Clay (%)	21.26
Silt (%)	13.17

Table 2. Chemical characteristics of simulated acid rain used in pot study.

Properties		Treatments					
		T0	T1	T2	T3	T4	T5
		pH					
		6.00 ± 0.2	3.5	4.0	4.5	5.0	5.5
EC	($\mu s\ cm^{-1}$)	15.62	130.53	46.87	26.67	11.95	7.14
Salinity	(ppt)	0.02	0.08	0.03	0.02	0.02	0.02
TDS	"	7.77	70.37	24	13.03	5.98	3.59
K^+	"	0.17	2.2	0.97	0.53	0.36	0.3
Ca^{2+}	"	1.74	1.51	0.7	1.01	0.72	0.62
Mg^{2+}	"	0.07	0.09	0.05	0.07	0.06	0.04
Na^+	"	0.33	0.75	0.27	1.06	0.39	0.2
NH_4^+	"	0.5	0.21	0.28	0.02	0.04	0.03
NO_3^-	"	1.03	5.3	2.2	3.2	2.43	1.03
PO_4^{3-}	"	0.32	0.22	0.22	0.2	0.25	0.31
S_2^-	"	0.005	0.03	0.02	0.03	0.01	0.01
Cl_2^-	"	0.5	0.5	1.0	1.0	1.0	0.5
NO_2^-	"	0.04	0.04	0.04	0.1	0.06	0.04

Note: " = represent mg L^{-1}.

at T3 (pH 4.5) (Table 4). However, T0 (pH 3.5) and T5 (pH 5.5) produced the highest biomass in the roots (4.34 g) and stems (7.00 g), respectively.

Leaf chlorosis and necrosis were observed at SAR pH (pH 3.5, 4.0 and 4.5) (Table 5). Three out of the 36 seedlings were affected. There was no significant

Table 3. Effect of SAR on height and number of leaves increment in *S. macroptera*.

Trt.	pH level	Height increment (cm)				
		Day 0	Day 30	Day 60	Day 90	Mean (pH)
T0	6.0 ± 0.2	48.57	4.40$^{abc\cdots}$	6.00$^{a\cdots}$	8.33$^{a\cdots}$	6.24$^{ab\cdots}$
T1	3.5	51.20	2.67$^{bc\cdots}$	4.17$^{a\cdots}$	4.67$^{a\cdots}$	3.83$^{b\cdots}$
T2	4.0	47.00	2.00$^{c\cdots}$	3.00$^{a\cdots}$	6.83$^{a\cdots}$	3.94$^{b\cdots}$
T3	4.5	50.25	3.33$^{bc\cdots}$	5.50$^{a\cdots}$	6.33$^{a\cdots}$	5.06$^{ab\cdots}$
T4	5.0	49.17	9.00$^{a\cdots}$	7.17$^{a\cdots}$	6.50$^{a\cdots}$	7.38$^{ab\cdots}$
T5	5.5	51.08	7.67$^{ab\cdots}$	5.75$^{a\cdots}$	10.50$^{a\cdots}$	8.25$^{a\cdots}$
Mean (days)		49.47	4.60$^{b\cdot}$	5.24$^{ab\cdot}$	7.19$^{a\cdot}$	

	pH level	Number of leaves				
T0	6.0 ± 0.2	7.86	6.00$^{a\cdots}$	5.00$^{ab\cdots}$	3.00$^{a\cdots}$	4.67$^{ab\cdots}$
T1	3.5	7.40	1.00$^{b\cdots}$	3.00$^{ab\cdots}$	6.00$^{a\cdots}$	3.33$^{ab\cdots}$
T2	4.0	7.33	0.67$^{b\cdots}$	3.33$^{b\cdots}$	4.00$^{a\cdots}$	2.44$^{b\cdots}$
T3	4.5	7.67	4.33$^{ab\cdots}$	6.67$^{a\cdots}$	6.33$^{a\cdots}$	5.78$^{a\cdots}$
T4	5.0	8.00	5.33$^{a\cdots}$	4.33$^{ab\cdots}$	5.33$^{a\cdots}$	5.00$^{ab\cdots}$
T5	5.5	8.83	2.67$^{ab\cdots}$	2.67$^{ab\cdots}$	4.67$^{a\cdots}$	3.56$^{ab\cdots}$
Mean (days)		7.86	3.33$^{a\cdot}$	4.17$^{ab\cdot}$	4.89$^{a\cdot}$	

Note: Mean with different letters was significantly different at $p \le 0.05$ (Tukey's test), ' = Mean comparison by day, " = Mean comparison by SAR level, "' = Interaction between day and SAR level.

Table 4. Effect of SAR on biomass of *S. macroptera*.

Trt.	pH level	Biomass (g)			
		Leaves	Stems	Roots	Total
T0	6.0 ± 0.2	7.33^{ab}	7.97^{ab}	4.34^{a}	19.65^{b}
T1	3.5	5.64^{b}	5.68^{bc}	2.54^{ab}	13.86^{f}
T2	4.0	5.94^{b}	6.20^{abc}	2.89^{ab}	15.03^{d}
T3	4.5	6.90^{ab}	5.10^{c}	2.46^{b}	14.46^{e}
T4	5.0	8.40^{a}	8.18^{abc}	3.48^{ab}	20.06^{a}
T5	5.5	7.02^{ab}	7.00^{a}	4.18^{ab}	18.20^{c}

Note: Mean within same column with different letters was significantly different at $p \le 0.05$ (Tukey's test).

difference at $p \le 0.05$ in foliar N, K, and Na concentrations (Table 6). Perhaps this could be the effects of nutrient leaching in foliar.

Higher acidity decreased P and Ca in the leaves, and increased Zn concentrations (Table 6). T3 (pH 4.5) showed the highest increase in Ca and P of the leaves while the lowest was in T1 (pH 3.5). As SAR increased from pH 3.5 to pH 4.5, P concentration in leaves increased from 0.085 to 0.231%. Similar results were obtained for Ca (Table 6). Nitrogen, P, K, Ca and Na accumulation were relatively lower at lower SAR pH, compared to that of the control. Increase in SAR pH from 3.5 to 6.0 ± 0.2 decreased accumulation of N in the leaves. The highest K concentration in the stems was in T4 (pH 5) and the lowest in T0 (Table 7). Other treatments showed no significant difference. The accumulation of N (Table 7) was highest in T1 (pH 3.5) and lowest in T2 (pH 4), T3 (pH 4.5), and T4 (pH 5).

For Zn and P accumulation, no significant difference was observed among treatments. T4 showed the highest uptake for K and Ca while T3 caused the lowest Na accumulation compared to other treatments. No significant difference was observed in N, K, Ca, Na and Zn concentrations in the roots (Table 8). T0 (pH 6 ± 0.02) caused the highest increase in P concentrations compared to other treatments (Table 8). Regardless of SAR pH, N, Ca and Na concentrations in the roots were not significantly different (Table 8). T0 (6.0 ± 0.2) pH showed the highest P while T5 (pH 5.5) caused the lowest. Similarly to P accumulation, T5 caused the highest accumulation of P and T1 caused the lowest. Highest concentration of Zn was recorded in T0 (6.0 ± 0.2) and lowest in T3 (pH 4.5).

Lack of phosphorus in leaves, roots and stems could

Table 5. Effect of SAR on the occurrence of chlorosis and necrosis in *S. macroptera*.

Trt.	pH level	Chlorosisper seedling				
		Day 0	Day 30	Day 60	Day 90	Average
T0	6.0 ± 0.2	0	0	0	0	0/6
T1	3.5	0	1	1#	1#	1/6
T2	4.0	0	0	0	1	1/6
T3	4.5	0	0	0	1	1/6
T4	5.0	0	0	0	0	0/6
T5	5.5	0	0	0	0	0/6
Average		0/36	1/36	0/36	2/36	3/36

	pH level	Necrosis per seedling				
T0	6.0 ± 0.2	0	0	0	0	0/6
T1	3.5	0	0	1	1#	1/6
T2	4.0	0	0	1	1*+1	2/6
T3	4.5	0	0	0	0	0/6
T4	5.0	0	0	0	0	0/6
T5	5.5	0	0	0	0	0/6
Average		36/36	0/36	2/36	1/36	3/36

Note: 36 = Total number of seedling, # = No increasing number of chlorosis and necrosis, * = Plant mortality.

Table 6. Effect of SAR on foliar nutrient concentration and accumulation in *S. macroptera* leaves.

Trt.	SAR pH level	Nutrient concentration (%)					
		N	P ($\times 10^{-2}$)	K ($\times 10^{-1}$)	Ca	Na ($\times 10^{-1}$)	Zn ($\times 10^{-3}$)
T0	6.0 ± 0.2	1.1[a]	5.5[b]	7.7[a]	2.2[ab]	3.0[a]	1.4[ab]
T1	3.5	1.5[a]	8.5[b]	8.6[a]	1.8[b]	3.1[a]	2.8[a]
T2	4.0	1.1[a]	12.0[ab]	6.1[a]	2.3[ab]	3.4[a]	1.1[ab]
T3	4.5	1.4[a]	23.1[a]	7.6[a]	2.6[a]	3.4[a]	1.2[ab]
T4	5.0	1.3[a]	8.1[b]	8.4[a]	2.2[ab]	3.9[a]	0.6[b]
T5	5.5	1.3[a]	4.0[b]	8.5[a]	2.5[ab]	4.3[a]	2.2[ab]

Trt.	SAR pH level	Nutrient accumulation (mg plant^{-1})					
		N	P	K	Ca	Na	Zn ($\times 10^{-2}$)
T0	6.0 ± 0.2	81[ab]	3.1[b]	56[ab]	159[ab]	21[ab]	8.0[b]
T1	3.5	85[ab]	4.1[b]	47[bc]	100[b]	17[b]	16.7[ab]
T2	4.0	66[b]	9.1[ab]	34[c]	137[ab]	19[ab]	5.7[b]
T3	4.5	98[ab]	15.1[a]	52[b]	183[a]	27[ab]	7.7[b]
T4	5.0	107[a]	6.1[b]	70[a]	184[a]	31[a]	5.0[b]
T5	5.5	91[ab]	2.1[b]	60[ab]	173[a]	30[a]	27.0[a]

Note: Mean within same column with different letters was significantly different at $p \leq 0.05$ (Tukey's test).

also be one of the reasons for the low biomass of the test plants (Table 4).

Effect of SAR on soil cultivated with *S. macroptera*

There was no significant difference in soil pH and available P cultivated with *S. macroptera* (Table 9). However, for NH_4, it was significant where T2 (pH 4.0) recorded the lowest (11.68 mg kg^{-1}) concentration whilst the highest (37.36 mg kg^{-1}) was in T5 and T0 (pH 5.5 and pH 6.0 ± 0.02). Treatment 2 (pH 4.0) produced the most significant available NO_3 with highest value (22.42 mg kg^{-1}) and T0 (pH 6.0 ± 0.02) produced the lowest value

Table 7. Effect of SAR on stems nutrient concentrations and accumulation in *S. macroptera*.

Trt.	SAR pH level	Nutrient concentration (%)					
		N ($\times 10^{-1}$)	P ($\times 10^{-2}$)	K ($\times 10^{-1}$)	Ca	Na ($\times 10^{-1}$)	Zn ($\times 10^{-4}$)
T0	6.0 ± 0.2	4ᵃ	9.0ᵃ	3.9ᵇ	2.5ᵃ	3.6ᵃ	6.5ᵃ
T1	3.5	8ᵃ	8.8ᵃ	5.2ᵃᵇ	2.5ᵃ	4.8ᵃ	6.3ᵃ
T2	4.0	5ᵃ	11.8ᵃ	5.9ᵃᵇ	2.7ᵃ	4.0ᵃ	10.0ᵃ
T3	4.5	6ᵃ	5.9ᵃ	5.9ᵃᵇ	2.7ᵃ	3.7ᵃ	9.7ᵃ
T4	5.0	5ᵃ	5.2ᵃ	7.7ᵃ	2.9ᵃ	5.2ᵃ	7.7ᵃ
T5	5.5	5ᵃ	8.2ᵃ	5.7ᵃᵇ	2.6ᵃ	4.3ᵃ	9.5ᵃ

Trt.	SAR pH level	Nutrient accumulation (mg plant^{-1})					
		N	P	K	Ca	Na	Zn ($\times 10^{-2}$)
T0	6.0 ± 0.2	30ᵃᵇ	7.3ᵃ	31ᵃᵇ	202ᵃ	28ᵃᵇ	4.5ᵃ
T1	3.5	60ᵃ	4.7ᵃ	23ᵇ	114ᶜ	33ᵃ	4.0ᵃ
T2	4.0	27ᵇ	6.7ᵃ	36ᵃᵇ	148ᵇᶜ	24ᵃᵇ	5.0ᵃ
T3	4.5	21ᵇ	2.8ᵃ	30ᵃᵇ	135ᵇᶜ	19ᵇ	4.3ᵃ
T4	5.0	28ᵇ	6.0ᵃ	47ᵃ	206ᵃ	33ᵃ	6.0ᵃ
T5	5.5	37ᵃᵇ	5.9ᵃ	40ᵃᵇ	184ᵃᵇ	30ᵃ	7.0ᵃ

Note: Mean within same column with different letters was significantly different at p ≤ 0.05 (Tukey's test).

Table 8. Effect of SAR on roots nutrient concentrations and accumulation in *S. macroptera*

Trt.	SAR pH level	Nutrient concentration (%)					
		N ($\times 10^{-1}$)	P ($\times 10^{-3}$)	K ($\times 10^{-1}$)	Ca	Na ($\times 10^{-1}$)	Zn ($\times 10^{-4}$)
T0	6.0 ± 0.2	5.1ᵃ	8ᵃ	6.3ᵃ	3.0ᵃ	4.1ᵃ	13.0ᵃ
T1	3.5	7.3ᵃ	6ᵃᵇ	7.1ᵃ	2.9ᵃ	4.0ᵃ	11.5ᵃ
T2	4.0	5.8ᵃ	8ᵃ	7.6ᵃ	3.9ᵃ	4.6ᵃ	16.5ᵃ
T3	4.5	7.7ᵃ	6ᵃᵇ	7.6ᵃ	2.9ᵃ	4.0ᵃ	9.0ᵃ
T4	5.0	7.4ᵃ	5ᵃᵇ	7.5ᵃ	2.9ᵃ	4.3ᵃ	10.5ᵃ
T5	5.5	6.1ᵃ	3ᵇ	7.3ᵃ	2.9ᵃ	4.0ᵃ	11.0ᵃ

Trt.	SAR pH level	Nutrient accumulation (mg plant^{-1})					
		N	P ($\times 10^{-1}$)	K	Ca	Na	Zn ($\times 10^{-2}$)
T0	6.0 ± 0.2	18ᵃ	4ᵃ	31ᵃ	97ᵃ	18ᵃ	8.0ᵃ
T1	3.5	17ᵃ	2ᵃᵇ	18ᵇ	74ᵃ	10ᵃ	3.7ᵃᵇ
T2	4.0	13ᵃ	2ᵃᵇ	22ᵃᵇ	87ᵃ	10ᵃ	3.7ᵃᵇ
T3	4.5	12ᵃ	2ᵃᵇ	19ᵃᵇ	72ᵃ	10ᵃ	2.0ᵇ
T4	5.0	21ᵃ	2ᵃᵇ	26ᵃᵇ	101ᵃ	15ᵃ	5.6ᵃᵇ
T5	5.5	23ᵃ	1ᵇ	30ᵃ	120ᵃ	17ᵃ	3.3ᵃᵇ

Note: Mean within same column with different letters was significantly different at p ≤ 0.05 (Tukey's test).

(12.61 mg kg^{-1}). The soils treated with T0 (pH 6.0 ± 0.02) were lower in terms of exchangeable acidity, Al, and H. Treatments 2 and T3 (pH 4.0 and pH 4.5) produced higher in exchangeable Al about 1.11 and 1.40 cmol kg^{-1}, respectively.

The highest H concentration in soils was in T1 (pH 3.5) with 1.29 cmol kg^{-1} whilst other treatments did not show any statistical difference. This treatment also produced the similar results in available SO_4^{2}. With the exception of Mg, no significant differences were recorded for K, Ca and Na (Figures 2 and 3). The lowest concentration of exchangeable Mg was produced in T3 (pH 4.5) (0.26 cmolkg^{-1}) and the highest in T5 (pH 5.0) (1.18 cmol kg^{-1}).

No significant differences among treatments were recorded for exchangeable Fe (Figure 4). Treatment 5 (pH 5.5) produced the lowest concentration for exchangeable Cu and Zn about 0.0018 and 0.0028 cmol kg^{-1} while the highest was in T2 (0.005 cmol kg^{-1}) and T1

Table 9. Effect of SAR on selected soil chemical properties in *S. macroptera*.

Soil chemical properties		Treatment					
		T0	T1	T2	T3	T4	T5
pH in water		4.43[a]	3.93[a]	4.17[a]	4.13[a]	4.10[a]	4.21[a]
pH in KCl		3.40[a]	3.23[a]	3.27[a]	3.45[a]	3.48[a]	3.60[a]
Exchangeable NH_4^+	(mg kg^{-1})	37.36[a]	28.02[ab]	11.68[c]	21.02[bc]	25.69[abc]	37.36[a]
Exchangeable NO_3^-	(mg kg^{-1})	12.61[b]	21.02[ab]	22.42[a]	16.35[ab]	18.68[ab]	18.68[ab]
Available P	(mg kg^{-1})	11.86[a]	8.87[a]	10.97[a]	12.66[a]	10.60[a]	9.95[a]
Exchangeable acidity	(cmol kg^{-1})	1.01[b]	2.21[ab]	2.18[ab]	2.53[a]	0.98[b]	1.22[ab]
Exchangeable Al	(cmol kg^{-1})	0.14[c]	0.92[ab]	1.11[a]	1.40[a]	0.35[bc]	0.38[bc]
Exchangeable H	(cmol kg^{-1})	0.87[ab]	1.29[a]	1.07[ab]	1.13[ab]	0.63[b]	0.84[ab]
Available SO_4^{2-}	(mg kg^{-1})	36.63[b]	90.47[a]	67.50[ab]	37.87[b]	66.27[ab]	41.70[b]

Note: Mean within same rows with different letters was significantly different at p ≤ 0.05 (Tukey's test).

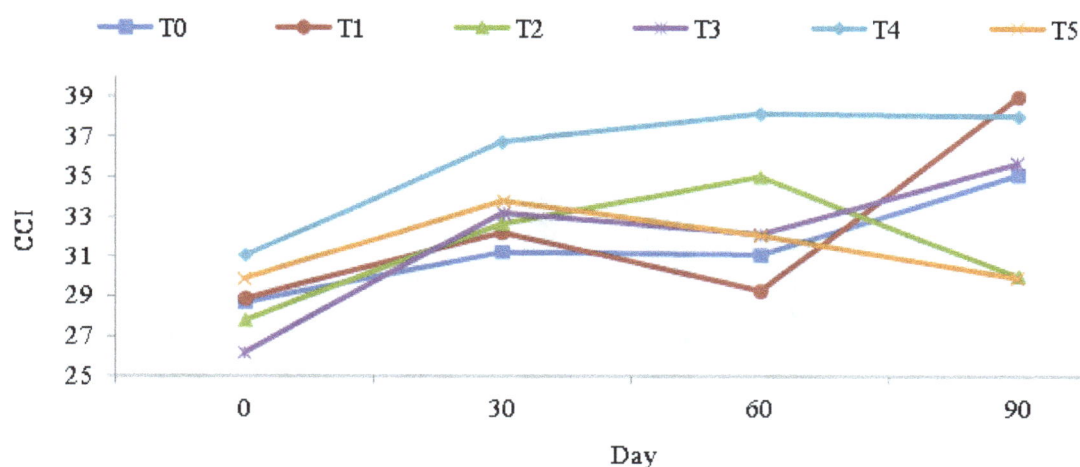

Figure 1. Effect of SAR on chlorophyll content in*Shoreamacroptera*. Note: No significant differences among T0, T1, T2, T3, T4, and T5 regardless of the time using Tukey's test at p ≥ 0.05.

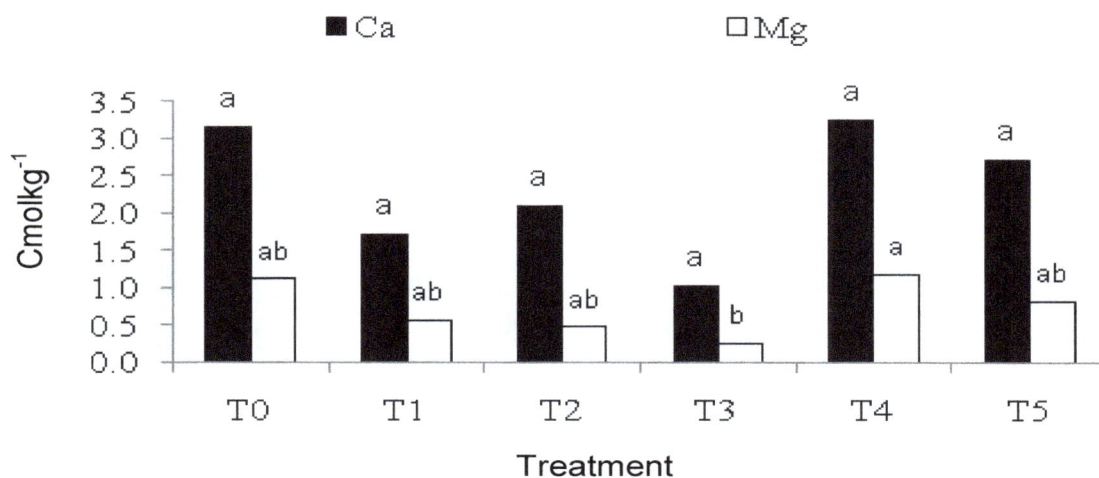

Figure 2. Effect of SAR on soil exchangeable Ca and Mg concentration. Note: Mean with different letters was significantly different at p ≤ 0.05 (Tukey's test).

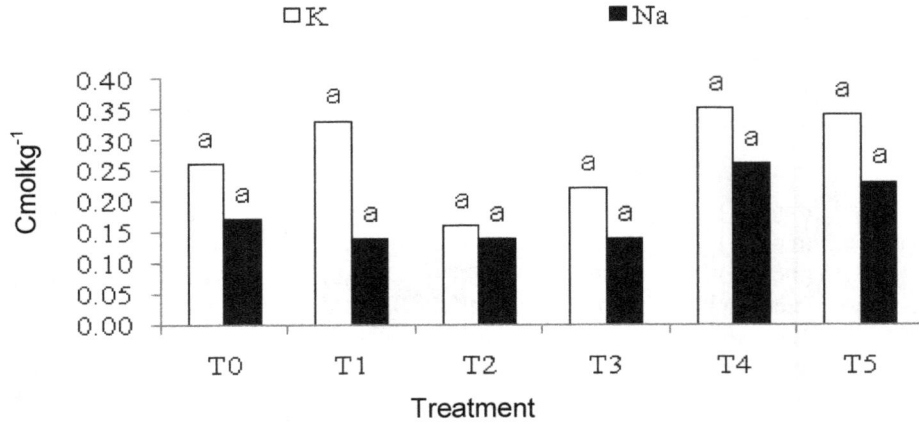

Figure 3. Effect of SAR on soil exchangeable Cu concentration.

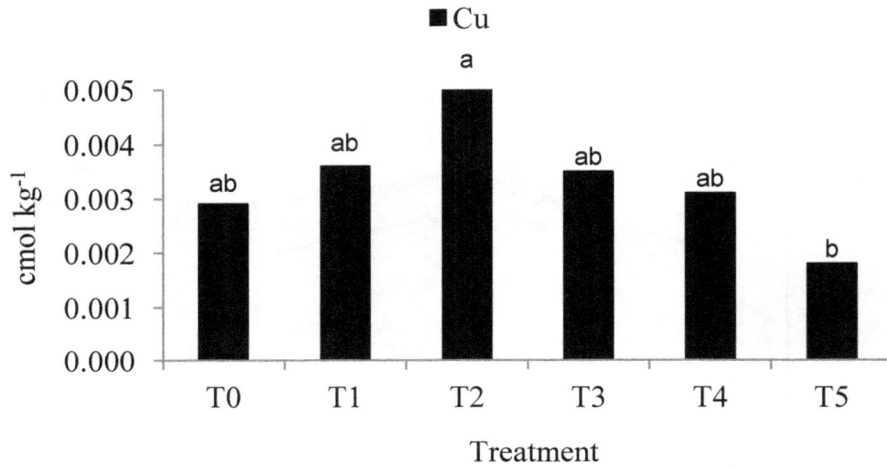

Figure 4. Effect of SAR on soil exchangeable Cu concentration. Note: Mean with different letters was significantly different at $p \leq 0.05$ (Tukey's test).

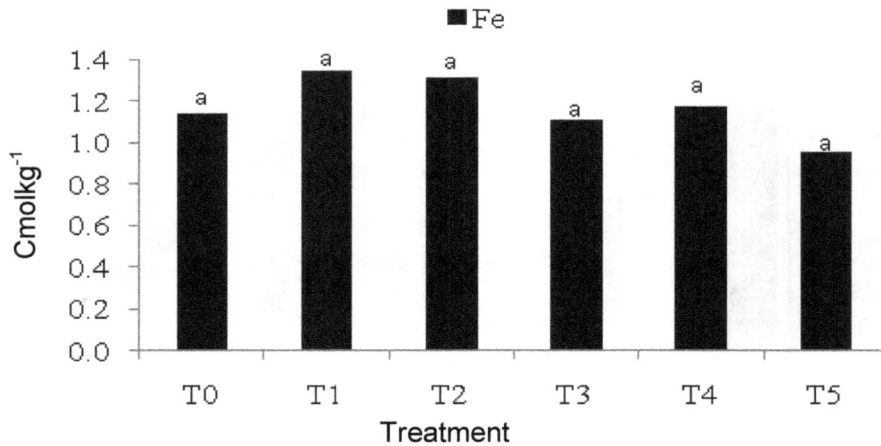

Figure 5. Effect of SAR on soil exchangeable Fe concentration. Note: Mean with different letters was significantly different at $p \leq 0.05$ (Tukey's test).

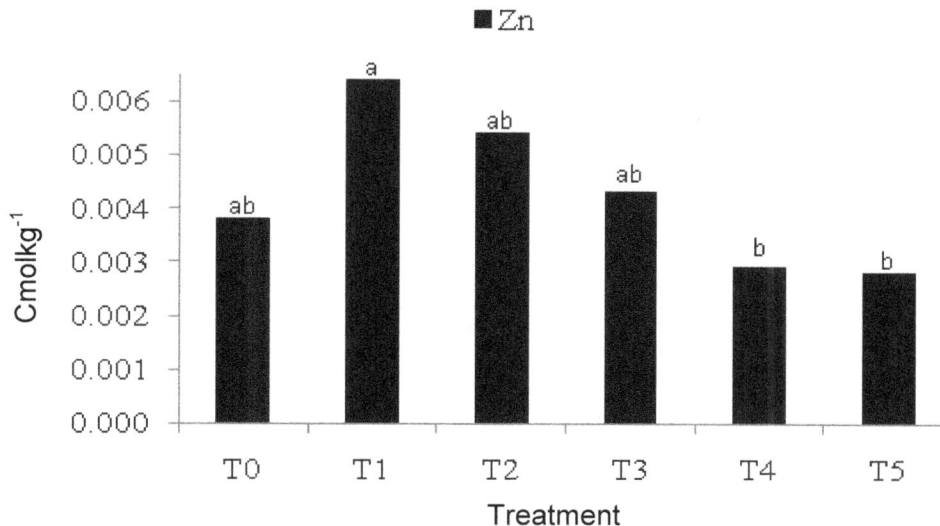

Figure 6. Effect of SAR on soil exchangeable Zn concentration. Note: Mean with different letters was significantly different at $p \leq 0.05$ (Tukey's test).

(0.006 cmol kg^{-1}), respectively (Figures 5 and 6).

DISCUSSION

In this study, increment of plant height of *S. macroptera* after 90 days exposed to SAR at T1 (pH 3.5) was 3.83 cm from initial day at 47.00 cm. It was 2.41 cm lower than T0 (pH 6.0 ± 0.2) at 6.24 cm or 38.61%. The value obviously showed role of different acidity level in reduction of plant height. A similar pattern on plant height reduction in forest and crop plant has been reported by Balasubramanian et al. (2007). This could probably be due to the cell division function which was retarded by lack of macronutrient in soil.

Exposure of *S. macroptera* to SAR had a definite impact on decreasing number of leaves. The decrease was mostly in T2 (pH 4.0) (2.44) as compared to T0 (pH 6.0 ± 0.2) (4.67). Perhaps, this effect comes from stress mechanism of leaves when exposed to different acidity. According to Sonia and Khan (1996), leaf growth is affected by simulated acid rain by reducing transpiring area with essential nutrient uptake.

Biomass was substantially reduced by the acidity of SAR. The result of the present studies was same with the findings of Malek (1995), in *Larix decidua* Mill. *S. macroptera* seedlings recorded reduced in leaf, stem and root weight with SAR of T1 (pH 3.5) compared with pH T0 (pH 6.0 ± 0.2). Another study by Sonia and Khan (1996) showed that acid rain stress caused significant reduction in stem weight by slowing cell division and expansion. They also report increased rainwater acidity and decreased redistribution of photosynthesis, which affects root elongation.

The effects of SAR pH acidity on chlorosis and necrosis are related to leaves of *S. macroptera*. The symptoms include the plant having deficient nitrogen, phosphorus, potassium and magnesium. This is true, from the result on plant accumulation in Table 6.

Increasing SAR acidity does not affect soil pH. It probably buffers soil capacity and series of soil type that was used. Nyalau soil series is considered as acidic soil (Paramananthan, 2000); whenever acidic solution fills the soil, it has a tendency to control H$^+$ ions. Fertility on soil mainly depends on the macronutrient available in soils. Exchanging NH$_4^+$ for the lower SAR treatments presents lower value compared to control. As stated by Walna et al. (2000), enzymatic activities in highest pH level are faster in influencing the growth of soil microorganism and mineralization activities.

Exchangeable H$^+$ recorded highest in lower pH than control. It showed more free hydrogen was deposited in soil surface. The H$^+$ ions come from the mixture of acid for preparing the SAR. This is also indicator for higher presence of Al and heavy metal on soil. The role of hydrogen ions is to hydrolyze water molecules to release macronutrient and the same time triggers the heavy metal to combine with Al, which is called oxylation formation process. This result will give toxicity to the soil directly (Cronan and Grigal, 1995).

Conclusion

S. macroptera is sensitive to simulated acid rain (SAR) and has the potential to be sensitive plant for acid deposition especially in Malaysia. However, soil chemical characteristics were slightly affected by SAR due to little changes in its characteristics. In order to validate this data, long term experiment is suggested.

ACKNOWLEDGEMENTS

The authors would like to thank the Ministry of Higher Education, Malaysia and Universiti Putra Malaysia, for the financial (RUGS: 9199765 and FRSG: 5523701) and technical support during the conduct of the research and preparation of this paper. We would also like to thank Asia Center for Air Pollution research, Japan (ACAP) for their technical support during the conduct of this research.

REFERENCES

Ang LH, Maruyama Y (1995). Survival andearly growth of Shoreaplatyclandos, Shoreamacroptera, Shoreaassamica, and Hopea nervosa in open planting. J. Trop. Forest Sci. 7(4):541-557.

Ayers GP, Leong CP, Gillett RW, Lim SF (2002). Rainwater composition and acidity at five sites in Malaysia, in 1996. Water Air Soil Poll. 133:15-30.

Balasubramanian G, Udayasoorian C, Prabu PC (2007). Effects of short term exposed of simulated acid rain on the growth of Acacia nilatica. J. Trop. Forest Sci. 19(4):198-206.

Bremner JM (1965). Total Nitrogen.In D.D. Black, L.E. Evans and Ensmingeret, C.A. Method of soil analysis. Madison, WI. Am. Soc. Agronomy. 2:1149-1178.

Cottenie A (1980). Soil testing and plant testing as a basis of fertilizer recommendation. FOA Soils Bull. 38:70-73.

Cronan CS, Grigal DF (1995). Use of calcium and aluminum ratios as indicators of stress in forest ecosystems. J. Environ. Qual. 24:209-226.

Ernst WHO (2012). The use of higher plants as bioindicators, In: B.A. Markert, A.M. Breure and H.G. Zechmeister, eds, Trace Metals and other Contaminants in the Environment. Elsevier. pp. 423-463.

Heij GJ, de Vries W, Posthumus AC, Mohren GMJ (1991). Effects of air pollution and acid deposition on forests and forest soils, In: G.J. Heij and T. Schneider, eds, Studies in Environmental Science, Elsevier. pp. 97-137.

John TVS, Delphis FL, Shreeram PI, Michelle LB, Myron JM (2012). The effects of phenoseason and storm characteristics on throughfall solute washoff and leaching dynamics from a temperate deciduous forest canopy. Sci. Total Environ. 430:48-58.

Keeney DR, Nelson DW (1982). Nitrogen-Inorganic Forms.In D.R. Keeney and D.E. Bakeret. Methods of Soil Analysis Part 2. Madison, WI: Argon Monogr. ASA and SSAA. pp. 19-33.

Liu J, Peng S, Faivre-Vuillin B, Xu Z, Zhang D, Zhou G (2008). Erigeron annuus (L.) pers., as a green manure for ameliorating soil exposed to acid rain in southern China. J. Soils Sediments. 8(6):452-460.

Liu X, Lei D, Mo J, Du E, Shen J, Lu X, Zhang Y, Zhou X, He C, Zhang F (2011). Nitrogen deposition and its ecological impact in China: An overview. Environ. Pollut. 159(10):2251-2264.

Malek S (1995). The effect of acid rain and mineral fertilization on the biometrical features of the Larix decidua Mill. seedling. Water Air Soil Pollut. 88:93-107.

Manokaran N, Kochummen KM (1993). Tree growth in primary lowland and hill dipterocarps forests. J. Trop. For Sci. 6(3):323-345.

Mehlich A (1953). Determination of P, K, Ca, Mg, and NH_4.USA: Soil test Division Mimeo, North Carolina Department of Agriculture.

Murphy J, Riley J (1962). A modified single solution for the determination of phosphate in natural waters. Anal. Chim. Acta. 27:31-36.

Paramananthan S (2000). Soils in Malaysia: Their Characteristics and Identification vol. 1. Kuala Lumpur: Akademi Sains Malaysia.

Pomares-Gracia F, Pratt PF (1987). Recovery of 15H-labelled fertilizer from manured and sludge-amended soils. Soil Sci. Soc. Am. J. 42:717-720.

Malaysia Meteorology Department (MMD) (2008). Report on the Status of Acid Deposition in Malaysia, Kuala Lumpur, Malaysia.

Rowell DL (1994). Soil science: Method and Applications. New Jersey: Pearson Education, Inc.

SAS (2001).SAS/STAT software. Cary. NY: SAS Institute.

Sonia R, Khan M (1996). Effect of simulated acid rain on Cicerarietinum var. Pant G-14. J. Environ. Pollut. 3(3):197-201.

Symington CF (2004). Ecological distribution of the Dipterocarps in the Malay peninsula. Forest Manual Dipterocarpaceae. Ashton P.S. and Appanah, eds, Forest Research Institute Malaysia and Malaysia Nature Society. Kuala Lumpur. pp. 14-32.

Tan KH (2005). Soil Sampling, Preparation, and Analysis 2nd. USA: CRC Press.

Walna B, Mala SD, Siepak J (2000). The impact of acid rain on potassium and sodium status in typical soils of the WielkoPolski national park (Poland). Water Air Soil Pollut. 121:31-41.

Wang ZH, Liu XJ, Ju XT, Zhang FS, Malhi SS (2004). Ammonia volatilization loss from surface-broadcast urea: Comparison of vented- and closed-chamber methods and loss in winter wheat-summer maize rotation in north China plain. Commun. Soil Sci. Plant Anal. 35(19):2917-2939.

20

Improving and sustaining soil fertility by use of enriched farmyard manure and inorganic fertilizers for hybrid maize (BH-140) production at West Hararghe zone, Oromia, Eastern Ethiopia

Zelalem Bekeko

Department of Plant Sciences, Haramaya University Chiro Campus, P. O. Box 335, Chiro, Ethiopia.

A study was conducted at the Haramaya University Chiro Campus to determine the effect of enriched FYM and inorganic fertilizers on grain yield of maize and soil chemical properties. FYM was used either alone or in combination with inorganic fertilizers as follows: Control (zero fertilizers and FYM), 10 tons/ha FYM, 8 tons/ha FYM and 25 kg/ha of N + 20 kg/ha P, 6 tons/ha FYM, 50 kg/haN + 40 kg/ha P, 4 tons/ha FYM, 75 kg/ha N + 60 kg/ha P, 2 tons/ha FYM, 100 kg/ha N+80 kg/ha P, 100 kg/ha N + 100 kg/ha P. The treatments were arranged in randomized complete block design with four replications from 2008 to 2011. Result showed that 59.60 Cmole/kg, 4.58%, 0.82%, 62.7 ppm, cation exchange capacity (CEC), %OC, total N, and available P, respectively were noted after FYM application over years indicating improved soil chemical properties. Similarly, combined analysis of variance on hybrid maize (BH-140) yield over years showed no significant difference among treatments in comparison with 10 tons/ha FYM and 100 kg/ha N + 100 kg P/ha (p < 0.05). But 4 tons/ha FYM and 75 kg/ha N+60 kg/ha P increased maize yield from 5.1 tons/ha in 2009 to 8.15 tons/ha in 2010. From this finding, it was noted that enriching FYM with inorganic fertilizers can boost hybrid maize grain yield significantly through improving the physicochemical properties of the soil. On the basis of these results, it can be concluded that enriched FYM should be used for hybrid maize production at Western Hararghe in order to get maximum grain yield and sustain soil productivity. Thus, it is recommended that application of 4 tons/ha FYM incorporated with 75 kg of N and 60 kg of P at Chiro can significantly increase hybrid maize (BH-140) yield and sustain its productivity over years. Besides, it also reduces the cost of inorganic fertilizers which is becoming a bottle neck to smallholder farmers of Eastern Ethiopia. However, profitability of this technology needs to be tested at different locations and in different seasons.

Key words: Soil fertility, farm yard manure, Hybrid maize (BH-140), cation exchange capacity (CEC) inorganic fertilizers

INTRODUCTION

Maize is one of the most important cereals broadly adapted worldwide (Christian et al., 2012). In Ethiopia, it is grown in the lowlands, the mid-altitudes and the highland regions. It is an important field crop in terms of area coverage, production and utilization for food and feed purposes. However, maize varieties mostly grown in the highlands (1,700 to 2,400 m.a.s.l) of Ethiopia are local cultivars with poor agronomic practices (Beyene et al., 2005; Soboksa et al., 2008). They are low yielding, vulnerable to biotic and a biotic constraint and also exhibit undesirable agronomic performances such as late maturity and susceptibility to root rot and stalk lodging (Legesse et al., 2007; Soboksa et al., 2008). Enhancement of maize production and productivity can be achieved through identification of potentially superior inbred line combinations in the form of hybrids along with proper supplementation of plant nutrition (Worku et al., 2001; Betran et al., 2003; Wonde et al., 2007; Shah et al., 2009; Achieng et al., 2010).

One of the major problems affecting food production in Africa including Ethiopia is the rapid depletion of nutrients in smallholder farms (Badiane and Delgado, 1995; Achieng et al., 2010). Soil nutrient replenishment is therefore a prerequisite for halting soil fertility decline. This may be accomplished through the application of mineral and organic fertilizers (Wakene et al., 2005). Animal manures are valuable sources of nutrients and the yield-increasing effect of manure is well established (Leonard, 1986; Wakene et al., 2005; Silvia et al., 2006). Organic matter in the soil improves soil physical conditions by improving soil structure, increases water-holding capacity, and improves soil structure and aeration, as well as regulating the soil temperature (Gachene and Gathiru, 2003; Wakene et al., 2005). Organic matter contains small varying amounts of plants nutrients, especially nitrogen, phosphorus and potassium which are slowly released into the soil for plant uptake (Gachene and Gathiru, 2003; Achieng et al., 2010).

Chemical fertilizers are used in modern agriculture to correct known plant nutrient deficiencies, to provide high levels of nutrition, which aid plants in withstanding stress conditions, to maintain optimum soil fertility conditions, and to improve crop quality. Adequate fertilization programs supply the amounts of plant nutrients needed to sustain maximum net returns (Leonard, 1986). The broad aim of integrated nutrient management is to utilize available organic and inorganic sources of nutrients in a judicious and efficient manner. Based on the evaluation of soil quality indicators, Dutta et al. (2003) reported that the use of organic fertilizers together with chemical fertilizers, compared to the addition of organic fertilizers alone, had a higher positive effect on microbial biomass and hence soil health. Sutanto et al. (1993) in their studies on acid soils for sustainable food crop production noted that farmyard manure (FYM) and mineral fertilizer produced excellent responses. Boateng and Oppong (1995) studied the effect of FYM and method of land clearing on soil properties on maize yield and reported that plots treated with poultry manure and NPK (20-20-0) gave the best yield results.

Soil fertility depletion on smallholder farms is one of the fundamental biophysical root causes responsible for declining food production in eastern part of Ethiopia (Heluf et al., 1999). Especially, in the highlands of Hararghe, where maize is grown among the major cereals in the high rainfall areas such as Chiro, Doba, Tullo, Mesela, Gemechis, Kuni, Boke Habro and Daro Labu, its productivity is severely constrained by poor soil fertility and poor crop management practices. Yield is too small usually less than 2 tons/ha as compared to a potential yield of over 5 tons/ha in the region (Zelalem, 2012). Particularly, nutrient deficiency is one of the major constraints to maize production and productivity in these areas. Intercropping is widely used in these areas by combining maize or sorghum with perennial crops like Khat (Chata edulis) which exposes the soil to rampant nutrient degradation leading to poor crop yield (Heluf et al., 1999; Fininsa, 2001; Ararsa, 2012).

Application of FYM and inorganic fertilizers N and P significantly increases grain yield of hybrid maize cultivars and improves some soil chemical and physical properties such as available P, cation exchange capacity (CEC), total nitrogen, the texture, structure and water holding capacity of the soil (Debelle et al., 2001; Wakene et al., 2005; Shah et al., 2009; Tesdale et al., 1993; Heluf et al., 1999; Asfaw et al., 1998; Achieng et al., 2010).

The recycling and the use of nutrients from organic manure have been given more consideration for insuring sustainable land use in agricultural production development (Ararsa, 2012). The positive influence of organic fertilizers on soil fertility, crop yield and quality has been demonstrated in the works of many researchers (Hoffman, 2001). Organic materials are a good source of plant nutrients and have a positive effect on improvement of the soil physical structure (Silvia et al., 2006; Zelalem, 2012). Application of animal manures to agricultural fields is a widely used method of increasing soil organic matter and fertility (Debelle et al., 2001; Wakene et al., 2005; Heluf et al., 1999; Khaliq et al., 2009). Most solid livestock manures can be applied directly to crop fields or piled for composting. In organic farming, nitrogen is supplied through organic amendments in the form of manure. Applying organic nitrogen fertilizer without prior knowledge of nitrogen mineralization and crop needs can result in nitrate-nitrogen (NO_3-N) leaching below the root and potential groundwater contamination(Debelle et al ., 2001).

The incorporated use of organic sources of nutrients not only supply essential nutrients but also has some positive interaction with chemical fertilizers to increase their efficiency and thereby to improve the soil structure (Elfstrand et al., 2007). Integrated use of chemical fertilizers and organic material may be a good approach for sustainable production of crops. Integrated use of organic matter and chemical fertilizers is beneficial in improving crop yield, soil pH, organic carbon and available N, P and K in sandy loam soil (Rautaray et al., 2003).

Khaliq et al. (2009) used partially decomposed cattle and chicken manure amended with wood ash and reported that higher plant yield of fodder maize was obtained by the use of chicken manure. In western part of Ethiopia, Wakene et al. (2005) and Debelle et al. (2001)reported the benefit of FYM in maize production and soil maintainace. But no investigation was made at eastern part of Ethiopia where the soil fertility is highly depleted using FYM and inorganic fertilizers for yield sustenance and soil health.

Hybrid maize variety (BH-140) was deployed to the farmers of Western Hararghe zone, Eastern Ethiopia during the last 5 years through the Oromia Bureau of Agriculture to boost the production and productivity of maize in the zone. However, the response of this hybrid maize (BH-140) cultivar to FYM and inorganic fertilizers and effect of these fertilizers on some soil properties at Western Hararghe zone is not studied. Therefore, the present investigation was carried out to evaluate the effects of FYM and inorganic fertilizers on grain yield of hybrid maize (BH-140) as well as soil chemical properties at Western Hararghe zone, Oromia Regional State, Ethiopia.

MATERIALS AND METHODS

Description of the study area

Western Hararghe is located between 7°55' N to 9°33' N latitude and 40°10' E to 41°39' E longitude. The major crops grown in the study area are sorghum, maize, chat, field beans, potato and tef. The area is characterized by Charcher Highlands having undulating slopes and mountainous in topography. The mean annual rainfall ranges from 850 to 1200 mm/year with minimum and maximum temperatures of 12 and 27°C, respectively.

Treatment details

The response of hybrid maize variety (BH-140) was used as test crop, to N and P fertilizers. FYM was used either alone or in combination with inorganic fertilizers as follows: control (zero fertilizers and FYM), 10 tons/ha FYM, 8 tons/ha FYM and 25 kg/ha of N + 20 kg/ha P, 6 tons/ha FYM, 50 kg/haN + 40 kg/ha P, 4 tons/ha FYM, 75 kg/ha N + 60 kg/ha P, 2 tons/ha FYM, 100 kg/ha N + 80 kg/ha P, 100 kg/ha N + 100 kg/ha P. The treatments were arranged in randomized complete block design with four replications at the Haramaya University Chiro Campus from 2008 to 2011 cropping seasons.

Experimental procedures

The experimental field was prepared by using local plough according to farmers' conventional farming practices. The field was ploughed four times each year during the experimental seasons. A plot size of 4 m length by 4.5 m width with six rows per plot was used. Spacing was 0.75 and 0.25 m between rows and plants, respectively. Planting was done in May 2008, 2009, 2010 and 2011 at a rate of 25 kg/ha. Enriched FYM was prepared by adding 10 kg of urea by pit method in 12 m^3 pit from cattle manure subjected to microbial fermentation for 90 days (Debelle et al., 2001; Achieng et al., 2010).

Urea (46-0-0) and triple superphosphate (TSP) (0-46-0) were used as sources of N and P, respectively. All P fertilizer and half

dose of N fertilizer as per treatment were applied as basal application at planting and the remainder N was top-dressed at 35 days after planting and FYM was applied each year 1 month before the sowing date. Seeds of hybrid maize (BH-140) were sown on 10th of May 2008, at 20th of May 2009 and 15th of May 2010 at the rate of 25 kg/ha. Sowing was completed on the same day. Then after, all necessary cultural practices were employed to raise a successful crop.

An area of 5.65 m^2, corresponding to 32 plants in the central four rows, was harvested immediately after physiological maturity for grain yield. During harvests, border plants at the ends of each row were excluded to avoid border effects. Grain moisture percent (MOI%) was estimated using a Dickey-John multi grain moisture tester. Grain yield (GY t ha^{-1}) was calculated using shelled grain and adjusted to 12.5% moisture (Mosisa et al., 2007).

Forty-six (46) surface soil samples (0 to 30 cm depth) were collected from representative spots of the entire experimental field after final plough and composited to two replicate samples for each analysis. These were analyzed for soil texture, pH, CEC, organic carbon, available P and total N. Similarly, surface soil samples at the same depth were collected at blooming stage (75 days after planting at the end of the final experimental seasons, 2011). One representative soil sample was taken from every plot, using auger to make composite sample per treatment for the analysis of total N and available P.

Soil texture was expressed by using Bouyoucos hydrometer method (Tesdale et al., 1993). Available P was extracted with a sodium bicarbonate solution at pH 8.5 following the procedure described by Olsen et al. (1954). The pH of the soil was measured potentiometrically in the supernatant suspension of a 1:2.5 soil: water mixture by using a pH meter, and organic carbon was determined by following Walkley and Black (1934) wet oxidation method as described by Jackson (1958). CEC was measured by using 1 M-neutral ammonium acetate. Total nitrogen was determined by using Kjedahl method as described by Jackson (1958).

In order to record the soil profile characteristics at the experimental site, a 2 m by 1.5 and 1.60 m deep pit was excavated adjacent to the experimental field and soil profile was described *in situ*. Soil samples were taken from all the identified horizons and pH, texture, organic matter content, total N, available P and CEC were analyzed using the same procedures. Bulk density, particle density and pore spaces were also determined.

Statistical analysis

The data recorded in this study were subjected to statistical analysis. Analyses of variance were carried out using MSTATC soft ware. Significant differences between and/or among treatments were delineated by least significant differences (LSD). Interpretations were made following the procedure described byGomez and Gomez (1984)

RESULTS

Grain yield

Combined analysis of variance on grain yield of hybrid maize (BH-140) over years showed no significant difference between Treatments 2 and 7 (10 t/ha FYM and 100 kg/ha N +100 kg/ha P) (Tables 2 and 3) and also the result indicated that all proportions of FYM and inorganic fertilizer treatments significantly increased maize grain yield as compared to the control treatment (Tables 1 and

Table 1. Effect of enriched FYM on grain yield (mean values) of hybrid maize (BH-140) at Chiro, Western Hararghe from 2008 to 2011.

Treatment	Mean grain yield of maize (kg/ha)					
	Rep_1	Rep_2	Rep_3	Rep_4	Total	Mean
Control (0 FYM and 0 N and P)	1563	1784	1586	1657	6590	1647.5
10t/ha FYM+0 N and P	6579	6934	6601	6496	26610	6652.5
8 t/ha FYM and 25 kg/ha N + 20 kg/ha P	5546	5955	6266	5661	23428	5857
6 t/ha FYM and 50 kg/ha N + 40 kg/ha P	5497	5353	4978	4854	20682	5170.5
4 t/ha FYM and 75 kg/ha N + 60 kg/ha P	7601	8155	8042	8836	32634	8158.5
2 t/ha FYM and 100 kg/ha N + 80 kg/ha P	7269	6837	6228	6340	26674	6668.5
100 kg/ha N + 100 kg/ha P	6568	6821	7343	7256	27988	6997
Total	4063	41839	41044	41100	164606	6858.58

Table 2. Analysis of variance (mean value) for the effect of enriched FYM and inorganic fertilizers on grain yield of hybrid maize (BH-140) at West Hararghe zone, Oromia, Eastern Ethiopia (2008 to 2011).

Sources of variation	DF	SS	MS	F_{cal}	F_{tab}
Treatment	6	104302819.7	17383803.28		
Replication	3	109453.84			
Error	18	2497295.16	138738.62	125.29*	3.26
Total	27	106909568.7			
CV (%)	5.43				
LSD = 0.05	571.53				

*Significant at 5%.

Table 3. Mean separation (Duncan's multiple range test) for the effect of enriched FYM and inorganic fertilizers on grain yield of hybrid maize (BH-140) at Western Hararghe zone, Oromia, Eastern Ethiopia (2008 to 2011).

	1647.5	6652.5	5857	5170.5	8158.5	6668.5	6997
1647.5	0	-5005*	-4209.5*	-3523*	-6511*	-5021*	-5349.5*
6652.5		0	795.5^{ns}	1482^{ns}	-1506*	-16*	-344.5*
5857			0	686.5^{ns}	-2301.5*	-811.5*	-1140*
5170.5				0	-2988*	-1498*	-1826.5*
8158.5					0	1490^{ns}	1161.5^{ns}
6668.5						0	-328.5*
6997							0

*Significant at 5%; ^{ns}, non significant at 5%.

2) and the highest grain yield (8158 kg/ha) was obtained in the Treatment 4 (4 ton/ha FYM + 75 kg/ha N and 60 kg/ha P) and the lowest grain yield (1647.5 kg/ha) was obtained in the control plots (Table 1).

The analysis of variance also elucidated no significant difference among Treatments 2, 6 and 7 (10 t/ha FYM + 0 N and P, 2 t/ha FYM and 100 kg/ha N + 80 kg/ha P and 100 kg/ha N + 100 kg/ha P), respectively (Table 3) at (p < 0.05) on grain yield of hybrid maize (BH-140). But 4 ton/ha FYM and 75 kg/ha N + 60 kg/ha P increased maize yield from 5.1 t/ha in 2009 to 8.15 t/ha in 2010 (Table 1).

Effect on chemical properties of soil

Result showed that 59.60 Cmole/kg, 4.58%, 0.82%, 62.7 ppm, CEC, %OC, total N, and available P, respectively over years indicating improved soil chemical properties after application of FYM (Table 5). From the analysis over 4 years, total N, available P and CEC increased by 2-fold as compared to their respective value before application of FYM and inorganic fertilizers in 2008 cropping season (Table 4). The percentage in OC did not significantly increased during the experimental seasons but the value of the soil pH showed slight decline (Tables 4 and 5).

Table 4. Chemical and physical properties of soil at Chiro, Western Hararghe before FYM and inorganic fertilizers application at 2008 cropping season

Depth (cm)	Horizon	Particle size distribution (%)			Textural class	PD G (cm³)	BD G (cm³)	PS (%)	pH	OC (%)	Total N (%)	Av.P (ppm)	CEC Cmol/kg
		Clay	Sand	Silt									
0 - 30	AP	50	38	12	Clay	2.38	0.99	57.9	8.01	4.04	0.35	37.9	39.71
30 - 90	Bt1	18	56	26	Sand	2.50	1.38	46.8	8.45	0.48	0.56	34.8	-
90 - 150	Bt2	20	54	26	Sand	2.50	1.34	46.4	8.78	0.45	0.34	33.8	-

Table 5. Effect of enriched FYM and inorganic fertilizers on Chemical properties of soil at Chiro, western Hararghe (2008 to 2011)

Depth (cm)	Horizon	Particle size distribution (%)			Textural class	PD g (cm³)	BD g(cm³)	PS (%)	pH	OC (%)	Total N(%)	Av.P (ppm)	CEC Cmol/kg
		Clay	Sand	Silt									
0 - 30	AP	50	38	12	Clay	2.38	0.99	57.9	7.31	4.58	0.82	62.3	59.6
30 - 90	Bt1	18	56	26	Sand	2.50	1.38	46.8	7.89	0.58	0.76	36.7	-
90 - 150	Bt2	20	54	26	Sand	2.50	1.34	46.4	8.20	0.40	0.47	34.0	
BAP										4.01	0.44	37.9	39.71
AP										4.58*	0.82*	62.7*	59.6*

*Significant at 5%

DISCUSSION

The long-term effects of the combined application of organic and inorganic fertilizers in improving soil fertility and crop yield have been demonstrated by many workers (Chen et al., 1988). Wang et al. (2001) reported that organic and inorganic fertilizers showed great benefits not only for the increase in the N uptake by the plant but also in the improvement of the fodder yield on maize.

Intensive cultivation of high yielding hybrid maize varieties requires application of plant nutrients in large quantities. Supplying these nutrients from chemical fertilizers has got certain limitations and inherent problems. Further, these chemical fertilizers can supply only a few plant nutrients like nitrogen, phosphorus and potash and also they are becoming very expensive for resource poor farmers. Silvia et al. (2006) reported that non-inclusion of organic manures such as FYM, compost, green manures, etc. in the manurial schedule have resulted in the depletion of fertility status of the arable soils and their consequent degradation. Debelle et al. (2001) also reported organic manures, especially FYM, have a significant role for maintaining and improving the chemical, physical and biological properties of soils and in sustaining maize yield in western part of Ethiopia. They also reported that 10 ton/ha of FYM is statistically at par with current agronomic recommendation of inorganic fertilizers N and P for maize. In the present finding, it is also observed that 10 tons/ha of FYM and 100 kg/ha N + 100 kg/ha P showed no significant difference on yield (Tables 1 and 3).

Wakene et al. (2005) indicated that the urgency of using organic manure has been gaining ground

in the wake of increasing cost of fertilizer with every passing year and certain other inherent limitations with the use of chemical fertilizers. FYM is the oldest organic manure used by man ever since he involved in farming. It has stood the test of time and is still very popular among the poor and marginal farmers. It consists of litter, waste products of crops mixed with animal dung and urine. Therefore, it contains all the nutrient elements present in the plant itself and returns these nutrients to the soil when it is applied to the field for the benefit of succeeding crop.

This study also confirms the role of FYM and chemical fertilizer combinations in increasing grain yield of maize and the results showed that manure and chemical fertilizer can increase grain yield of maize but a combination of them has more effect on increase in grain yield (Tables 1 and 2). In a recent evaluation of the direct effects of cattle manure on corn, it was verified by Silvia et al. (2006) that manure increased green ear yield and grain yield in two corn cultivars in Brazil. Dutta et al. (2003) also reported the benefit of organic matter application on soil in enhancing the soil microbial population and soil health.

Cattle manure also increased water retention and availability, phosphorus, potassium and sodium contents in the soil layer from 0 to 20 cm (Silva et al., 2006). Therefore, the present finding is in agreement with their report in which the total N, available P and CEC of the soil increased in the soil layer from 0 to 30 cm depth (Table 5). In the present finding also, the pH of the soil has declined after application of FYM from 8.01 (Table 4) in 2008 to 7.31 (Table 5) in 2010 which is in agreement with the reports of Khaliq et al. (2004). Similarly, Wakene et al. (2005) also reported addition of organic matter especially FYM into tropical soils enhances the development of soil acidity from the release of organic acids into the soils over years. Thus, the present finding is also inline with their report. The residual effect of organic fertilizers on yield, including FYM manure, has been found to be positive in sorghum (Partidar and Mali, 2002), corn (Ramamurthy and Shivashankar, 1996) and *Brassica juncea* (L.) (Rao and Shaktawat, 2002). Therefore, there was a direct effect of FYM manure on grain yield of maize (Khaliq et al., 2004; Onasanya et al., 2009). Similarly, Rautaray et al. (2003) reported integrated use of organic matter and inorganic fertilizers is beneficial in improving crop yield, soil pH, organic carbon and available N, P and K in sandy loam soils which is in agreement with this finding.

Conclusion

From this finding, it is concluded that enriching FYM with inorganic fertilizers boost hybrid maize (BH-140) grain yield significantly through improving the physicochemical properties of the soil. Thus, it is recommended that application of 4 t/ha FYM incorporated with 75 kg of N and 60 kg of P at Chiro can significantly increase hybrid maize (BH-140) yield and sustain its productivity over years. However, profitability of this technology needs tobe tested at different locations and in different seasons in eastern part of Ethiopia.

ACKNOWLEDGEMENTS

The author is grateful to the Ministry of Agriculture and Rural Development of the Federal Democratic Republic of Ethiopia Chiro ATVET College for funding this research project, the Ethiopian National Soil Laboratory for Soil analysis, Professor Temam Hussein for his technical advice during the field experiment, Nemomsa Beyene for his moral support and sound comments on the article, Belay Dinsa, Tedila Mulu, Bacha Reda, and Tadesse Bedada for their assistance during the field data collection.

REFERENCES

Achieng JO, Ouma G, Odhiambo G,. Muyekho F (2010). Effect of farmyard manure and inorganic fertilizers on maize production on Alfisols and Ultisols in Kakamega, Western Kenya. Agric. Biol. J. N. Am. 1(4):430-439.

Ararsa G (2012).GIS based land suitability evaluation for sustainable agricultural development at Kuni (Sebale) watershade West Hararghe Zone, Oromia. M.Sc. thesis Mekele University, Ethiopia.

Asfaw B, Heluf G, Yohannes U (1998). Effect of tied ridges on grain yield response of Maize (*Zea mays* L.) to application of crop residue and residual NP on two soil types at Alemaya, Ethiopia; South Afri. J. Plant Soil. 15:123-129.

Badiane O, Delgado CL (1995). A 2020 vision for food, agriculture and the environment, Discussion Paper 4. International Food Policy Research Institute, Washington, DC.

Betran FJ, Beck D, Banziger M, Edmeades GO (2003). Genetic analysis of inbred and hybrids grain yield under stress and non stress environments in tropical maize. Crop Sci. 43:807-817.

Beyene Y, Anna-Maria B, Alexander A (2005). A comparative study of molecular and morphological methods of describing genetic relationships in traditional Ethiopian highland maize. Afr. J. Biotechnol. 4 (7):586-59.

Boateng JK, Oppong J (1995). Proceedings of Seminar on organic and sedentary agriculture held at the Science and Technology Policy Research Institute (C.S.I.R) Accra. 1-3 Nov. p. 85.

Chen LZ, Xia ZL, Au SJ (1988). The integrated use of organic and chemical fertilizer in China. SFCAAS (Ed.). Proceedings of International Symposium on Balanced Fertilization. Chinese Agric. Press. pp. 390-396.

Christian R,Angelika CE,Christoph G, Jan L,Frank T, Albrecht EM (2012). Genomic and metabolic prediction of complex heterotic traits in hybrid maize. Nature Genet. 44:217–220.

Debelle T, Friessen DK (2001). Effect of enriching farmyard manure with mineral fertilizer on grain yield of mazie at Bako, western Ethiopia. Seventh Eastern and Southern African Maize conference 11th-15th February. pp. 335-337.

Dutta S, Pal R, Chakeraborty A, Chakrabarti K (2003). Influence of integrated plant nutrient phosphorus and sugarcane and sugar yields. Field Crop Res. 77:43-49.

Elfstrand S, Bath B, Martensson A (2007). Influence of various forms of green manure amendment on soil microbial community composition, enzyme activity and nutrient levels in leek. Appl. Soil Ecol. 36:70-82.

Fininsa C (2001). Association of maize rust (*Puccinia sorghi*) and leaf blight epidemics with cropping systems in Hararghe highlands, eastern Ethiopia. Crop Prot. 20:669-678.

Gachene GKK, Kimaru G (2003). Soil fertility and land productivity, Nairobi, Kenya. 1:57-70.

Gomez KA, Gomez AA (1984). Statistical Procedures for Agircultural Research.2nd Edition.John and Wiley and Sons, New York. P. 680.

Heluf G, Asfaw B, Yohannes U, Eylachew Z (1999). Yield response of maize (*Zea mays* L.) to crop residue management on two major soil types of Alemaya, eastern Ethiopia: I. Effects of varying rates of applied and residual NP fertilizers. Nutr. Cycling Agro. 54:65-71.

Hoffman J (2001). Assessment of the long-term effect of organic and mineral fertilization on soil fertility. 12th WFC-Fertilization in the third millennium, Beijing 10:27-34.

Jackson ML (1958). Soil Chemical Analysis Practice Hall of India. New Delhi. p. 29.

Khaliq T, Mahmood T, Kamal J, Masood A (2004). Effectiveness of farmyard manure, poultry manure and nitrogen for corn (Zea mays L..) Productivity. Int. J. Agric. Biol. 2:260-263.

Khaliq T, Mahmood T, Kamal J, Masood A(2009). Effectiveness of farmyard manure, poultry manure and nitrogen for corn (Zea mays L.). Productivity Int. J. Agric. Biol. 2:260-263.

Legesse BW Myburg AA, Tumasi A (2007).Genetic diversity of maize inbred lines revaled by AFLP markers. Afr. Crop Sci. Conf. Proc. 8:649-653.

Leonard D (1986). Soil, Crop, and Fertilizer Use: A Field Manual for

Development Workers. Under contract with Peace Corps. 4th edition revised and expanded. United State Peace Corps. Information collection and exchange. Reprint R0008.

Mosisa W Wonde A Berhanu T Legesse W Alpha D, Tumassi A (2007). Performance of CIMMTY maize germplasm under low nitrogen soil conditions in the mid altitude sub humid agro ecology of Ethiopia. Afr. J. Sci. Conf. Proc. 18:15-18.

Onasanya OP, Aiyelari A, Onasanya S, Oikeh FE, Nwilene-Oyelakin (2009). Growth and Yield Response of Maize (*Zea mays* L.) to Different Rates of Nitrogen and Phosphorus Fertilizers in Southern Nigeria. World J. Agric. Sci. 5(4):400-407.

Olsen SR Cole CV, Watanabe FS, Dean LA (1954). Estimation of available phosphorus in soils by extraction with sodium bicarbonate. USA Circular. 939:1-19.

Partidar M, Mali AL (2002). Residual effect of farmyard manure, fertilizer and biofertilizer on succeeding wheat (*Triticum aestivum*). Indian J. Agron. 47:26-32.

Ramamurthy V, Shivashankar SW (1996). Residual effect of organic matter and phosphorus on growth, yield and quality of maize (*Zea mays*). Indian J. Agron. 41:247-251.

Rao SS, Shaktawat MS (2002). Residual effect of organic manure, phosphorus and gypsum application in preceding groundnut (*Arachis hypogea*) on soil fertility and productivity of Indian mustard (*Brassica juncea*). Indian J. Agron. 47:487-494.

Rautaray SK, Ghosh BC, Mittra BN (2003). Effect of fly ash, organic wastes and chemical fertilizers on yield, nutrient uptake, heavy metal content and residual fertility in a rice-mustard cropping sequence under acid lateritic soils. Bioresour. Technol. 90:275-283.

Shah STH, Zamir MSI, Waseem M, Ali A, Tahir M, Khalid WA (2009). Growth and yield response of maize (*Zea mays* L.) to organic and inorganic sources of nitrogen. Pak. J. Life Soc. Sci. 7(2):108-111.

Silva PSL, Silva J, Olivera FHT, Sousa AKF, Duda GP (2006). Residual effect of cattle manure application on green ear yield and corn grain yield. Horticultura Brasileira. 24: 166-169.

Soboksa G Mandefro N, Gezahegne B (2008). Stability of drought tolerant maize genotypes in the drought stressed areas of Ethiopia. Seventh Eastern and Southern African Regional Maize Conference. pp. 301-304.

Sutanto R, Suproyo A, Mass A (1993). The management of upland acids Soils for Sustainable food crop production in Indonesia. Rwanda Agriculture research Institute. Soil Manag. Abstracts 5(3):1576.

Tesdale SL, Nelson WL, Beaton JD, Havlin JL (1993). Soil Fertility and Fertilizers (5th ed.). Macmillan Publishing Company, USA. Provide page number(s).

Wonde A, Mosisa W, Berhanu T, Legesse W, Alpha D, Tumassi A (2007). Performance of CIMMTY maize germplasm under low nitrogen soil conditions in the mid altitude sub humid agro ecology of Ethiopia. Afr. J. Sci. Conf. Proc. 18:15-18.

Worku M, Habtamu Z, Grima T, Benti T, Legesse W, Wonde A, Aschalew G, Haji T (2001). Yield stability of maize (*Zea mays*) genotypes across locations. Seventh Eastern and Southern African Regional Maize Conference. pp. 139-142.

Wakene N (2001). Assessment of important physicochemical properties of Alfisols under different management systems in Bako area, Western Ethiopia; M.Sc. Thesis, School of Graduate Studies, Alemeya University, Ethiopia. P. 93.

Wakene N, Heluf G, Friesen DK (2005). Integrated Use of Farmyard Manure and NP fertilizers for Maize on Farmers' Fields. J. Agric. Rural Dev. Trop. Subtropics 106(2):131-141.

Walkley A, Black CA (1934). An examination of the Degtjareff method for determining soil organic matter and a proposed modification of the chromic acid titration method. Soil Sci. 37:29-38.

Wang XB, Cia DX, Hang JZZ (2001). Land application of organic and inorganic fertilizers for corn in dry land farming in a region of north China sustaining global farm. D.E Ston. R.I.I.I. Montar and G.C. Steinhardt (Eds.). pp. 419-422.

Zelalem B (2012). Effect of Nitrogen and Phosphorus fertilizers on Some Soil Properties and Grain Yield of Maize (BH-140) at Chiro, Western Hararghe, Ethiopia. Afr. J. Agric. Res. (Submitted).

21

Zinc deficiency in Indian soils with special focus to enrich zinc in peanut

P. Arunachalam[1], P. Kannan[1], G. Prabukumar[1] and M. Govindaraj[2]

[1]Dryland Agricultural Research Station, Chettinad-630 102, Tamil Nadu Agricultural University, Tamil Nadu, India.
[2]Centre for Plant Breeding and Genetics, Coimbatore-641 003, Tamil Nadu Agricultural University, Tamil Nadu, India.

In India, zinc (Zn) is now considered as fourth most important yield limiting nutrient in agricultural crops. Zn deficiency in Indian soils is likely to increase from 49 to 63% by 2025. India is leading in groundnut acreage but behind the China in production due to less productivity. Apart from rain-dependant cultivation and mineral nutrition play a vital role in groundnut productivity. Among the nutrients, Zn deficiency cause yield loss to the maximum of 40% in groundnut. The average response of groundnut to zinc fertilization ranged from 210 to 470 kg ha[-1]. Hence, it is ideal to follow suitable crop improvement and agronomic management strategies to enhance the uptake and availability of Zn in peanut. There are reports emerging that genetic variability exists among the peanut genotypes for zinc response and accumulation in kernel. This implies that high zinc dense confectionary peanut genotypes can be exploited for the further breeding programmes. In addition, Zn fertilization strategies viz., soil application of enriched Zn, seed coating and foliar application can be suitably adapted with available sources of Zn fertilizer to enhance Zn availability and uptake by peanut under changing climate. This article attempts to examine the status of Zn deficiency in semiarid tropics and approaches to enhance Zn content in peanut kernel through crop improvement and agronomic manipulation.

Key words: Zinc deficiency, peanut, biofortification, zinc rich genotype.

INTRODUCTION

Globally, India is leading in peanut acreage, but behind China in production due to low productivity. India's peanut average productivity is 938 kg ha[-1], which is far behind the most of the peanut growing countries with the highest productivity of 3540 kg ha[-1] in USA and the world mean productivity is 1348 kg ha[-1] (Thamaraikannan et al., 2009). The low productivity in India is mainly due to rain-dependent cultivation, poor soil fertility and mismanagement of plant nutrients especially micronutrients.

Peanut is a poorman's nut due to its high-energy, protein and minerals at a comparatively low cost, is consumed by a large number of people world-wide, and is also a rich source of micronutrients including Zn which makes the crop more important. The 100 g peanut contains 567 Kcal of energy with carbohydrate of 16.13 g, protein of 25.8 g, total fat of 49.24 g, dietary fibre of 8.5 g and Cholesterol free. Among the vitamins and minerals, peanut has high folic acid content (240 μg) and 3.27 mg of zinc, respectively (USDA National Nutrient data base). Peanut is a good source of zinc (Singh, 2007); the kernels are eaten after roasting, frying, salting or boiling and in many preparations and confectionery products. However, in India, due to low productivity the per capita availability of peanut is less, that is, 10 kg peanut per capita are available for domestic consumption. Fat and oil consumption averages less than 5 kg per capita per year.

Zinc deficiency severely affects growth and yield of oilseed crop. Among oilseeds, peanut in particular, suffers from Zn deficiency. As peanut is a good source of Zn, consumption of high Zn density peanut genotypes may be a solution to ensure adequate level of zinc in Indian population (Singh and Lal, 2007). Zinc is an important micronutrient, plant response to Zn deficiency occurs in terms of decrease in membrane integrity, susceptibility to heat stress, decreased synthesis of carbohydrates, cytochromes nucleotide auxin and chlorophyll. Further, Zn-containing enzymes are also inhibited, which include alcohol dehydrogenase, carbonic anhydrase, Copper-zinc superoxide dismutase, alkaline phoshatase, phosphosipase, carboxypeptidase and RNA polymerase (Marschner, 1993). Depending on the zinc level, zinc deficiency status of plants can be classified, that is less than 10 mg kg^{-1} is definite zinc deficiency and more than 20 mg kg^{-1} is sufficient Zn. This article attempts to examine critically the scanty and scattered reports available on the status of Zn deficiency in Semiarid Tropics and approaches to improve zinc use efficiency in terms of pod yield and seed zinc content in peanut.

SOIL VERSUS ZINC NUTRITION

Zinc deficiency: An India concern

In India, zinc is now considered the fourth most important yield-limiting nutrient after, nitrogen, phosphorus and potassium, respectively. Analysis of 256,000 soils and 25,000 plant samples from all over India showed that 48.5% of the soils and 44% of the plant samples were potentially zinc-deficient and that this was the most common micronutrient problem affecting crop yields in India. Deficiency of zinc has increased in Southern States due to extensive use of NPK without micronutrients. Periodic assessment of soil test data also suggests that zinc deficiency in soils of India is likely to increase from 49 to 63% by the year 2025 as most of the marginal soils brought under cultivation are showing zinc deficiency (Singh, 2006). Farming families consuming their zinc deficient crop produce leads to low zinc in their blood plasma compared to those which were fed on produce received from farms fertilized with zinc regularly. Zinc supplementation is therefore essential for maintaining high zinc content in soil, seed and blood plasma of human and animals (Singh et al., 2009).

Soil factors affecting availability of zinc

Although genotypic factors are important in determining either tolerance or susceptibility of a crop cultivar to zinc deficiency, it is soil factors which are responsible for low available zinc supply. In general, the soils most commonly associated with zinc deficiency problems in plants mainly due to the factors like neutral to alkaline in reaction, especially where the pH is above 7.4, high calcium carbonate content in topsoil or in subsoil exposed by removal of the topsoil during field leveling or by erosion, coarse texture (sandy soil) with a low organic matter status, permanently or intermittently waterlogged soil, high available phosphate status, high bicarbonate or magnesium concentrations in soil or irrigation water and acid soil of low zinc status developed on highly weathered parent material (Figure 1).

Calcareous soils with a high content of calcium carbonate (>15%) are typical soils of semi-arid and arid climates. Presence of calcium carbonate decreases the availability of zinc due to higher soil pH. The main types of salt affected soils are the saline soils (Solonchaks), sodic soils (Solonetz) and both mainly occur in arid and semi-arid regions. Saline soils contain high concentrations of soluble salts which restrict the types of crops which can be grown and reduces the availability of zinc. The poor availability of zinc caused by water logging can be due to a relatively high pH, zinc being present as the insoluble sulphide (ZnS) and elevated concentrations of ferrous, bicarbonate, and phosphate ions (Doberman and Fairhurst, 2000).

Interactions between zinc and other plant nutrients

High soil phosphate levels are one of the most common causes of zinc deficiency in crops by cations added with phosphate salts can inhibit zinc absorption from solution, H$^+$ ions generated by phosphate salts inhibit zinc absorption from solution and phosphorus enhances the adsorption of zinc into soil constituents. Nitrogen appears to affect the zinc status of crops by both promoting plant growth and by changing the pH of the root environment. In many soils, nitrogen is the chief factor limiting growth and yield and therefore, not surprisingly, improvements in yield have been found through positive interactions by applying nitrogen and zinc fertilizers. Several macronutrient elements, including calcium, magnesium, potassium and sodium are known to inhibit the absorption of zinc by plant.

Interactions of zinc with other micronutrients

Zinc interact with copper, iron, manganese and boron influence their concentration in plants by a) zinc-copper interactions occur due to copper and zinc sharing a common site for root absorption or copper nutrition affects the redistribution of zinc within plants, b) iron-zinc interactions study resulted increasing zinc supplies to plants have been observed to increase the iron status, to decrease it and to have no effect on it (Loneragan and

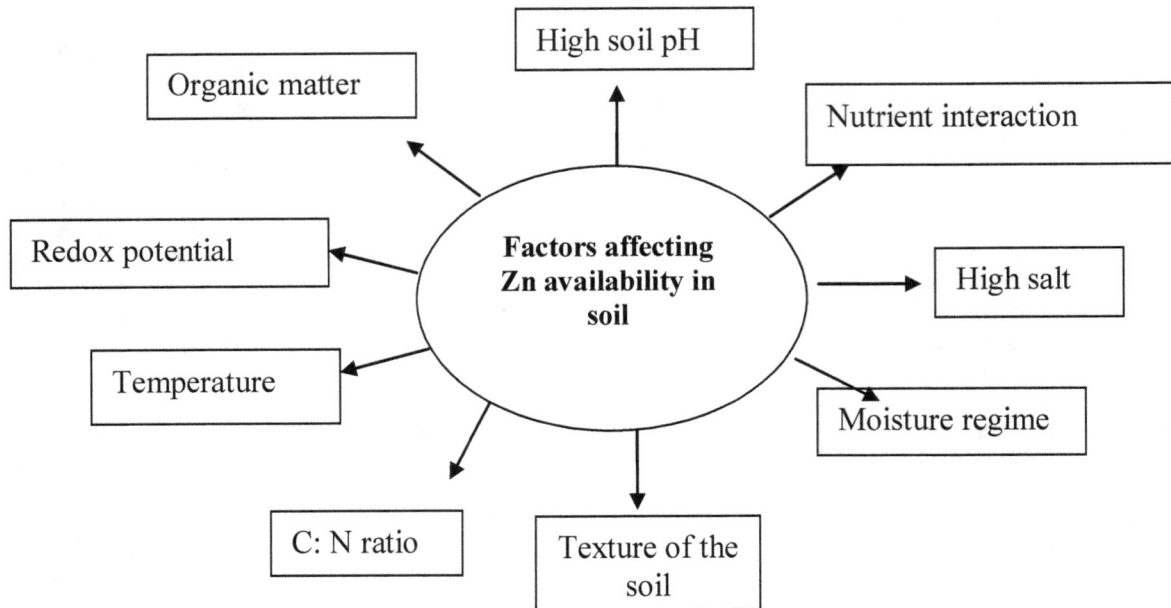

Figure 1. Important soil chemical and physical factors affecting availability of Zn to roots.

Webb, 1993). In contrary to the previous report, Zn application had adverse effect on Fe concentration and Fe uptake in plants (Imtiaz et al., 2003). Mn and boron interaction resulted positive and negative response in availability in soil and uptake and distribution in the plant.

PEANUT

An Indian scenario

In India, peanut is being grown to an area of 8 million ha, production of 7.5 million tonnes, with an average productivity of 938 kg ha^{-1} during 2006 to 2007 (Thamaraikannan et al., 2009). The South West monsoon decides the fate of peanut in India, because around 75% of peanut crop grown under rainfed conditions during Kharif season (June - September). The peanut crop is mainly grown in the States of Gujarat, Tamil Nadu, Andhra Pradesh, Karnataka and Maharashtra, which accounts for 89% of area and production in India. Though Gujarat is leading in area and its production is not appreciable due to low productivity compared to other States (Figure 2).

In Central India, multi-nutrient deficiencies are widely causing poor crop yields (Singh, 2009). The yield reduction reported in peanut due to zinc deficiency is 30 to 40%. The average response of peanut to Zn ranged from 210 to 470 kg ha^{-1} (Takkar and Nayyar, 1984). An increase in energy value as well as total lipids and crude protein in peanut was registered with zinc application (Nayyar, 1990).

Zinc deficiency and its associated consequence in peanut

In India, Zn deficiency was recorded about 50% of the peanut growing soils causing considerable yield losses (Singh, 1999; Singh et al., 2004). Zinc is the one of the eight trace element needed for the normal plant growth and reproduction. Zinc is needed for peanut as a tracer and aids in the use of other trace elements by the plants. The Zn deficiency in peanut caused irregular mottling and yellow-ivory interveinal chlorosis in the upper leaves. Under severe deficiency, the entire leaflets became chlorotic. The main symptoms of zinc deficiency are decreased internodal length and restricted development of new leaves. Deficient plants accumulate reddish pigment in stems, petioles and leaf veins (Alloway, 2008). In zinc deficient soils, fertilization of Zn increase the nodulation, chlorophyll content and pod yield. ILZRO (1975) reported that deficiency causes reduced pegging but no distinct leaf symptoms. Zinc deficiency in peanuts is often associated with high soil pH, high soil calcium contents and high soil phosphorus concentrations. Chahal and Ahluwalia (1977) studied the zinc uptake at various growth stages of peanut in non-calcareous typic Torripsamment, low in available Zn. Highest amount of Zn was accumulated in shoot portion at mid-flowering stage and declined severely at 75 days of plant growth. Zinc concentration in shoot portion increased again at maturity. The maximum zinc translocation from the shoot portion to fruits occurred between 50 and 75 days growth period. Phosphorous application showed an antagonistic effect on zinc uptake. In peanut, 35 days after pod

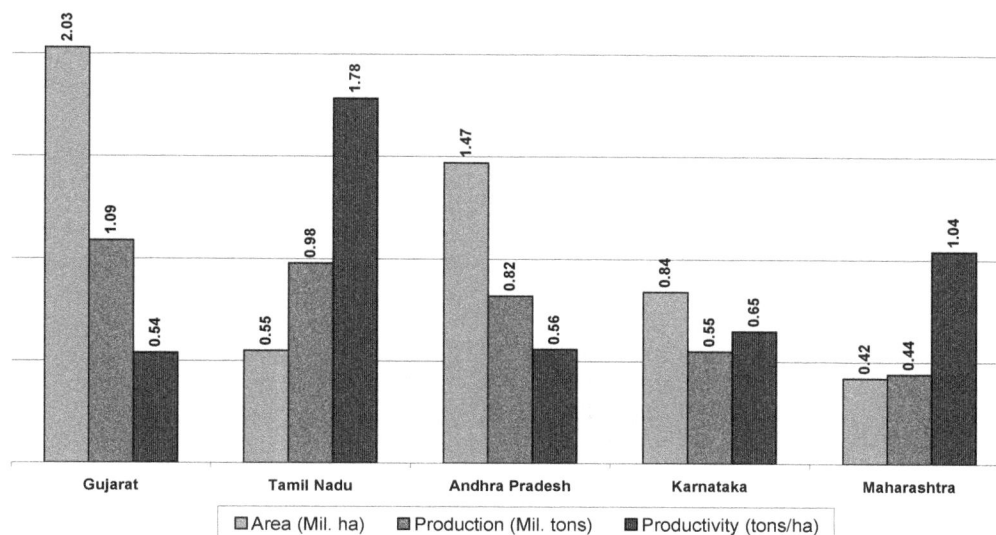

Figure 2. Area, production and productivity of major peanut producing States of India during 2006-2007.

formation is an active period for oil-filling.

The active oil-filling stage was associated with a decrease in the starch, soluble sugars and proteins so as to make available energy and carbon skeleton for the synthesis of oil. The oil content in the matured kernels was decreased by 11, 12 and 25% with Zn, S and Zn+S deficiency, respectively (Sukhija et al., 1987). The blade of the young fully expanded leaf is recommended for diagnosis of Zn deficiency in peanuts and 8-10 mg Zn/kg of drymatter is the critical value. When diagnosing the leaf sample and whole plant of peanut critical concentration is varied between 15 to 25 mg Zn kg^{-1} of drymatter (Bell et al., 1990).

STRATEGIES TO ENHANCE ZINC IN PEANUT KERNEL

Exploiting genetic variability for zinc in peanut

Selecting and breeding of food crops which are more efficient in the uptake of trace minerals from the soil and load more trace minerals into their seeds have benefits agricultural productivity and human nutrition. The micronutrient use efficiency by crop plants is genetically controlled and the physiological and molecular mechanisms of micronutrient efficiency are just beginning to understand (Khoshgoftarmanesh et al., 2010). There was limited attempts have been made in this direction peanut. Few attempts were made to exploit the inherent variability available in peanut for zinc accumulation potential in peanut.

The study conducted at Directorate of Groundnut Research, Junagath to assess the variability in zinc

among the seventy peanut genotypes. The Zn concentration in seeds of peanut genotypes ranged from 11 to 77 mg kg^{-1} with a mean value of 45 mg kg^{-1}. Nineteen genotypes with more than 50 mg kg^{-1} of Zn in their seeds were categorized as high zinc-density genotypes (Lal and Singh, 2007). The genotypes recorded with maximum zinc concentrations are NRCG-4659 (77 mg kg^{-1}), PBS-14032 (76 mg kg^{-1}) and NRCG-6820 (73 mg kg^{-1}) with the pod yield of 1273 kg, 1247 kg and 1313 kg per hectare, respectively. The data of various parameters indicated that seeds from most of the high Zn-density genotypes were also rich in P, Ca and Fe, which are also required in a daily diet.

Singh et al. (2007) reported that the pod yield ranged from 857 to 1527 kg ha^{-1} in the peanut genotypes studied with GG-5 and ICGV-86590 identified as high yielding commercial peanut cultivars and good sources of zinc. Arunachalam et al. (2012) estimated the kernel zinc content in 21 peanut genotypes by basal application of 25 kg ha^{-1} of $ZnSO_4$. The variability for zinc content in peanut kernel ranges from 28.7 mg kg^{-1} (ICGV 07219) to 70.2 mg kg^{-1} (ICGV 07225) with a mean of 56.3 mg kg^{-1}. The genotypes have recorded high zinc content were ICGV 07225 (70.2 mg kg^{-1}), ICGV 07220 (69.8 mg kg^{-1}), ICGV 07222 (69.4 mg kg^{-1}), Narayani (66.4 mg kg^{-1}), ICGV 07247(65.4 mg kg^{-1}), TLG 45 (62.6 mg kg^{-1}) and JL 24 (60.6 mg kg^{-1}).

Agronomic management to enhance zinc in peanut

Zinc fertilization is also required on a regular basis on many alluvial soils and soils of peanut belt in India, and also for high yielding areas in the southern regions.

Soil application

Peanut responded with higher pod yield wherever soil application of zinc was practiced in alluvial soils. The response of peanut to zinc fertilization ranged from 210 to 470 kg ha^{-1}. Application of zinc varies with soil type and it ranged from 0.5 to 1.5 kg per hectare (Takkar and Nayyar, 1984; Takkar et al., 1989). Zinc plays as an activator of many enzymes in plants and is directly involved in the biosynthesis of growth substances like auxin which produce more cells and dry matter that in turn will be stored in seeds as a sink. Thus, the increase seed yield is more expected (Devlin and Withan, 1983).

Patil et al. (2003) observed significant increase in peanut yield by soil application of Fe and Zn along with recommended dose of fertilizer in black soils. Chitdeshwari and Poongothai (2003) reported that, the response of peanut to the soil application of zinc 5 kg ha^{-1} + boron 1 kg ha^{-1} + sulphur 40 kg ha^{-1} significantly increased the pod yield to the tune of 24.2% for TMV 7 and 14.8% for JL 24 over control. The micronutrient response of peanut studied in different States of India under rainfed conditions concluded that an application of Zn, B and S along with N+P was economical. Application of Zn through drip irrigation increased chlorophyll content, pod numbers and yield. It also increased fertilizer-use efficiency and kept the soil loose for peg penetration and pod development. The drip irrigation application was superior over the other soil and foliar Zn applications by precise application at appropriate times with desired concentration, uniform distribution, less damage to crop and soil and ultimately higher yield (Singh, 2007).

The soil application of 5, 1, 0.5 kg ha^{-1} Zn, B, and molybdenum (Mo), respectively along with NPK increased the peanut yield to 30% (Nayak et al., 2009). Muthukumararaja and Sriramachandrasekharan (2012) observed increased in yield with zinc fertilization in zinc deficient soil. The soil application of 10 kg ha^{-1} Fe and 5 kg ha^{-1} Zn was recommended by All India Coordinated Research Project on micronutrients (Singh, 2010).

The requirement and response to zinc application varies with the soil types. Application of 40 kg Zn ha^{-1} recorded high pod yield, protein content and zinc uptake by peanut, whereas application of 20 kg Zn ha^{-1} recorded the highest oil content and sulphur uptake in vertisol (Tathe et al., 2008). An application of 10 kg of zinc in swell-shrink soils, 5 kg in zinc to alluvial red and lateritic soils per hectare was found optimum in ameliorating zinc deficiency (Singh, 2008). Basal application of 25 kg ha^{-1} of ZnSO$_4$ in alfisol increased the pod yield per plant in peanut varieties viz., TMV 7 from 19.2 to 21.4 g, TMV (Gn) 13 from 18.4 to 22.5 g and VRI (Gn) 6 from 35.7 to 38.6 g (Arunachalam et al., 2012).

Foliar application

There are reports that positive response of peanut yield to the foliar application. But the quantum and time of zinc spray were varied in different studies. Gobarah et al. (2006) reported highest peanut seed yield, oil and protein with the application of P$_2$O$_5$ along with foliar spray of zinc. Sixty peanut cultivars were evaluated for foliar application of zinc, the 0.2% ZnSO$_4$ solution was applied on the peanut foliage, thrice at 40, 55 and 70 days after emergence at 500, 1000 and 1000 L ha^{-1}, respectively. On an average of 16% increase in the seed Zn concentration of peanut cultivars was recorded due to foliar application of Zn (Singh et al., 2007). Deficiency of Zn can also be effectively controlled with 2 to 4 spray of 0.5% zinc sulphate salt solution on standing crops (Singh, 2008). Foliar spraying of 1 g L^{-1} zinc recorded highest seed yield (2910 kg ha^{-1}) with increase in pod yield, plant height, 100 seed weight, seed length and seed width. The application of 80 kg ha^{-1} nitrogen along with zinc foliar application enhanced the seed yield to 3742 Kg ha^{-1} (Pendashtek et al., 2011).

Seed treatment and nano-coating of zinc

Seed treatment with Teprosyn Zn + P at the recommended level (8 ml kg^{-1} seed) increased the pod yield of peanut over NPK control. Seed treatment gave higher zinc use efficiency than soil application of zinc sulphate at the rate of 5.5 kg Zn ha^{-1} or two foliar spray of Zn at 30 and 45 days after flowering, also increased the peanut pod yield at similar magnitude (Singh et al., 2003).

An attempt was made by the Prasad et al. (2012) to study the effect of nanoscale zinc oxide on peanut. Peanut seeds treated with 1000 ppm ZnO showed better germination than the seeds treated with bulk ZnSO$_4$. Peanut seeds treated with nanoscale ZnO showed the maximum seedling vigour index at 1000 ppm and increased concentration of ZnO to 2000 ppm has decreased the vigour index. Nanoscale ZnO showed large root growth of seedling compared to bulk ZnSO$_4$ and control. Prasad and coworkers suggested that the growth promoting effect of nanoscale ZnO at optimum concentrations and inhibitory effect at high concentrations on root and shoot growth and pod yield in peanut. Seed treated with nanoscale ZnO enhanced the zinc level in seeds, which increased the germination, root growth, shoot growth, dry weight and pod yield. Significant zinc uptake by the leaf and kernel was observed with the foliar application of nanoscale ZnO compared to chelated zinc sulfate.

ZINC FERTILIZERS

There are different forms and source of zinc available as listed in Table 1. The solubility of several zinc minerals decreases in the following order namely Zn (OH)$_2$ (amorphous) > Zn(OH)$_2$ > ZnCO$_3$ (smithsonite) > ZnO

Table 1. Commonly used zinc fertilizers.

Compound	Formula	Zn content (%)
Inorganic zinc fertilizers		
Zinc sulphate monohydrate	$ZnSO_4 . H_2O$	36
Zinc sulphate heptahydrate	$ZnSO_4 . 7H_2O$	22
Zinc oxysulphate	$ZnSO_4 \, xZnO$	20-50
Basic zinc sulphate	$ZnSO_4 . 4Zn(OH)_2$	55
Ammoniated zinc sulphate	$Zn(NH_3)4SO_4$	10
Zinc oxide	ZnO	50-80
Zinc carbonate	$ZnCO_3$	50-56
Zinc chloride	$ZnCl_2$	50
Zinc nitrate	$Zn(NO_3)_2 . 3H_2O$	23
Chelated zinc fertilizers		
Disodium zinc EDTA	Na_2Zn EDTA	8-14
Sodium zinc EDTA	$NaZn$ EDTA	9-13
Sodium zinc HEDTA	$NaZnH$ EDTA	6-10
Zinc polyflavonoid	-	5-10
Zinc lignosulphonate	-	5-8

(zincite) > Zn $(PO_4)_2 . 4H_2O$) (willemite) > soil Zn > Zn Fe_2O_4 (franklinite). All of the Zn $(OH)_2$ minerals, ZnO and $ZnCO_3$ are about 105 times more soluble than soil zinc (adsorbed to solid surfaces) and would therefore makes highly suitable fertilizer sources of zinc.

Zinc forms soluble complexes with chloride, phosphate, nitrate and sulphate ions, but the neutral sulphate ($ZnSO_4$) and phosphate ($ZnHPO_4$) species are the most important and contribute to the total concentration of zinc in solution. The $ZnSO_4$ complex may increase the solubility of Zn^{2+} in soils and accounts for the increased availability of zinc when acidifying fertilisers, such as ammonium sulphate [$(NH_4 (SO_4)_2)$] are used.

The soil and seed application of $ZnCl_2$ and $ZnSO_4$ showed a positive response with germination, pod number, pod yield and oil content. However, both fertilizers were detrimental to peanut seedlings when applied as seed dressing (Singh, 2007). Application FYM enriched Zn observed improvement in the available zinc status of deficient soils is probably the result of an increase in soluble, organically-complexed forms of zinc. Low molecular weight organic acids namely, humic acid and fulvic acid form soluble complexes with zinc and contribute to the total soluble concentration in a soil. Barrow (1993) reported that organic ligands reduced the amounts of zinc adsorbed in soil and the effect was most pronounced with those ligands, including humic acids which complexed zinc most strongly. Soluble forms of organically-complexed zinc can result in zinc becoming increasingly mobile and plant available in soils.

In many cases, complexation of organic zinc with organic ligands will result in decreased adsorption onto mineral surfaces. Enriching the $ZnSO_4$ with organic sources such as farmyard manure, composted coirpith and poultry manure atleast 20 to 30 days. Application of enriched $ZnSO_4$ as a basal was found to be economically viable and sustainable. In case of severe deficiency of zinc foliar application of $ZnSO_4$ at 0.2 to 0.5% is recommended to temporarily arrest the deficiency.

CONCERNS AND FUTURE DIRECTIONS

Soil and foliar application of zinc meet out only 30 to 40% of zinc requirement of crop plants while the remainder gets absorbed in clay colloids and become immobile and some parts may goes out to the environment due to soil and edaphic factors. As only a small quantum of applied zinc is utilized by the crop plants. To enhance zinc uptake by plants and to increase the seed zinc concentration are depends on use of zinc responsive peanut varieties along with better agronomic management. These kinds of approaches are economic way to combat zinc malnutrition especially developing country like India. To reduce zinc malnutrition, the crop fortification is important aspect which not only enhances the zinc content in kernel and also increases the productivity of peanut. The future strategies to be adopted for enhancing the zinc in peanut kernels are narrated below:

i) Screening programme to explore the inherent genetic variability in peanut for zinc both in kernel and foliage, since the fodder is also very much desired. This would enable peanut genotypes to be matched to soils and reduce the requirement for zinc fertilizers. List of peanut varieties which are highly susceptible to zinc deficiency would also be useful for farmers and extension workers so that they can be extra vigilant with those most at risk of deficiency. Zinc efficient confectionary type peanut strains could be used in breeding programmes to develop high dense zinc varieties to grow in areas where people are affected by zinc deficiency,

ii) Studying the solublization and mobilization of zinc in soil and interaction with other nutrients in various types of soils are essential for the precise recommendation and use of zinc. The majority of peanut crop is grown under rain-fed conditions. Hence, there is a need to develop the comprehensive nutrient management practices to be developed by taking into account of macro, micro and soil amendment (gypsum) requirement of peanut. Development of easy and economic, field-based biochemical test kits for assessing the zinc status of crops without relying on analytical laboratories is a key in the efficient use of zinc in agriculture near future,

iii) The peanut is a zinc responsive crop, but most of the peanut growing soils are deficient in zinc that fails to support the zinc dense cultivars. Hence, GIS/remote sensing strategy may be used to delineate the zinc deficient peanut area for site specific recommendation,

iv) Better understanding of molecular and physiological mechanisms of zinc uptake, mobility and partitioning in peanut is required to design the suitable source, method of application and the stage of application. Investigation is needed on plant anatomical and rhizosphere changes responsible for the variability in absorption, translocation and uptake of zinc in peanut,

v) Creating mass awareness about zinc nutrient application in peanut or food crops, prioritization of government policy for emphasizing micro-nutrient usage in agriculture is very much desired.

At last to enrich micronutrient in food crops, the collaborative multidisciplinary research involving soil scientists, agronomists, plant breeders, human nutritionists and food technologists on the bio-fortification with zinc in a form which is bio-available to consumers is need of the hour. Further, the positive approach of government in their policies to enhance the micro-nutrients in stable food crops and their use in regular foods are very much desired to address the hidden hunger problem.

REFERENCES

Alloway BJ (2008). Zinc in soils and crop nutrition, Second edition, published by IZA and IFA Brussels, Belgium and Paris, France.

Arunachalam P, Kannan P, Balasubramaniyan P, Prabukumar G, Prabhaharan J (2012). Response of groundnut (Arachis hypogaea L.) genotypes to biofortification through soil fertilization of micronutrients in Alfisol conditions. Annual Report of Dryland Agricultural Research Station, Chettinad, India.

Barrow NJ (1993). Mechanisms of reaction of zinc with soil and soil components. In: Robson, A.D. (ed.) Zinc in Soils and Plants, Kluwer Academic Publishers, Dordrecht, pp. 15-32.

Bell RW, Kirk G, Placket, Loneragan JF (1990). Diagnosis of zinc deficiency in peanut (Arachis hypogaea L.) by plant analysis. Comm. Soil Sci. Plant Anal. 21:273-285

Chahal RS, Ahluwalia, SPS (1977). Neutroperiodism in different varieties of groundnut with respect to zinc and its uptake as affected by phosphorus application. Plant Soil 47(3):541-546

Chitdeshwari T, Poongothai S (2003). Yield of groundnut and its nutrient uptake as influenced by Zinc, Boron and Sulphur. Agric. Sci. Dig. 23(4):263-266.

Devlin RM, Withan FH (1983). Plant physiology. Wadsworth publishing company: California.

Doberman A, Fairhurst T (2000). Rice: Nutrient disorders and nutrient management. Potash and Phosphate Institute of Canada and International Rice Research Institute, Los Baños, Philippines.

Gobarah ME, Mohamed MH Tawfik MM (2006). Effect of phosphorus fertilizer and foliar spraying with zinc on growth, yield and quality of groundnut under reclaimed sandy soils. J. Appl. Sci. Res. 2(8):491-496.

ILZRO (1975). Zinc in Crop Nutrition, International Lead Zinc Research Organisation Inc, Research Triangle Park.

Imtiaz M, Alloway BJ, Shah KH, Siddiqui, SH, Memon MY, Aslam M, Khan P (2003). Zinc nutrition of wheat II: Interaction of zinc with other trace elements. J. Asian Plant Sci. 2:156-160.

Khoshgoftarmanesh AH, Schulin R, Chaney RL, Daneshbakhsh B Afyuni M (2010). Micronutrient-efficient genotypes for crop yield and nutritional quality in sustainable agriculture: A review. Agron. Sust. Develop. 30:83–107.

Lal C, Singh AL (2007). Screening for high zinc density groundnut genotypes in India, National Research Centre for Groundnut (ICAR), Junagadh, India.

Loneragan JF, Webb MJ (1993). Interactions between zinc and other nutrients affecting the growth of plants. In: Robson, A.D. (ed.) Zinc in Soils and Plants. Kluwer Academic Publishers, Dordrecht, pp. 119-134.

Marschner H (1993). Zinc uptake from soils, In: Robson, A.D. (ed.) Zinc in Soils and Plants. Kluwer Academic Publishers, Dordrecht, pp. 59-78.

Muthukumararaja TM, Sriramachandrasekharan MV (2012). Effect of zinc on yield, zinc nutrition and zinc use efficiency of low land rice. J. Agric. Tech. 8(2):551-561.

Nayak SC, Sarangi D, Mishra GC, Rout DP (2009). Response of groundnut to secondary and micronutrients. SAT eJournal- 7.

Nayyar VK (1990). Micronutrient in soils and crops of Punjab. Research Bulletin, Department of Soils, PAU, Ludhiana. p. 148.

Patil CV, Yaledahalli NA, Prakash SS (2003). Integrated nutrient management for sustainable productivity of groundnut in India. Paper presented at the National Workshop on Groundnut Seed Technology, Raichur, 6-7 February, 2003.

Pendashtek M, Tarighi F, Doustan HZ (2011). Effect of foliar zinc spraying and nitrogen fertilization on seed yield and several attributes of groundnut (Arachis hypogaea L.). World Appl. Sci. J. 13(5):1209-1217.

Prasad TNV, Sudhakar KVP, Sreenivasulu Y, Latha P, Munaswamy V, Raja Reddy K, Sreeprasad TS, Sajanlal PR, Pradeep T (2012). Effect of nanoscale zinc oxide particles on the germination, growth and yield of peanut. J. Plant Nutr. 35(6):905-927.

Singh AL (1999). Mineral nutrition of groundnut. In: Helantaranjan, A. (Ed.) Advances in Plant Physiology. Scientific Publishers (India) Jodhpur, pp. 161-200.

Singh AL (2007). Prevention and correction of zinc deficiency of groundnut in India. In: Proceeding of Zinc Crops 2007 Conference for improving crop production and human health, Istanbul, Turkey 24-26th May 2007.

Singh AL, Basu MS, Singh NB (2004). Mineral disorders of groundnut. National Research center for groundnut (ICAR), Junagadh India. p. 85.

Singh AL, Chaudhari V, Misra JB (2007). Zinc fortification in groundnut and identification of Zn responsive cultivars of India. Directorate of Groundnut Research, Junagadh, India.

Singh AL, Lal C (2007). Screening for high density groundnut genotypes in India. In: Proceeding of Zinc Crops 2007 Conference for improving crop production and human health, Istanbul, Turkey 24-26th May 2007.

Singh MV (2006). Micronutrients in crops and in soils of India. In: Alloway BJ (ed.) Micronutrients for global crop production. Springer. Business.

Singh MV (2008). Micronutrient deficiencies in crops in India. In:Micronutrients in global crops. (ed. Alloway Brown) Springer, New York.

Singh MV (2009). Micro nutritional problem in soils of India and improvement for human and animal health. Indian J. Fert. 5(4):11-16.

Singh MV (2010). Micronutrient deficiency in Indian soils and field usable practices for their correction. AICRP (micronutrient) Annual Report, Indian Institute of Soil Science, Bhopal, India.

Singh MV, Narwal RP, Bhupal RG, Patel KP, Sadana US (2009). Changing scenario of micronutrient deficiencies in India during four decades and its impact on crop responses and nutritional health of human and animals. The Proceedings of the International Plant Nutrition Colloquium XVI. Department of Plant Sciences, UC Davis.

Singh MV, Patel KP, Ramani VP (2003). Crop responses to secondary and micronutrients in swell-shrink soils. Fert. News 48(4):63-66.

Sukhija PS, Randhawa V, Dhillon KS, Munshi SK (1987). The influence of zinc and sulphur deficiency on oil-filling in peanut (Arachis hypogaea L.) kernels. Plant Soil 103(2):261 267.

Takkar PN, Nayyar UK (1984). Crop response to micronutrient application. Proceedings of FAI-NRC seminar, Jaipur, pp. 95-123

Takkar PN, Chhibba IM, Mehta SK (1989). Bulletin No. 1. Indian Institute of Soil Science Bhopal, India.

Tathe AS, Patil GD, Khilari JM (2008). Effects of sulphur and zinc on groundnut in vertisols. Asian J. Soil Sci. 3(1):178-180.

Thamaraikannan M, Palaniappan G, Dharmalingam S (2009). Groundnut: The King of Oilseeds. Market Survey, India.

22

Using universal soil loss equation and soil erodibility factor to assess soil erosion in Tshesebe village, north east Botswana

Trust Manyiwa and Oagile Dikinya

Department of Environmental Science, University of Botswana, Private Bag 00704, Gaborone, Botswana.

Soil erodibility (K) factor in the Universal Soil Loss Equation (USLE), defines the resistance of soil to detachment by rainfall impact and/or surface flow force. Whilst there are a number of factors of erosion, this study aims to use erodibility factor and related length slope factor to assess soil erosional loss at field scale. To quantify soil erodibility the following properties were measured; texture, organic matter content and structural properties of the soil samples in eroded and non-eroded sites. Sub sampling was conducted in both eroded and non-eroded site and a total of six samples were collected in each site. In addition, slope length and slope angle were determined to evaluate the slope effect on the degree of soil loss associated with the K-factor. The measured or estimated K-factor value compared with the USLE K-based nomograph. The average soil erodibility (K-factor) was 0.031 and (t ha h ha^{-1} MJ^{-1}mm^{-1}) for eroded and non-eroded area, respectively. The high K-factor value in eroded area (almost doubled) was associated with low organic matter content (0.75%) compared to high organic matter in non-eroded (1.18%) as well as the significant slope (3°) in eroded than non-eroded areas (1°). The results also show that K-factor significantly (P<0.05) correlates with soil texture and organic matter due to their strong binding effect on aggregate stability and water infiltration hence enhanced particles' resistant to detachment. Interestingly there was no significant difference in K- factor values between eroded and non-eroded areas. Further, the K-factor based nomograph over-predicted the measured K-factor value by 10 times in eroded and 19 times in non-eroded soil, with a strong correlation in eroded (r^2=0.77) than in non-eroded (r^2=0.10).

Key words: Universal soil loss equation (USLE), soil erodibility, soil erosion, soil properties, eroded and non-eroded areas.

INTRODUCTION

Soil erosion is a major soil degradation threat in most vulnerable ecological systems especially in the fragile semi-arid environments like Botswana (where there is less biomass to sustain soil structural integrity). It is a serious problem associated with land use (Morgan, 1996). Soil erodibility (K-factor) has been used recently as an indicator of erosion (Parysow et al., 2003; Tejada and Gonzalez, 2006; Zhang et al., 2007) because of its susceptibility to particulate detachment and transport by erosion agents such as wind and water. In practice, K represents an integrated average annual value of the total soil and soil profile reaction to a large number of erosion and hydrological processes (Bonilla and Johnson, 2012). The K factor is one of the key parameters required for soil erosion prediction across the world (Zhang et al., 2007). Therefore assessment of erosional soil losses is the basis for effective conservation planning and management of the vulnerable ecosystems. There exist several models to predict the extent of water induced erosion (Brady and Weil, 2002, 2008) such as WEPP (La°en et al., 1991), EUROSEM (Morgan et al., 1992), and GUEST (Ciesiolka et al., 1995; Rose et al., 1997). EUROSEM and GUEST models have been developed to describe and quantify soil erosion processes and are particularly suitable for adaptation across arrange of scales in the landscape.The model deals with: the interception of rainfall by the plant cover; the volume and kinetic energy of the rainfall reaching the ground surface as direct through fall and leaf drainage; the volume of stremflow; the volume of surface depression storage; the detachment of soil particles by raindrop impact and by runoff; sediment deposition; and the transport capacity of the runoff (Morgan et al., 1992). On the other hand, WEPP is an American model based on a continuous simulation approach in which changing soil moisture conditions are modelled from daily calculations of the soil water balance. In this way, the conditions at the start of each rainstorm are predicted. The problems with continuous simulation models are that they require a large amount of input data on changing climatic and land use conditions over a year. These continuous simulation models are highly sensitive to the modelling of evapotranspiration and dynamic properties of the soils and they yield predictions for a large number of events that produce only small amounts of runoff and soil loss (Morgan et al., 1992).

However, the Universal Soil Loss Equation (USLE) has been useful in predicting the average rate of soil loss due to water erosion from agricultural lands (Wischmeier and Smith, 1978). In the early 1990s the basic USLE was updated and computerized to create an erosion prediction tool called the Revised Universal Soil Loss Equation (RUSLE) (Renard et al., 1997). The USLE/RUSLE soil loss prediction is dependent upon soil properties including texture, organic matter content and structure of the soil. The RUSLE uses the same basic factors of the USLE although some are modified and better defined. The predicted soil loss A is estimated using the following equation: A= RKLSCP, where; R=rainfall erosivity; K= soil erodibility; L= slope length; S = slope gradient or steepness; C= cover and management and P= erosion control practices.

Amongst the USLE factors,soil erodibility (K) factor is applicable to most tropical soils (El-Swaify and Dangler, 1976; Roose, 1977; Angima et al., 2003) and was found to strongly correlate with soil loss (Tejada and Gonzalez, 2006). The erodibility (K) factor reflects the ease with which the soil is detached by splash during rainfall and/or by surface flow especially on sloping areas (Angima et al., 2003). The two most significant and closely related soil characteristics influencing soil erodibility are infiltration capacity and structural stability (Millward and Mersey, 1999). These are largely influenced by soil texture, organic matter and soil plasticity. High infiltration capacity means that less water will be available for runoff and the surface is less likely to be ponded and more susceptible to splashing. In particular, soils which are highly permeable have high infiltration capacities (e.g. sandy soils)and are more prone to water erosion since the soil easily allows water to penetrate and therefore easily washed away (Zachar, 1982). On the other hand, stable aggregates resist the beating action of rain and thereby save soil even though runoff may occur. The factors that determine aggregate stability include bulk density, Atterberg limits as well as texture and organic matter content of soils (Toy et al., 2002).

Moreover, soils with larger sand and silt proportions are more vulnerable to water erosion due to lack of stability of soil particles (Toy et al., 2002). Similarly, soils with relatively low organic matter content are very vulnerable to water erosion (Brady and Weil, 2002) since organic matter increases the stability of soil. A 36% decrease in K-factor value was observed in organic matter amended soil in respect to the control (Tejada and Gonzalez, 2006). Furthermore, the susceptibility of soil to water erosion also depends on slope length (Toy et al., 2002) and is most prevalent in sloping areas (Angima et al., 2003). Liu et al. (2000), in their studies on 'slope length effect on soil loss for steep slopes' also reported the greater sensitive of slope effect to soil loss due to differences in rainfall. Whilst there are a number of factors of erosion, this study does not intend to cover all the factors of soil erosion. Rather it focuses on the erodibility or (K) factor and related factors of slope length (LS) factor in assessing soil erosion in typical tropical soil in fragile semi-arid environment Botswana. Thus the objective of the study was to use or apply erodibility K-factor as an indicator of erosion to assess erosion in Tshesebe village, north east Botswana. The village used as a case study is an agricultural area and was observed to be vulnerable to erosional losses as evidenced by gully formations in the area.

MATERIALS AND METHODS

Description of study area

The study area is located in Tshesebe village(20°45'0" N and 27°34'0" E, with an elevation of about 1170 m) in the North East District of Botswana (Figure 1). The area receives about 506 mm of rainfall, with the highest rainfall in December and January and receives nil rainfall on June and July. Generally the daily maximum temperatures range between 27.3 and 35°C, while the mean temperatures range between 6.1 and19.7°C (Radcliffe,et al 1990). The village lies in the ecological zone known as hardveld, characterized by predominance of tree Savanna and acacia scrub. The vegetation is thick along Ntsheriver and streams found in the area. Mophane trees (colospermummophane) and terminaliasericiaare also very common in the area. The soils are predominantly imperfectly drained Luvisols and Arenosols

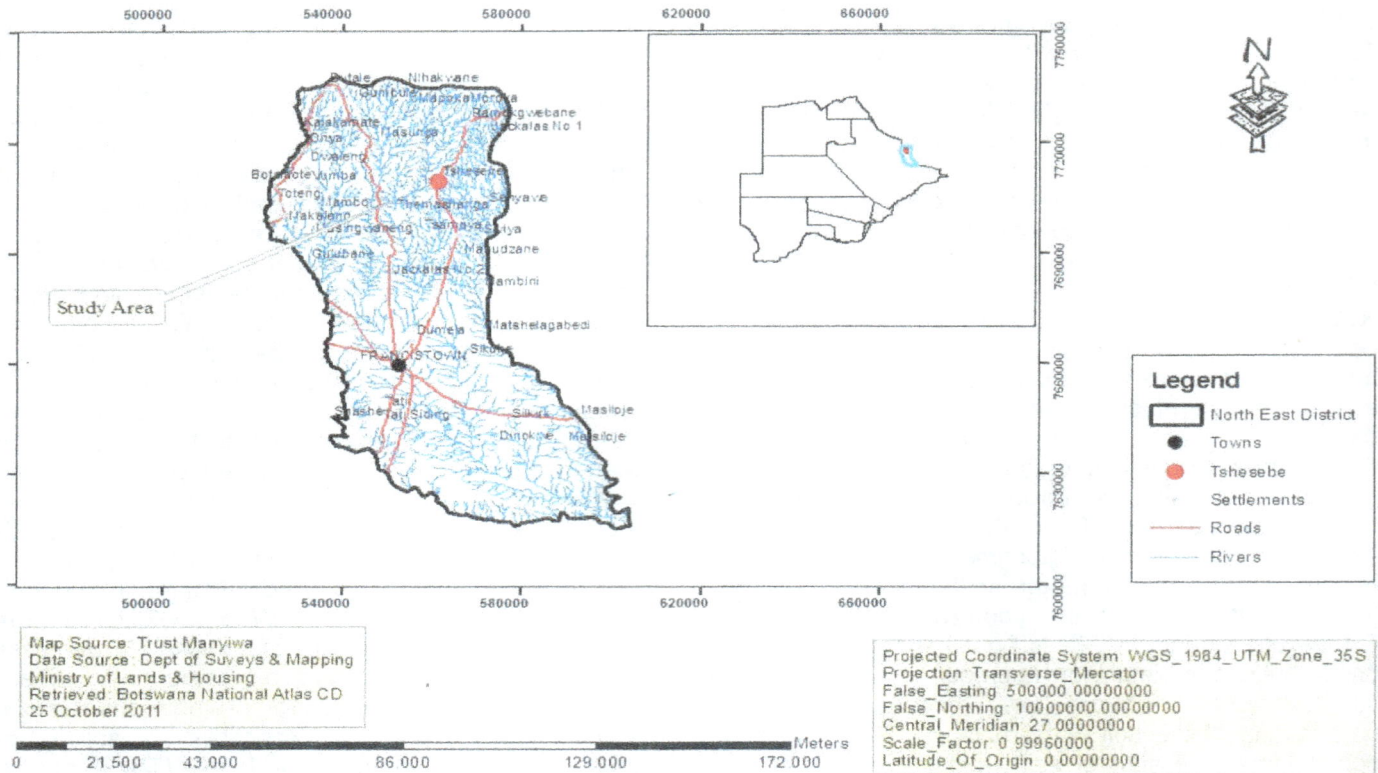

Figure 1. The study area.

(Anon, 1991). The geological parent material is gneiss (Radcliffe et al., 1990). Soil erosion is prevalent as evidenced by gully formations in the area.

Sampling and soil morphological properties measurements

Soil samples were collected from eroded and non-eroded surface soils that is, from sampling points; A1, B1, C1, D1, E1 and F1 and A2, B2, C2, D2, E2 and F2 from eroded area and non-eroded area (control sites), respectively. Sampling depth was 0 to 15 cm and 15 to 30 cm and the samples were mixed to form a composite sample. Samples from the eroded sites were collected on a line parallel to the slope direction. Samples were then passed on a 2 mm sieve for laboratory analysis. Soil morphological properties including soil structure type, class and permeability class were also collected based on FAO (2006), (Table 1).

The K-factor parameter determinations

Selected physical properties related to texture and structure of soils were measured including particle size analysis, soil bulk density, plastic limit and liquid limit and soil organic matter to quantify soil erodibility factor.

Particle size analysis

Soil texture was determined using the Hydrometer or the Bouyoucos method for mechanical analysis or particle size analysis by measuring the proportion of different sized particles in a soil and

hence it's textural class. This is because for most agricultural purposes, the Bouyoucos method is sufficiently precise (Hanks and Ashcroft, 1970).

Bulk density and porosity

The core method was used to determine the soil bulk density and porosity. A cylindrical tube (5 cm long, 5 cm diameter) was driven into the soil to collect the samples. The bulk density and porosity of soil samples were estimated according to Rowell (1994).

Atterberg limits determination

Atterberg limits were measured using standard American Society for Testing and Materials (ASTM) devices (Faniran and Areola, 1978). Atterberg limits refers to the water content of fine grained soils at different states of consistency and are based on plastic limit (PL) and liquid limit (LL) and more importantly on plasticity index (PI). The plastic limit is the water content (in %), at which soil can no longer be deformed by rolling into 3.2 mm diameter without crumbling. While liquid limit is water content at which a soil changes from plastic to liquid behavior. The plasticity index is a measure of plasticity or the difference between the liquid limit and the plastic limit (that is, PI = LL-PL). The Casagrande Method was used to determine attaerberg limits (McBride, 1993).

Soil organic matter

Soil organic matter was determined using the Walkley-Black

Table 1. Surface soil structure, slope angle and length for both eroded and non-eroded areas.

Parrameter	Eroded	Non-eroded
*Soil structure	Granular	Crumb
*Soil structure class	1	1
*Permeability Class	2 = moderate to rapid	3 = moderate
Slope length (m)	20 m	9 m
Slope angle (°)	3°	1°

*Defined according to (FAO, 2006).

Method (Tiessen and Moir, 1993).

Statistically analysis

Statistically data analysis was done using methodology by Wheater and Cook (2003) for the t-test (paired and unpaired) and to check if there is any significant difference between eroded and non-eroded areas at significance level of P<0.05 (or 95% confidence limit). The t-test was computed according to the following equation:

$$t = \frac{\overline{x}_1 - \overline{x}_2}{S\overline{x}_1 - \overline{x}_2} \quad \dots \tag{1}$$

$$S\overline{x}_1 - \overline{x}_2 = \sqrt{\frac{S_1^2}{n_1} + \frac{S_2^2}{n_2}}$$

Where, and $\overline{x} =$ mean of samples, $n =$ sample size and $S =$ variance.

RESULTS AND DISCUSSION

Parameterization of erodibility factor

Soil particle size distribution, organic matter, structure and slope effects

Particle size analysis and respective soil textural class are presented in Table 2. Generally, the results show that sand content is generally high in all samples (Table 2). Soil textural class for soils in the eroded area is mainly sandy (at least 68%) characterized by weak structure and granular type (Table 1). The weak structure (granular) as evidenced by the relatively low organic matter (0.75% for eroded and 1.18% for non-eroded areas, Table 2) makes the soil susceptible to erosion in eroded areas. This was supported by Ball (1990) who reported an increase of erosion with decreasing organic matter. On the other hand, soil samples in the non-eroded area have more clay (27%) and are less susceptible to erosion. Similarly, the slope lengths and slope angles of the eroded area are high (20 m and 3°, respectively) as compared to those in non-eroded area (9 m and 1°). Slope length and slope angle contribute to the erodibility of soil as slope leads to colluvial deposited materials (Brady and Weil,

2002) and hence more erosional loses. This is because slope leads to materials being transported by mass movement, while in the non-eroded areas the slope is relatively flat and less material is transported hence less erosional losses as manifested by low K-factor values [0.031 (t ha h ha^{-1} MJ^{-1}mm^{-1})].

Similarly, Table 2 shows that organic matter content is higher for non-eroded areas (1.18%) than eroded (0.75%) and it is in agreement with Charman and Murphy (1991) who stated that when organic matter is high, the soil will be less susceptible to erosion because of the binding effect of organic matter and therefore less vulnerability to particle detachment. The high organic matter in non-eroded area is high probably because of undisturbed litter as evidenced by presence of vegetation. This litter leads to the formation of humus which contributes to more organic matter in the non-eroded area (Brady and Weil, 2002).

Soil bulk density and porosity

Generally bulk density was higher in non-eroded than eroded sites because of surface structural loss (Figure 2a) whereas the porosity (which is indirectly proportionally to bulk density) was lower in non-eroded than eroded sites (Figure 2b). For instance, the average bulk density is 1.52 and 1.25 g/cm^3 for non-eroded and eroded area, respectively. The average porosity is 28 and 44% for non-eroded and eroded areas, respectively. This is because as particles are eroded soil material becomes loose, therefore reducing the bulk density and increasing soil porosity (Abu-Hamdeh and Al-jalil, 1999). The higher density is also attributed to high clay content (binding effect) in the non-eroded soil thus making it less vulnerable to erosion.

Plastic limit, liquid limit and plastic index

Plastic limit is the moisture content that defines where the soil changes from a semi solid to a plastic (flexible) state while the liquid limit is the moisture content that defines where soil changes from a plastic to viscous fluid state (Reddy, 1999). Plastic limit of the soil samples ranged 2.7

Table 2. Particle size distribution for eroded and non-eroded areas.

| Sampling point | % clay | % silt | % sand | Textural class | Organic matter | |
					%OM	%OC
Eroded area						
A1	15	4	77	Sandy loam	0.50	0.29
B1	16	3	81	Sandy loam	0.74	0.43
C1	30	5	65	Sandy clay loam	0.58	0.34
D1	32	5	65	Sandy clay loam	0.52	0.30
E1	28	5	68	Sandy clay loam	0.80	0.46
F1	27	5	64	Sandy clay loam	1.35	0.78
Mean ($\bar{x}$)	25	5	77	Xxxxxxx	0.75	0.43
Non-eroded area						
A2	36	4	60	Sandy clay loam	0.80	0.46
B2	19	8	73	Sandy loam	1.49	0.86
C2	16	4	80	Sandy clay loam	1.13	0.65
D2	23	9	68	Sandy clay	1.10	0.64
E2	32	3	64	Sandy clay loam	1.40	0.81
F2	36	5	60	Sandy clay loam	1.21	0.70
Mean ($\bar{x}$)	27	6	68	Xxxxxxxxx	1.18	0.69

Where OM and OC is Organic matter and Organic carbon respectively.

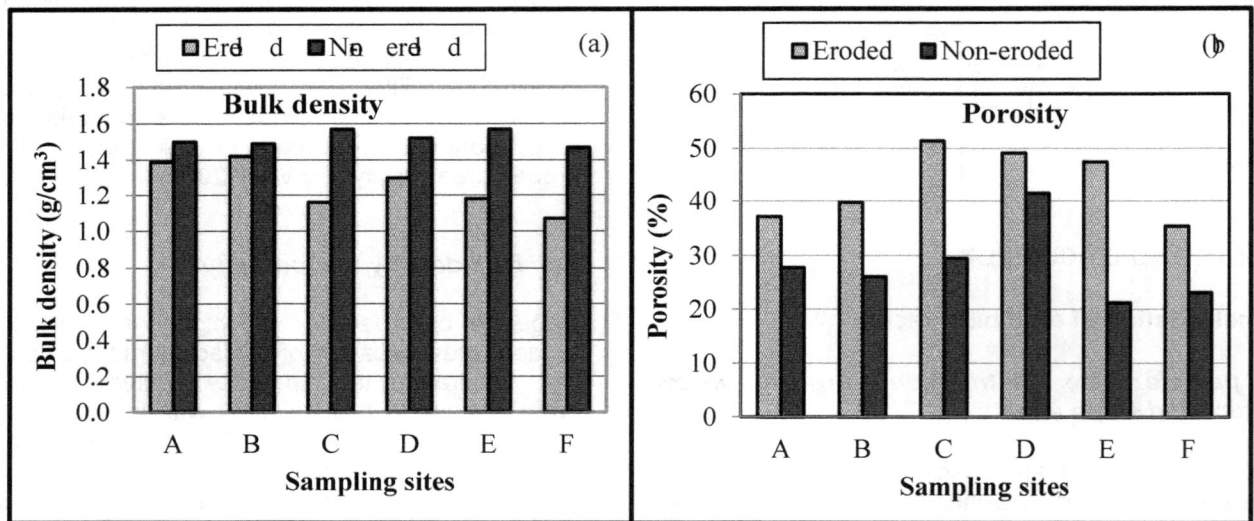

Figure 2. Soil bulk density (a) and porosity (b) for eroded and non-eroded sites.

to 6.8% in eroded and 3.1 to 6.8% in non-eroded. In general this reflects high structural stable soil material or high resistance to detachment in non-eroded sites and hence less vulnerability. In most cases % plastic limit and liquid limit are high in eroded than non-eroded areas and similar observation were reported by Nandi and Luffman (2012). The plastic index is high in eroded areas with an average of 16.3% (Figure 3) probably due the sandy nature of the soil (Table 2). This is in contrast withReddy (1999) who reported thatwhen the plastic index of a soil is high it will not be easily eroded. Other factors like slope

angle also contribute to the erodibility of the soil. The results also indicate that non-eroded areas have low plastic limit but they are not easily eroded due to a flat area and some vegetation cover thus preventing erosion even though its plastic limit is low. Similarly the average plasticity index was 16.3 and 14.4%, respectively for eroded and non-eroded areas and this has an influence on soil erodibility. For instance, soils with low plastic limit have high organic matter Ball (1990) thus explaining large erodibility in eroded area (with relatively high low organic matter). Soils with high content of clay particles

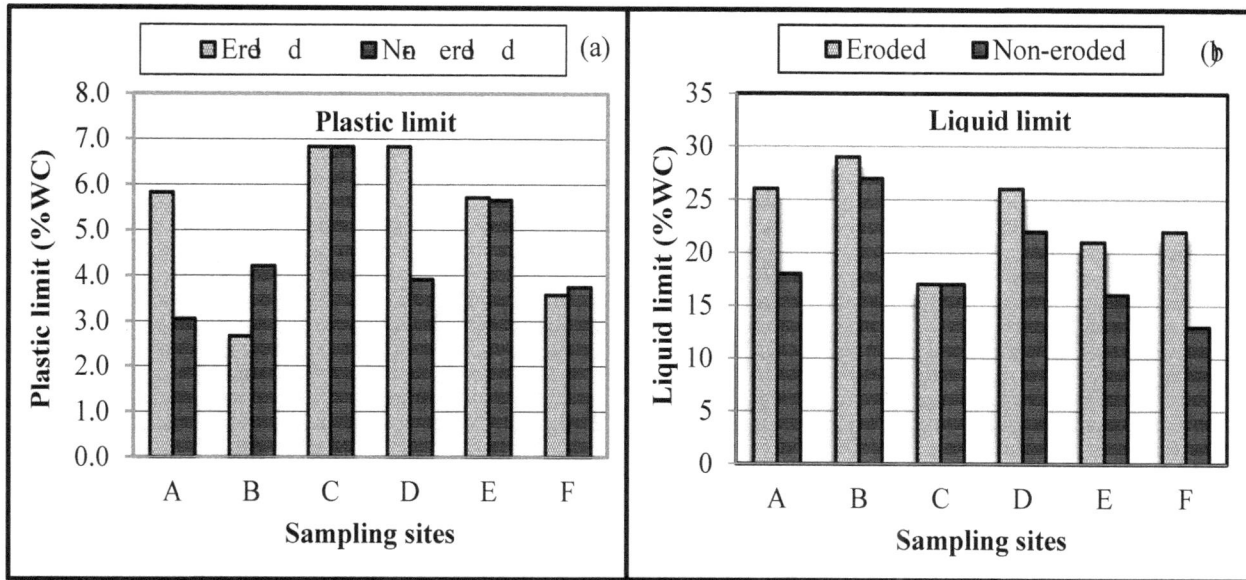

Figure 3. The plastic limit (a) and liquid limit (b) for both eroded and non-eroded areas.

normally have high liquid limit and plastic limit because of the binding potential of clay particles with instant retardation of detachment. While sand particles are easily merged together (because of no binding) hence they easily come together and lead to low plastic and liquid limit.

Table 3 shows that the eroded and non-eroded areas differ significantly in soil porosity, soil organic matter, plasticity limit, soil porosity, percentage sand, percentage clay and slope angle and slope length. This is attributed to that as soil is eroded top soil is washed away which is rich in humus and organic matter thus leading to a significant difference in soil organic matter between eroded and non-eroded areas. Erosion also washes away finer soil particles leaving sand particles in the eroded area compared to non-eroded area with less sand particles. Soil porosity differed due to difference in particle size distribution and therefore soil porosity influences plastic limit.

Quantifying and estimating soil erodibility factor

Direct measurement of erodibility (K) factor requires long-term data and time consuming, there exists few techniques developed to estimate the K factorvalues from readily available data on soil properties (Römkens et al., 1997; Zhang et al., 2008). Several monographs have been also developed to quantitatively estimate soil erosion based on soil properties at field or farm scale (Wischmeier et al., 1971; Wischmeier and Smith, 1978; Vaezi et al., 2011). To quantify the effects of the parameteric erodibility factors on erosion, the following equation was used (Williams et al., 1984):

$$K = \left\{0.2 + 0.3e^{\left[-\frac{0.286SAN(1-SIL)}{100}\right]}\right\} \left(\frac{SIL}{CLA} + SIL\right)^{0.3} x \left(1.0 - \left(\frac{0.25C}{C+e}\right)\left(\frac{0.25C}{C+e}\right)\right) (3.72 - 2.95)(1.0 - 0.7SN1 + e(-5.51 + 22.9SN1)$$

(2)

Where; SAN, SIL and CLA are %sand, %silt and %clay fractions, respectively, and C is the soil %organic carbon content (%) and SN1=(1-SAN/100).

Based on the computations from Equation 2, the K-factor value ranged from 0.013 to 0.055 (t ha h ha^{-1} MJ^{-1}mm^{-1}) for eroded and 0.012 to 0.026 (t ha h ha^{-1} MJ^{-1}mm^{-1}) for non-eroded soils. The results generally shows that the eroded areas have average higher K- factor values [0.031 (t ha h ha^{-1} MJ^{-1}mm^{-1})] while non-eroded areas have lower K- factor values [0.018 (t ha h ha^{-1} MJ^{-1}mm^{-1})]. According to Wawer et al. (2005), soils high in clay content have low K- factor values because they are resistant to detachment because of the binding effect of clay. This is evidenced by the high clay content for non-eroded (28.2%). Similarly, non-eroded soils are less affected by erosion due to high organic matter (1.2%) while eroded areas have relatively low organic matter content (0.75%). This organic matter binds the soil particles together and creates forces between particles and thus creating stability (Brady and Weil, 2008)(Figure 3).

Although the difference in the K-factor between eroded and non-eroded areas is not statistically significant, it is still worth to note that the two areas differ significantly in soil porosity, soil organic matter, plasticity limit, soil porosity, percentage sand, percentage clay and slope angle and slope length (Table 4).

The higher K- factor values for eroded areas indicate that the soils are more prone or susceptible to erosion (Wawer et al., 2005), probably because of high %sand in eroded areas (67.6%). The high sand content contributes

Table 3. Significant difference of parameterized erodibility (K) factor properties for eroded and non-eroded areas at P>0.05.

Property	Eroded ($\overline{x}$), (n=6)	Variance (S_1^2)	Non-eroded ($\overline{x}$), (n=6)	Variance (S_2^2)	Computed t-value	Significant difference (t-test)
Bulk density	1.25	0.16	1.52	0.03	1.79	Not significant
Porosity	43.30	5.71	28.20	1.22	2.51	Significant
Plastic limit	7.17	4.32	4.57	5.25	3.13	Significant
Liquid limit	23.50	6.84	18.80	5.23	0.44	Not significant
Plasticity index	16.33	5.57	14.43	6.99	2.09	Not significant
%Organic matter	0.75	0.32	1.18	0.25	2.68	Significant
% Sand	67.60	10.14	66.40	8.12	4.58	Significant
% Silt	5.30	2.25	5.45	2.30	0.14	Not significant
% Clay	26.90	8.49	28.20	8.95	3.75	Significant
Slope angle and slope length	9°and 20 m	12.02	1° 9 m	5.66	14.57	Significant

Table 4. Significant difference of K-values for eroded and non-eroded areas at P>0.05.

Values	Eroded	Non-eroded	t-value	Significant difference (t-test
($\overline{x}$), (n=6	0.031	0.018	1.91	Not significant
Variance (S_1^2)	0.015	0.005		

to less binding of aggregates hence easily eroded. Similarly the low clay content (27%) resulted in increased K- factor value of eroded soils since clay particles hold soil particles together and make them resistant to detachment (Zhang et al., 2007). On the other hand the high K- factor value in eroded soils was primarily due to low organic matter content (0.75%) because organic matter has the capacity to bind soil particles together (Brady and Weil, 2008). Other than relatively low clay and organic matter in non-eroded areas, the high K- factor value is a result of granular soil structure since it is generally more stable than and crumb structure (Daum, 1996).

To evaluate the effectiveness of USLE-K model, comparison between the measured (Williams et al., 1984) erodibility data with the (Wischmeier et al., 1971) nomograph data was done. The nomograph (which relates K to soil properties) was developed by Wischmeier et al. (1971) with the following equation form:

$$100K = 2.1 \times 10^4 \times (2 - OM) \times M^{1.14} + 3.25 \times (St\text{-}2) + 2.5 \times (Pt - 3) \quad (3)$$

Where, OM = Organic matter content (%), M = Silt plus fine sand content (%), St = Soil structure code (very fine granular = 1, fine granular = 2, coarse granular = 3, blocky, platy or massive = 4), Pt= Permeability class (rapid = 1, moderate to rapid = 2, moderate = 3, slow to moderate = 4, slow = 5, very slow = 6).

The equation was chosen because the K-factor is a lumped parameter that represents an integrated average annual value of the soil profile reaction to the processes of soil detachment and transport by raindrop impact and surface flow (Renard et al., 1997). Consequently K-factor is best obtained from direct measurements on natural plots (Kinnell, 2010). However, this is an infeasible task on national or continental scale. To overcome this problem measured K-factor values have been related to soil properties. The most widely used relationship is the soil-erodibility nomograph of Wischmeier and Smith (1978) (Table 5).

Conclusion

The results have shown that erodibility factor K significantly correlates with slope length, organic matter and %clay fractions as well as with structural properties including plastic limit, plastic index, bulk density and soil porosity. Generally the erodibility K-factor values were high in eroded than non-eroded areas. The average K-factor value in eroded area was 0.031 (t ha h ha^{-1} MJ^{-1}mm^{-1}) with a range of 0.013 to 0.055 t h (t ha h ha^{-1} MJ^{-1}mm^{-1}). Similarly the average K-value in non-eroded area was 0.018 t h (MJmm)$^{-1}$ with a range of 0.012 to 0.026 (t ha h ha^{-1} MJ^{-1}mm^{-1}). Soils in the eroded areas with limited organic matter and subsequently high erodibility values hence large proportionately erosional losses. Interestingly there was no significant difference in K- factor values between eroded and non-eroded areas at P<0.05. Further, the K-based nomograph over-predicted the measured K-factorvalue by 10 times in eroded and 19 times in non-eroded soil, with a stronger correlation in eroded (r^2=0.77) than in non-eroded (r^2=0.10).

Table 5. The comparison of nomograph-based estimates of erodibility factor (K_{nom}) and measured erodibility factor (K_{meas}) for eroded and non-eroded.

Sampling point	Eroded			Non-eroded		
	K_{meas}	K_{Nom}	K_{Nom}/K_{Meas}	K_{meas}	K_{Nom}	K_{Nom}/K_{Meas}
A	0.027	0.30	11.1	0.017	0.30	17.6
B	0.013	0.24	18.5	0.019	0.41	21.6
C	0.021	0.31	14.8	0.026	0.30	11.5
D	0.043	0.31	7.2	0.012	0.45	37.5
E	0.055	0.45	8.2	0.013	0.27	20.8
F	0.026	0.31	11.9	0.020	0.31	15.5
Mean	0.031	0.32	10.3	0.018	0.34	19.1

K_{meas} – computed from the Williams 1984 equation and KNo_m – nomogaph (Wischmeier et al., 1971).

ACKNOWLEDGEMENTS

This work was funded by the Department of Tertiary Education Financing under the Ministry of Education Skills and Development, Botswana. The first author acknowledges Prof. O. Areola for his invaluable and useful guidance of the research project which forms part of this manuscript. The Department of Environmental Science also provided facilities to enable laboratory analysis.

REFERENCES

Abu-Hamdeh N, Al-jalil HF (1999). Hydraulically powered soil core sampler and its application to soil density and porosity estimation. Soil. Tillage Res. 52:(1-2):113-120.

Angima SD, Stott DE, O'Neill MK, Ong MK, Weesies GA (2003). Soilerosion prediction using RUSLE for central Kenyan highland conditions. Agric. Ecosyst. Environ. 97(1-3):295-308.

Anon A (1991). Soil Map of the Republic of Botswana.Soil Mapping and Advisory Services.FAO/UNDP/Botswana Government, Gaborone.

Ball A (1990). Soil Properties and Their Uses.John Wiley and Sons. London.

Bonilla CA, Johnson OI (2012). Soil erodibility mapping and its correlation with soil properties in Central Chile. Geoderma: 189–190, 116–123.

Brady CN, Weil RR (2008). The Nature and Properties of Soils. 14th Edition, Prentice Hall: New Jersey.

Brady CN, Weil RR (2002). The Nature and Properties of Soils. 13th Edition, Prentice Hall: New Jersey.

Ciesiolka CAA, Coughlan KJ, Rose CW, Escalante MC, Hashim GM, Paningbatan EP, Sombatpanit S (1995). Methodology for a multi-country study of soil erosion Manage. Soil Technol. 8:179-92.

Charman PE, Murphy BW (1991). Soils Their Properties and Management. Oxford University Press: New York.

Daum DR (1996). Soil compaction and conservation tillage. Conservation Tillage Series 3: U.S. Department of Agriculture: Agricultural Research Service.

El-Swaify SA, Dangler EW (1976). Erodibilities of selected tropical soils in relation to structural and hydrologic parameters. In: Foster, G.R. (Ed.), SoilErosion Prediction and Control. Soil and Water Conservation Society, Ankeny, IA, USA, pp. 105–114.

Faniran AO, Areola O (1978). Essentials of Soil Study (With Special Reference to Tropical Areas) London: Heinemann. Appendix: Laboratory Techniques of Soil Analysis, pp. 237-266.

FAO (2006). Guidelines of soils description, 4th ed. Rome. United Nations.

Hanks RJ, Ashcroft GL (1970).Physical properties of soil, Logan, Utah.

Kinnell PIA (2010). Event soil loss, runoff and the Universal Soil Loss Equation family of models: a review. J. Hydrol. 385:384e397.

La°en JM, Lane LJ, Foster GR (1991). WEPP: A new generation of erosion prediction technology. J. Soil Water Conser. 46(34):8.

McBride RA (1993). Soil consistency limits. In: M.R Carter (ed) "Soil Sampling Methods of Analysis," Lewis Publishers, Boca Ratan, FR: pp. 519-529.

Millward AA, Mersey JE (1999). Adapting the RUSLE to model soil erosion potential in a mountainous tropical watershed. Catena 38(2):109-129.

Morgan RPC, Quenton JN, Rickson RJ (1992). EUROSEM: Documentation Manual.' (Silsoe College: Silsoe, UK).

Morgan RPC (1996). Soil erosion and conservation. Addison-Wesley Boston MA.

Nandi A, Luffman E (2012). Erosion Related Changes to Physicochemical Properties of Ultisols Distributed on Calcareous Sedimentary Rocks. J. Sustain. Develop. 5:8.

Parysow P, Wang G, Gertner G, Anderson A (2003). Spatial uncertainty analysis for mapping soil erodibility on joint sequential simulation. Catena 53:65–78.

Radcliffe DJ, Venema JH, De Wit PV (1990). Soils of North Eastern Botswana. Government of Botswana.

Reddy K (1999). Engineering Properties of Soils Based on Laboratory. UIC.

Renard KG, Foster GA, Weesies GA, McCool DK (1997). Predicting Soil Erosion by Water: A Guide to Conservation Planning with the Revised Universal Soil Loss Equation (RUSLE).'Agriculture Handbook No: 703: USDA: Washington, DC.)

Römkens MJM, Young RA, Poesen JWA, McCool DK, El-Swaify SA, Bradford JM (1997). Soil erodibility factor (K), in: Renard, K.G., Foster GR, Weesies GA, McCool DK, Yoder DC (Compilers), Predicting Soil Erosion by Water: A Guide to Conservation Planning With the Revised Universal Soil Loss Equation (RUSLE). Agric. HB No. 703, USDA, Washington, DC, USA, pp. 65–99.

Roose EJ (1977). Application of the Universal Soil Loss Equation of Wischmeier and Smith in West Africa. In: Greenland, J., Lal, R. (Eds.). Conservation and Soil Management in the Humid Tropics. Wiley, Chichester, England, pp. 177–187.

Rose CW, Coughlan KJ, Ciesiolka CAA, Fentie B (1997). Program GUEST. (Grifth University Erosion System Template).In `A New Soil Conservation Methodology and Application to Cropping Systems in Tropical Steeplands'. (Eds K. J. Coughlan and C. W. Rose.) ACIAR Technical Report.

Rowell DL (1994). Soil Science: Methods and Applications, Longman Scientific and technical, England.

Tejada M, Gonzalez JL (2006). The relationships between erodibility and erosion in a soil treated with two organic amendments. Soil. Tillage. Res. 91:186–198.

Tiessen HJ, Moir O (1993). Total and organic carbon. In: Soil Sampling and Methods of Analysis, M.E. Carter, Ed. Lewis Publishers, Ann Arbor, MI. pp. 187-211.

Toy JT, George RF, Kenneth GR (2002). Soil Erosion.John Wiley & Sons Inc. New York.

Vaezi AR, Bahrami HA, Sadeghi SHR, Mahdian MHS (2011). Developing a nomograph for estimating erodibility factor of Calcareous soils in north west of Iran. Int. J. Geol. 4(5):93–100.

Wawer R, Nowocieñ E, Podolski B (2005). Real Calculated K-USLE Erodibiliity Factor for Selected Polish Soils. Polish J. Environ. Stud. 14(5):655-658.

Williams JR, Jones CA, Dyke PT (1984). A Modeling Approach to Determining the Relationship between erosion and productivity. Transactions of the ASAE 27(1):129-144.

Wheater CP, Cook PA (2003).Using Statistics to Understand the Environment. Routledge: New York.

Wischmeier WH, Smith DD (1978). `Predicting Rainfall Erosion Losses: A Guide to Conservation Planning.'Agriculture Handbook No. 537. (USDA: Washington, DC.).

Wischmeier WH, Johnson CB, Cross BV (1971).A soil erodibilitynomograph for farmland and construction sites. Soil Water Conser. 26:189–193.

Zachar D (1982). Soil Erosion. Prentice Hall, Upper Saddle River: New Jersey.

Zhang KL, Shu AP, Xu XL, Young QK, Yu B (2008). Soil erodibility and its estimation for agricultural soils in China. J. Arid Environ.11:1-10

Zhang ZG, Fan BE, Bai WJ, Jiao JY (2007). Soil anti-erodibility of plant communities on the removal lands in hilly-gully region of the Loess Plateau. Science Soil Water Conser. 5:7-13.

23

Long-term behavior of Cu and Zn in soil and leachate of an intensive no-tillage system under swine wastewater and mineral fertilization

Shaiane D.M. Lucas*, Silvio C. Sampaio, Miguel A. Uribe-Opazo, Simone D. Gomes, Nathalie C. H. Kessler and Naimara V. Prado

Department of Water Resources and Environmental Sanitation of the Western Paraná State University - University Street, 2069, 85819-110, Cascavel - PR, Brazil.

The fertilization of crops with swine wastewater is a common practice and attractive for the reduction of natural resources and environmental pollution control. The feasibility of such use is due to the large volume of waste generated and the amount of nutrients that are easily mineralized when applied in soil. However, copper and zinc, added in high concentrations in swine rations, can accumulate on soil-water-plant system after successive applications. Thus, this study aimed at evaluating the effects of doses of swine wastewater associated with mineral fertilization in the levels of copper and zinc in an intensive no-tillage system (maize, black oats and soybeans) for four years. The experiments were carried out in area with twenty-four lysimeters of drainage, in typical oxisol. Swine wastewater treated in stabilization ponds and biodigester was applied in doses of 0, 100, 200 and 300 m^3 ha^{-1} associated with presence and absence of mineral fertilization, resulting in eight treatments with three repetitions each. Increases of 15.60% copper and 188% zinc in soil were found after four years of swine wastewater application. In the presence of mineral fertilization, zinc was 57% higher than in its absence. Copper and zinc levels detected were within the recommended range of mineral nutrition for maize, black oats and soybeans.

Key words: Fertilization, heavy metals, leaching, pig slurry, water reuse.

INTRODUCTION

Urban, agribusiness and animal wastewater have been used in several places in the world as a viable alternative to reduce the use of natural resources, to control the pollution of water bodies, provide water and fertilizers for crops, recycle nutrients and to increase agricultural production (Ceretta et al., 2005; Freitas et al., 2005; Hespanhol, 2003; Toze, 2006). Among these types of wastewater, swine wastewater stands out because of the significant economic and social importance of such activity, as well as because of its chemical composition, which offers a large input of nutrients that are easily

mineralized when applied to soil, partly replacing the use of mineral fertilizers (Scherer et al., 2007). Many studies have been carried out to examine such properties in different crops, such as maize (Freitas et al., 2005; Prior et al., 2009; Sampaio et al., 2010; Scherer et al., 2010), beans (Doblinski et al., 2010), soybean (Dal Bosco et al., 2008; Caovilla et al., 2010; Smanhotto et al., 2010), black oats and ryegrass (Assmann et al., 2007), grassland (Queiroz et al., 2004). However, other authors have warned about the dangers of successive applications of swine wastewater, which may lead to the accumulation of heavy metals in soil and leaching to water bodies (Girotto et al., 2010). Copper and zinc stand out because pig feed have high concentrations of these elements to improve animal food converting rate (Berenguer et al., 2008; Girotto et al., 2010; Graber et al., 2005; Scherer et al.,

*Corresponding author. E-mail: shaianelucas@gmail.com.

Table 1. Soil chemical characterization of the experimental area before the performance of treatments.

pH	OM	CEC	Al + H	P	K	Ca	Mg	Cu	Fe	Mn	Zn
CaCl$_2$	g kg^{-1}	mmol$_c$ kg^{-1}					mg kg^{-1}				
5.47	14.00	95.70	38.80	3.00	1.50	34.10	21.40	8.62	69.20	39.26	0.82

Table 2. Description of treatments applied to the experimental area.

Treatment	SW doses (m^{3}ha^{-1})	Mineral fertilization
0 SW-A	0	Absent
0 SW-P	0	Present
100 SW-A	100	Absent
100 SW- P	100	Present
200 SW-A	200	Absent
200 SW- P	200	Present
300 SW-A	300	Absent
300 SW- P	300	Present

Table 3. History of crops managed under no-tillage in the experimental area from 2006 to 2009.

Accumulated days	Seeding	Crops
0	-	Natural soil
200	19/03/2006	Maize (*Zea mays*)
318	02/12/2006	Soybean (*Glycine max*)
418	19/07/2007	Black oats (*Avena strigosa*)
535	13/12/2007	Soybean (*Glycine max*)
594	05/07/2008	Black oats (*Avena strigosa*)
666	13/10/2008	Baby maize (*Zea mays*)
791	17/02/2009	Maize (*Zea mays*)
903	02/07/2009	Black oats (*Avena strigosa*)
1015	04/12/2009	Soybean (*Glycine max*)

2010). Therefore, high concentrations of copper and zinc in the soil are liable to be absorbed in large quantities by the crops and thus enter the food chain, or leach to underground water bodies (Berenguer et al., 2008; Queiroz et al., 2004).

The cumulative effect of heavy metals after long periods of application from domestic and industrial effluents has been studied (Dére et al., 2007; Katanda et al., 2007; Nyamangara and Mzezewa, 1999; Mapanda et al., 2005; Rattan et al., 2005; Rusan et al., 2007; Xu et al., 2010). However, research on long-term swine wastewater is still poor in the literature, mainly, associated with mineral fertilizers application.

In this context, this study aims to assess the effects of doses of swine wastewater associated with mineral fertilization on the concentrations of copper and zinc in intensive no-tillage system (maize, black oats, and soybeans) for over four years.

MATERIALS AND METHODS

The experiments were carried out in the Agricultural Engineering Experimental Center of the Western Paraná State University - Cascavel, Paraná, Brazil. The geographical coordinates 24º 54' south latitude and 53° 32' west longitude, at an altitude of 760 m. The climate, according to the classification of Köeppen, is super humid mesothermal subtropical, with average annual rainfall of 1800 mm, hot summers, infrequent frosts, and a trend of rainfall occurrence during the summer. However, there is no definite dry season. The mean temperature is 20ºC and relative humidity is 75% (Iapar, 1998).

The experimental area is composed of twenty-four drainage lysimeters, which represent experimental plots, with a volume of 1 m^3 and area of 1.60 m^2. The soil is classified as typical oxisol. Soil chemical characterization of the experimental area before treatment application was performed to determine pH, OM, CEC, Al+H, P, K, Ca, Mg, Cu, Fe, Mg and Zn by methods potentiometric, Walkley Black, extraction with Ca ethyl, ascorbic acid and espectometria (Mehlich), respectively, according to the protocols of Embrapa (1997) (Table 1).

Over four years, nine consecutive experiments were conducted with the same application doses of swine wastewater - SW (0, 100, 200 and 300 m^3 ha^{-1}) during crop cycles. In addition to the doses of application, the effects of mineral fertilization (MF) were assessed, having its presence and absence as factors. Thus, treatments applied to the experimental plots are presented in Table 2.

The succession of crops intensively managed under no-till is shown in Table 3. In all cultures, where necessary, cultivation was carried out using the recommended dosages and products. It is noteworthy that the initial characterization of the area in 2006, was identified as 0 (zero) days.

During the first three years, the collection of SW was made in the outlet of the first facultative pond; these values are presented in (Table 4). In the fourth year, in order to assess higher concentrations of elements present in the SW, collection was then made at the outlet of the biodigester (Table 5). Applications of SW were performed only once, usually seven days before the sowing of each crop.

Soil samples were collected before the seeding and after crop management, in a layer from 0 to 0.60 m depth to determine pH, OM, CEC and Cu and Zn content, according to the methods of Embrapa (1997). The collections of the leachate occurred on two occasions, after the first and last precipitation of crop cycle. For each sample, Cu and Zn were determined by methods 3111 in accordance with Apha (2005). The analyses of Cu and Zn content in plant tissue were determined according to Malavolta et al. (1997).

The experimental design was randomized blocks in factorial threefold with four doses of SW (0, 100, 200 and 300 m^3 ha^{-1}), two levels of mineral fertilization (absence and presence), and six time periods (200, 318, 418, 791, 903 and 1015 days), with three replications. The transformations of the data were carried out in order to meet the longitudinal analysis assumptions (Gomes, 2000). The effects of the treatments over time were assessed using longitudinal study (Singer et al., 2010). Once significance was proven in the longitudinal study, tests were performed to compare the means using Tukey test at 5% probability. The Pearson linear correlation and linear regression were used to assess the effects of

Table 4. Total rates of nutrient application, via MF and SW, from 2006 to 2008.

Time (days)	Treatments	MF (kg ha^{-1})			Application of SW for each element (Kg ha^{-1})							
		N	P	K	N	P	K	pH	EC	COD	Cu	Zn
200	0 SW-A	15	60	30	0	0	0	0	0	0	0	0
200	0 SW-P	22.50	90	45	0	0	0	0	0	0	0	0
200	100 SW-A	15	60	30	85	19.23	16.87	7.70	6.77	304.80	0.01	0.04
200	100 SW-P	22.50	90	45	77.50	19.23	16.87	7.70	6.77	304.80	0.01	0.04
200	200 SW-A	15	60	30	185	38.47	33.75	7.70	6.77	609.60	0.01	0.09
200	200 SW-P	22.50	90	45	177.50	38.47	33.75	7.70	6.77	609.60	0.01	0.09
200	300 SW-A	15	60	30	285	57.71	50.62	7.70	6.77	914.40	0.02	0.13
200	300 SW-P	22.50	90	45	277.50	57.17	50.62	7.70	6.77	914.40	0.02	0.13
318	0 SW-A	0	0	0	0	0	0	0	0	0	0	0
318	0 SW-P	0	50	50	0	0	0	0	0	0	0	0
318	100 SW-A	0	0	0	81.17	9.33	55.01	7.73	4.89	144.41	0.02	0.12
318	100 SW-P	0	50	50	81.17	9.33	55.01	7.73	4.89	144.41	0.02	0.12
318	200 SW-A	0	0	0	159.33	18.32	107.99	7.73	4.89	288.81	0.04	0.23
318	200 SW-P	0	50	50	159.33	18.32	107.99	7.73	4.89	288.81	0.04	0.23
318	300 SW-A	0	0	0	240.50	27.66	163	7.73	4.89	433.22	0.06	0.35
318	300 SW-P	0	50	50	240.50	27.66	163	7.73	4.89	433.22	0.06	0.35
418	0 SW-A	0	0	0	0	0	0	0	0	0	0	0
418	0 SW-P	0	0	0	0	0	0	0	0	0	0	0
418	100 SW-A	0	0	0	80.17	9.22	54.33	7.69	4.64	132	0.02	0.12
418	100 SW-P	0	0	0	80.17	9.22	54.33	7.69	4.64	132	0.02	0.12
418	200 SW-A	0	0	0	160.33	18.44	108.67	7.69	4.64	264	0.04	0.24
418	200 SW-P	0	0	0	160.33	18.44	108.67	7.69	4.64	264	0.04	0.24
418	300 SW-A	0	0	0	240.50	27.66	163	7.69	4.64	396	0.06	0.35
418	300 SW-P	0	0	0	240.50	27.66	163	7.69	4.64	396	0.06	0.35
535	0 SW-A	0	0	0	0	0	0	0	0	0	0	0
535	0 SW-P	0	80	40	0	0	0	0	0	0	0	0
535	100 SW-A	0	0	0	88.70	10.86	46.21	7.70	5.43	132.21	0.03	0.02
535	100 SW-P	0	80	40	88.70	10.86	46.21	7.70	5.43	132.21	0.03	0.02
535	200 SW-A	0	0	0	177.40	21.72	92.42	7.70	5.43	264.40	0.05	0.04
535	200 SW-P	0	80	40	177.40	21.72	92.42	7.70	5.43	264.40	0.05	0.04
535	300 SW-A	0	0	0	266.10	32.59	138.63	7.70	5.43	396.60	0.08	0.06
535	300 SW-P	0	80	40	266.10	32.59	138.63	7.70	5.43	396.60	0.08	0.06
594	0 SW-A	0	0	0	0	0	0	0	0	0	0	0
594	0 SW-P	0	0	0	0	0	0	0	0	0	0	0
594	100 SW-A	0	0	0	33.88	21.19	44	7.90	2.10	145	1.25	7.65
594	100 SW-P	0	0	0	33.88	21.19	44	7.90	2.10	145	1.25	7.65
594	200 SW-A	0	0	0	67.76	42.38	88	7.90	2.10	290	2.50	15.30
594	200 SW-P	0	0	0	67.76	42.38	88	7.90	2.10	290	2.50	15.30
594	300 SW-A	0	0	0	101.64	63.57	132	7.90	2.10	435	3.75	22.95
594	300 SW-P	0	0	0	101.64	63.57	132	7.90	2.10	435	3.75	22.95
666	0 SW-A	0	0	0	0	0	0	0	0	0	0	0
666	0 SW-P	45	0	0	0	0	0	0	0	0	0	0
666	100 SW-A	0	0	0	33.88	21.19	44	7.90	2.10	145	1.25	7.65
666	100 SW-P	45	0	0	33.88	21.19	44	7.90	2.10	145	1.25	7.65
666	200 SW-A	0	0	0	67.76	42.38	88	7.90	2.10	290	2.50	15.30
666	200 SW-P	45	0	0	67.76	42.38	88	7.90	2.10	290,00	2.50	15.30
666	300 SW-A	0	0	0	101.64	63.57	132	7.90	2.10	435	3.75	22.95
666	300 SW-P	45	0	0	101.64	63.57	132	7.90	2.10	435	3.75	22.95

Table 5. Total rates of nutrient application, via MF and SW, from 2009.

Time (days)	Treatments	MF (kg ha^{-1})			Application of SW for each element (Kg ha^{-1})							
		N	P	K	N	P	K	pH	EC	COD	Cu	Zn
791	0 SW-A	0	0	0	0	0	0	0	0	0	0	0
791	0 SW-P	120	80	90	0	0	0	0	0	0	0	0
791	100 SW-A	0	0	0	26.51	6.94	8.55	7.57	1.29	137.80	0.07	0.65
791	100 SW-P	120	80	90	26.51	6.94	8.55	7.57	1.29	137.80	0.07	0.65
791	200 SW-A	0	0	0	53.02	13.88	17.10	7.57	1.29	275.60	0.14	1.30
791	200 SW-P	120	80	90	53.02	13.88	17.10	7.57	1.29	275.60	0.14	1.30
791	300 SW-A	0	0	0	79.53	20.83	25.65	7.57	1.29	413.40	0.22	1.95
791	300 SW-P	120	80	90	79.53	20.83	25.65	7.57	1.29	413.40	0.22	1.95
903	0 SW-A	0	0	0	0	0	0	0	0	0	0	0
903	0 SW-P	0	0	0	0	0	0	0	0	0	0	0
903	100 SW-A	0	0	0	127.87	14.51	44.55	7.08	6.62	574	0.51	3.50
903	100 SW-P	0	0	0	127.87	14.51	44.55	7.08	6.62	574	0.51	3.50
903	200 SW-A	0	0	0	255.74	29.02	89.10	7.08	6.62	1148	1.01	7
903	200 SW-P	0	0	0	255.74	29.02	89.10	7.08	6.62	1148	1.01	7
903	300 SW-A	0	0	0	383.61	43.53	133.65	7.08	6.62	1722	1.52	10.50
903	300 SW-P	0	0	0	383.61	43.53	133.65	7.08	6.62	1722	1.52	10.50
1015	0 SW-A	0	0	0	0	0	0	0	0	0	0	0
1015	0 SW-P	0	50	50	0	0	0	0	0	0	0	0
1015	100 SW-A	0	0	0	60.48	10.74	22.45	7.09	3.95	576.70	0.20	1.45
1015	100 SW-P	0	50	50	60.48	10.74	22.45	7.09	3.95	576.70	0.20	1.45
1015	200 SW-A	0	0	0	120.96	21.48	44.90	7.09	3.95	1153.40	0.39	2.90
1015	200 SW-P	0	50	50	120.96	21.48	44.90	7.09	3.95	1153.40	0.39	2.90
1015	300 SW-A	0	0	0	181.44	32.22	67.35	7.09	3.95	1730.10	0.59	4.35
1015	300 SW-P	0	50	50	181.44	32.22	67.35	7.09	3.95	1730.10	0.59	4.35

SW, MF, and soil chemical properties (pH, OM, CEC) on Cu and Zn in the soil.

RESULTS AND DISCUSSION

Soil parameters

The simultaneous longitudinal study of the factors SW, MF and time of application (T) indicated that only the isolated T provided significant differences in pH, OM, CEC, Cu and Zn.

The pH increased from 5.47 to 6.77 in the soil (0 to 0.60 m) at 1015 days of SW application (Table 1 and 6). Increased pH values of soil were consistent with the high pH values of SW used, ranging from 7.08 to 7.70 (Table 4 and 5), which according to Ayers and Westcot (1991) conforms to the interval from 6.50 to 8.40 recommended for irrigation. Such increased pH in the soil is due to the alkaline characteristics of SW (Chantigny et al., 2004), and the extracts of cultures kept in the soil surface (Diehl et al., 2008). The organic matter generated by the decomposition of this material adsorbs H and Al increasing soil pH (Amaral et al., 2000; Anami et al., 2008; Miyazawa et al., 1993). The major means found for

OM refer to the 318 and 1015 days after the application of SW (Table 6), and that is associated with the concentrations of biodegradable organic matter and inert something is missing here present in the wastewater of anaerobic pond and biodigester (Table 4 and 5). Increased soil OM found at 903 days can be due to a change in the collection point of SW, as well as due to the migration of organic compounds of low molecular weight over the soil profile (Novais et al., 2007). Furthermore, increases in OM with corresponding increase in depth of soil observed could be due to the presence of roots at deeper layers. Xu et al. (2010) found increases in OM in soil layer 0.10 m, at 53, 89 and 185%, in areas with applications of household sewage in periods of 3, 8 and 20 years, respectively.

The highest CEC was found at 200 days, differing from the other periods. The applications of SW increased the initial values of CEC at 200 and 318 days (Table 6) as a function of K, Ca and Mg (Smanhotto et al., 2010; Queiroz et al., 2004). After a period of 418 days, CEC showed some stabilization, with a mean value of 120 mmol$_c$ dm^{-3}. This conforms with the findings of Katanda et al. (2007) which reported that high CEC in soils irrigated with sewage water can be attributed to the OM presence in these waters.

Table 6. Means for the parameters of the soil in the layer from 0 to 0.60 m.

Time (days)	pH (CaCl$_2$)	OM (g kg^{-1})	CEC(mmol$_c$ kg^{-1})	Cu (mg kg^{-1})	Zn (mg kg^{-1})
200	6.40^b	23.69^b	165.32^a	5.39^d	0.68^d
318	6.46^b	28.17^a	144.25^b	3.98^e	1.29^{bc}
418	6.66^{ab}	20.28^c	128.15^b	7.95^{ab}	1.43^{bc}
791	6.68^{ab}	20.08^c	109.90^c	8.47^a	1.59^{bc}
903	6.89^a	23.99^b	128.16^b	7.02^{bc}	2.06^b
1015	6.77^{ab}	25.77^{ab}	116.39^{bc}	6.44^{cd}	3.29^a

Transformation of Johnson for CEC and Cu; Box and Cox for Zn; same letters in column indicate means equal to the 5% level of significance by Tukey test.

Cu increased from 418 to 1015 days compared to 200 and 318 days (Table 6), indicating that Cu may accumulate with successive applications of SW. At 791 days, Cu approached the initial characterization of 8.62 mg kg^{-1} (Table 1). Cu concentrations were high in the initial characterization of soil, and they significantly reduced after the first cultivation.This indicates that this micronutrient, the part extracted from the culture, becomes less available. This fact could be due to the formation of stable organo-metallic complexes of low solubility, which, in addition to complexing with organic substances, they bind to non-exchangeable fractions of the soil, such as iron and manganese oxides (Oliveira and Mattiazzo, 2001; Queiroz et al., 2004). Cu decreased significantly in periods of soybean cultivation, confirming that soybean crop is more sensitive to this micro-nutrient than maize and wheat (Lavado et al., 2001).

The highest content of Zn was found at 1015 days, statistically differing from the other periods. Increased Zn (Table 6) indicates the existence of a positive linear relationship with successive applications of SW over time, also shown in the longitudinal study. Smanhotto et al. (2010) found an increase in Zn at rates of 200 and 300 m^3 ha^{-1} in soybean and Freitas et al. (2005) also found increase in Zn in the soil with the application of raw drained SW and found concentrations ranging from 13.10 to 16.30 mg kg^{-1}. Works of Queiroz et al. (2004) and Dal Bosco et al. (2008) also reported an increase in Zn in the surface layers (0 to 0.20 m) of soil with the application of SW.

Accumulation of copper and zinc for each treatment compared to controls (0 SW-A, 0 SW-P) at 1015 days and the initial values of these metals prior to application of the treatments are shown in Table 7.

The accumulation of Cu and Zn were more significant for the treatments of 300 SW-A and 300 SW-P compared to controls (0 SW-A and 0 SW-P). The same occurred to the 100 SW-P treatments for Cu. Increases of 15.60% for Cu in the 300 SW-A treatment and 188% for Zn in the 300 SW-P treatments were found in relation to respective controls at 1015 days. Cavanagh et al. (2011) found increases of 31% for Cu and 64% for Zn in the application of sewage in relation to controls.

The national standard determined that good quality soils must have concentrations below the limits of 200 and 450 mg kg^{-1} of soil, for Cu and Zn respectively (CONAMA, 2009). Guidelines values for agricultural soils are 140 mg kg^{-1} for Cu and 300 mg kg^{-1} for Zn, according to the international standard (CEC, 1986). Although Cu and Zn (Table 7) are below the national and international recommended limits, the results may indicate possible contamination of surface and groundwater.

Berenguer et al. (2008) have assessed for over six years the effect of SW in rates of 29 and 51 m^3 ha^{-1} per year in sandy soil under maize cultivation and found increases of 32% for Cu and 11% for Zn. Girotto et al. (2010) have found concentrations of Cu and Zn, of 85.70 and 70 mg kg^{-1} respectively in the upper soil layer (0 to 0.02 m) under no-tillage when dosing maximum swine manure (80 kg ha^{-1}) for 78 months.

Mapanda et al. (2005) found much higher values of Cu in soils irrigated for more than two decades with household sewage (from 21 to 145 mg kg^{-1}) in relation to irrigated areas for a period of 5 to 15 years using household sewage and industrial wastewater (7 to 44 mg kg^{-1}). Similar results were found in the works of Nyamangara and Mzezewa (1999), Rattan et al. (2005) and Khan et al. (2008).

According to Graber et al. (2005), Cu and Zn can form complexes with humic substances that influence respective mobility in the soil. They also form complexes with phosphates, and may increase the solubility of these elements in the soil solution. Scherer et al. (2007) and Girotto et al. (2010) have reported that Cu and Zn are accumulated in the soil, especially in bioavailable forms, and the highest Cu is found in the organic and mineral soil and Zn in mineral form.

According to Novais et al. (2007), in acidic environments, Cu has great mobility, which is inversely proportional to the uptake of the element to the solid fraction. In this sense, it can be inferred, considering the increase in values of pH in this study compared with initial soil (Table 6), associated with clay soil and high concentrations of OM, that Cu has low mobility in the soil, which favors its retention in it, for a long period of time. To assess which parameters most affect the accumulation of Cu and Zn, a Pearson linear correlation was carried out (Table 8).

Strong and positive correlations were found for pH and

Table 7. Accumulation of Cu and Zn (mg kg^{-1}) in each treatment compared to controls, at 1015 days of application of SW.

Treatments	Cu	Accumulation of Cu	Zn	Accumulation of Zn
Natural soil	8.62		0.82	
0 SW-A	5.96		2	
0 SW-P	6.03		1.92	
100 SW-A	6.07	0.11	2.54	0.54
100 SW-P	6.80	0.77	3.01	1.09
200 SW-A	6.44	0.48	3.07	1.07
200 SW-P	6.56	0.53	3.91	1.99
300 SW-A	6.89	0.93	4.40	2.40
300 SW-P	6.80	0.84	5.53	3.61

Table 8. Pearson's linear correlation of soil parameters assessed for each treatment.

	Cu	Zn	pH	OM	Cu	Zn	pH	OM
		0 SW-A				0 SW-P		
Zn	0.15				-0.24			
pH	0.03	0.75*			-0.25	0.74**		
OM	-0.74*	0.12	0.49		-0.69**	0.46	0.67**	
CEC	-0.23	-0.14	0.07	0.56	-0.74**	-0.15	0.16	0.72**
		100 SW-A				100 SW-P		
Zn	0.13				0.11			
pH	-0.34	0.73*			-0.14	0.34		
OM	-0.69**	0.47	0.89*		-0.60	0.40	0.63	
CEC	-0.85*	-0.11	0.27	0.66	-0.38	-0.10	0.73**	0.28
		200 SW-A				200 SW-P		
Zn	0.05				0			
pH	-0.32	0.64			-0.23	0.34		
OM	-0.69**	0.34	0.87*		-0.49	0.61	0.81*	
CEC	-0.76*	-0.43	0.18	0.60	-0.52	0.03	0.64	0.65
		300 SW-A				300 SW-P		
Zn	-0.51				0.01			
pH	-0.30	0.44			-0.33	0.43		
OM	-0.63	0.63	0.91*		-0.58	0.43	0.94*	
CEC	-0.70**	-0.09	0.38	0.48	-0.76	-0.27	0.43	0.62

*p<0.05;**p<0.01

Zn in the 0 SW-A, 0 SW-P, and 100 SW-A treatments. Thus, Zn is more closely related to soil pH. Positive correlations were found for OM, pH and CEC, indicating that higher levels of OM improve soil buffer capacity. According to Novais et al. (2007), OM increases from 20 to 90% of CEC of the upper layers of soil minerals and virtually the entire CEC of organic soils.

Cu negatively correlated with the other soil parameters. In treatments 0 SW-A, 0 SW-P, 100 SW-A and 200 SW-A strong correlations was found for Cu and OM, and in treatments 0 SW-P, 100 SW-A, 200 SW-A and 300 SW-A also strong correlations existed for Cu and CEC. These results indicate that Cu content is inversely related to OM and CEC; higher OM and CEC, the higher the complexation and adsorption of such element in the organic and mineral fractions were obtained, and thus less available in the soil (Scherer et al., 2010).

Changes in pH and Zn over four years of study were

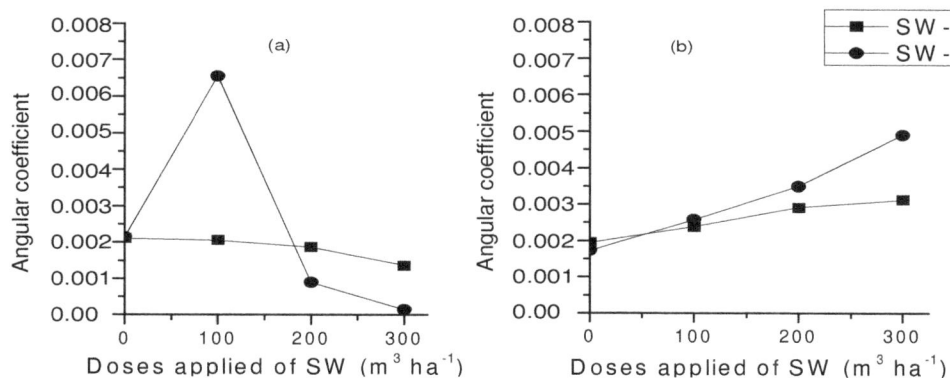

Figure 1. Angular coefficient of the linear model of pH (a) and Zn (b) as a function of the application doses of SW.

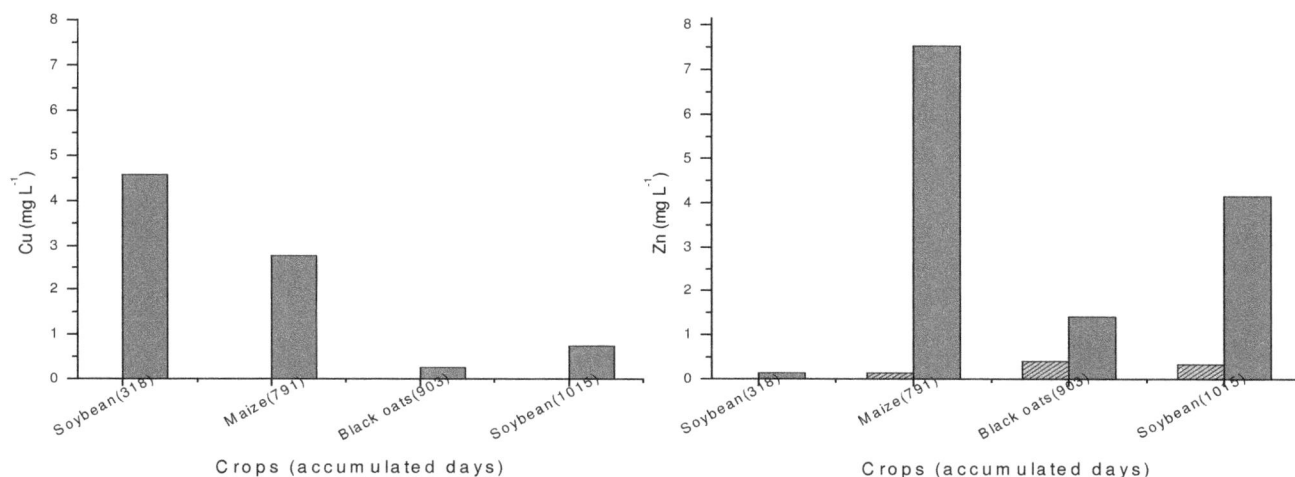

Figure 2. Minimum and maximum concentrations of Cu and Zn in leachate.

explained by simple linear models (Figure 1). For pH, an almost constant behavior is noted when MF is not used. Increased Zn trend over time is mainly influenced by application rate, and different behaviors are noted in the presence and absence of MF; in the presence of MF the slope coefficient increases significantly, that is, the daily rate of Zn accumulation in the soil. According Girotto et al. (2010) the Zn is strongly adsorbed to the functional groups, especially the mineral fraction of the soil, which explains the Zn increase with mineral fertilizer addition in our study.

Leachate parameters

Cu at 318 and 791 days is over the maximum limit of 2 mg L^{-1} for groundwater determined by Conama (2008) (Figure 2). The high value of Cu at 318 days is consistent with the lowest value found in the soil (Table 6), showing loss of the element by leaching. Zn to 791 days is also

above the recommended maximum limit of 5 mg L^{-1} (Conama, 2008). Similarly to the soil, Zn indicates tendency to increase the leached over time. The amount of SW associated with mineral fertilization highlights the leaching risk of Cu and Zn in groundwater.

Agronomic parameters

Maize, black oats and soybeans crops have Cu and Zn values within the range recommended by Malavolta et al. (1997). This range recommended to maize and black oats is 6 at 20 mg Cu kg^{-1} and 10 at 30 mg Zn kg^{-1}, while, to the soybean crop is 15 at 50 mg Cu kg^{-1} and 21 to 50 mg Zn kg^{-1} (Figure 3). At 903 days, black oats culture absorbed Cu near to the maximum limit of good plant nutrition, providing low concentrations in the leachate Figure 2). In general, soybean accumulated more Cu than maize and oats, thereby reducing Cu in the soil (Table 6) and in leachate (Figure 2). Zn increasing trend

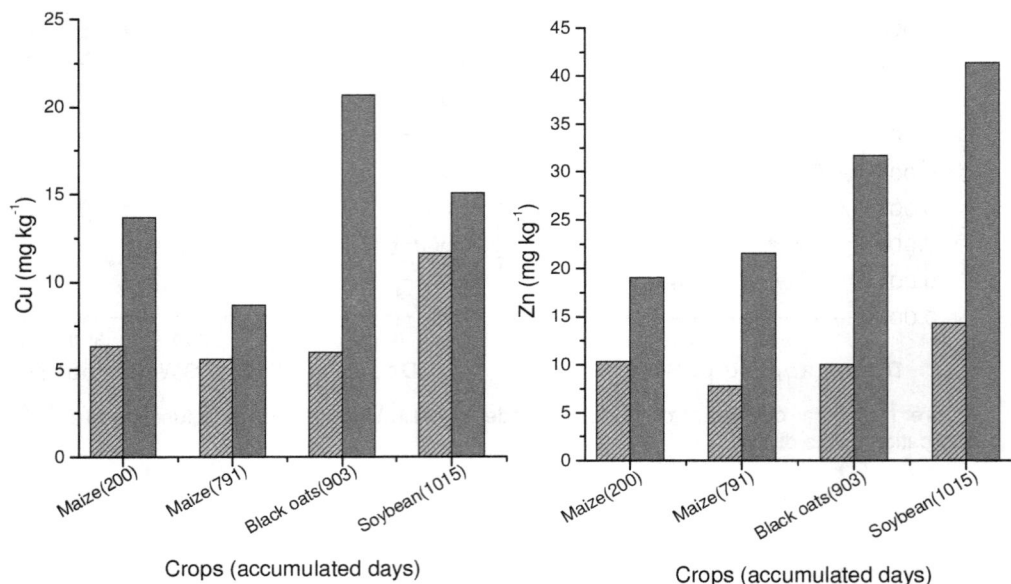

Figure 3. Minimum and maximum concentrations Cu and Zn in the plant tissue over the time of SW and MF application.

similar to that found in the soil (Table 6) was also detected in the plant tissue of cultures (Figure 3). The accumulation of Zn in soil-plant system favors higher losses of this micro-nutrient in leachate (Figure 2). Berenguer et al. (2008) have assessed Cu and Zn in maize for after six and seven years of SW application in rates of 29 and 51 m^3 ha^{-1} per year, and have found that the amounts absorbed by cultures were in accordance with the limits of phytotoxicity. The authors found at the end of the seventh year of experiment the values 2.11 and 17.5 mg kg^{-1} for Cu and Zn, respectively.

Conclusions

Finally, the applications of swine wastewater, for maize, black oats and soybeans treated with biodigester and stabilization pond, in the long term (four years), favored the accumulation of copper (15%) and zinc (188%) in the soil and therefore provides higher concentrations that standard limits determined for leachate. Considering controls as reference, zinc in the presence of mineral fertilization has shown soil accumulation potential 57% higher than in the absence of MF. In maize, black oats and soybeans, Cu and Zn are in the recommended range of good nutrition.

ACKNOWLEDGEMENTS

The authors would like to thank the Western Paraná State University, Conselho Nacional de Desenvolvimento Científico e Tecnológico (CNPQ) and the Coordenação de Aperfeiçoamento de Pessoal de Nível Superior (CAPES) for supporting this study.

ABBREVIATIONS

SW, Swine wastewater; **T,** time of application; **pH,** hydrogen potential; **OM,** organic matter; **CEC,** cation exchange capacity; **P,** phosphorus, **Al+H,** acidity; **K,** potassium, **Ca,** calcium, **Mg,** magnesium; **Cu,** copper; **Fe,** iron; **Mn,** manganese; **Zn,** zinc; **EC,** electric conductivity; **N,** nitrogen; **COD,** chemical oxygen demand; **MF,** mineral fertilization.

REFERENCES

Amaral AS, Spader V, Anghinoni I, Meurer EJ (2000). Resíduos vegetais na superfície do solo afetam a acidez do solo e a eficiência do herbicida flumetsulam. Cienc. Rural. 30(5):789-794.

Apha (2005). American Public Health Association. Standard methods for the examination of water and wastewater. New York: American Public Health Association, p.1600.

Anami MH, Sampaio SC, Suszek M, Damasceno S, Queiroz MMF (2008). Deslocamento miscível de nitrato e fosfato proveniente de água residuária da suinocultura em colunas de solo. Rev. Bras. Eng. Agríc. Ambient. 12(1):75-80.

Assmann TS, Assmann JM, Cassol LC, Diehl RC, Mantelo C, Magiero E (2007). Desempenho da mistura forrageira de aveia preta mais azevém e atributos químicos do solo em função da aplicação de esterco líquido de suínos. Rev. Bras. Cienc. Solo. 31:1515-1523.

Ayers RS, Westcot DW (1991). A qualidade da água na agricultura. Campina Grande: UFPB. p. 218.

Berenguer P, Cela S, Santiveri F, Boixadera J, Lloveras J (2008). Copper and zinc soil accumulation and plant concentration in irrigated maize fertilized with liquidi swine manure. Agron. J. 100(4):1056-1061.

Conama (2008). Conselho Nacional do Meio Ambiente (Brasil). Resolução nº 396, de 03 de abril de 2008. Dispõe sobre aclassificação dos corpos de água e diretrizes ambientais para o seu enquadramento das águas subterrâneas e dá outras providências. Diário oficial da União, 07 de abril de 2008.

Conama (2009). Conselho Nacional do Meio Ambiente (Brasil). Resolução n. 420, de 28 de dezembro de 2009. Critérios e valores orientadores de qualidade do solo quanto à presença de substâncias químicas. Diário Oficial da União. Brasília, 20 de Dezembro de 2009.

Ceretta CA, Basso CJ, Vieira FCB, Herbes MG, Moreira ICL, Bergwanger AL (2005). Dejeto de suínos: I – perdas de nitrogênio e fósforo na solução escoada na superfície do solo, sob plantio direto. Cienc. Rural. 35(6):1296-1304.

Caovilla FA, Sampaio SC, Smanhotto A, Nobrega LHP, Queiroz MHF, Gomes BM (2010). Características químicas de solo cultivado com soja e irrigado com água residuária da suinocultura. Rev. Bras. Eng. Agríc. Ambient 14(7):692–697.

Cavanagh A, Gasser MO, Labrecque M (2011). Swine wastewater as fertilizer on willow plantation. Biomass Bioenergy 35(10):4165-4173.

CEC (1986). Council Directive of 12 June 1986 on the protection of the environment, and in particular of the soil, when sewage sludge is used in agriculture. Official J. European Communities L181:6-12.

Chantigny MH, Rochette P, Argers DA, Massé D, Côté D (2004). Ammonia volatilization and selected soil characteristics following application of anaerobically digested swine wastewater. Soil Sci. Soc. Am. J. 68(1):306-312.

Dal Bosco TC, Sampaio SC, Uribe-Opazo MA, Gomes S.D, Nóbrega LHP (2008). Aplicação de água residuária de suinocultura em solo cultivado com soja: cobre e zinco no material escoado e no solo. Eng. Agric.28(4):699-709.

Diehl RC, Miyazana M, Takahashi HW (2008). Compostos orgânicos hidrossolúveis de resíduos vegetais e seus efeitos nos atributos químicos do solo. Rev. Bras. Cienc. Solo. 32: 2653-2659.

Dére C, LamyY I, Jaulin A, Maizeu S (2007). Long-term fate of exogenous metals in a sandy Luvisol subjected to intensive irrigation with raw wastewater. Environ. Pollut. 145(1):31-40.

Doblinski AF, Sampaio SC, Nóbrega LHP, Gomes SD, Dal Bosco TC (2010). Nonpoint source pollution by swine farming wastewater in bean crop. Eng. Agric. Ambient. 14(1):87–93.

Embrapa (Empresa Brasileira de Pesquisa Agropecuária) (1997). Manual de Métodos de análise de solo. 2. ed. Rio de Janeiro. Embrapa Solos. p. 212.

Freitas WS, Oliveira RA, Cecon PR, Pinto FA, Galvão JCC (2005). Efeito da aplicação de águas residuárias de suinocultura em solo cultivado com milho. Eng. Agric. 13(2):95-102.

Girotto E, Ceretta CA, Brunetto G, Santos DR, Silva LS, Lourenzi CR, Lorensini F, Vieira RCB, Shumatz R (2010). Acúmulo e formas de cobre e zinco no solo após aplicações sucessivas de dejeto líquido de suínos. Rev. Bras. Cienc. Solo. 34(3):955-965.

Gomes FP (2000). Curso de estatística experimental. 14. ed. Piracicaba: Degaspari. p. 477.

Graber I, Hansen JF, Olesen SE, Pettersen J, Ostergaard HS, Krogh L (2005). Accumulation of copper and zinc in Danish agricultural soils in intensive pig production areas. Danish J. Geog. 105(2):15-22.

Hespanhol I (2003). Potencial de reúso de água no Brasil: agricultura, indústria, município e recarga de aqüíferos. Barueri: Manole. pp. 37-96.

Iapar (Instituto Agronômico do Paraná). Cartas Climáticas do Estado do Paraná. Londrina: IAPAR, 1998.

Katanda Y, Mushonga C, Banganayi F, Nyamangara J (2007). Effects of heavy metals contained in soil irrigated with a mixture of sewage sludge and effluent for thirty years on soil microbial biomass and plant growth. Phys. Chem. Earth. 32:1185-1194.

Lavado RS, Porcelli CA, Alvarez R (2001). Nutrient and heavy metal concentration and distribution in maize, soybean and wheat as affected by different tillage systems in the Argentine Pampas. Soil Tillage Res. 62(1-2):55-60.

Khan S, Cao Q, Zheng YM, Huang YZ, Zhu YG (2008). Health risks of heavy metals in contaminated soils and food crops irrigated with wastewater in Beijing, China. Environ. Pollut. 152(1):686-692.

Malavolta E, Vitti GC, Oliveira S (1997). Avaliação do estado nutricional das plantas.2. ed. Piracicaba: Potafós. p. 319.

Mapanda F, Mangwayana EN, Nyamangara J, Giller KE (2005).The effect of long-term irrigation using wastewater on heavy metal contents of soils under vegetables in Harare, Zimbabwe. Agric. Ecosyst. Environ. 107:151-165.

Miyazawa M, Pavan MA, Calegari A (1993). Efeito de material vegetal na acidez do solo. Rev. Bras. Cienc. Solo. 17: 411- 416.

Novais RF, Alvarez VH, Barros NF, Fontes RLF, Cantarutti RB, Neves JCL (2007). Fertilidade do Solo. Viçosa: Sociedade Brasileira de Ciência do Solo. p.1015.

Nyamangara J, Mzezewa J (1999). The effect of long-term sewage sludge application on Zn, Cu, Ni and Pb levels in a clay loam soil under pasture grass in Zimbabwe. Agriculture. Ecosyst. Environ. 73(3):199-204.

Oliveira FC, Mattiazzo ME (2001). Mobilidade de metais pesados em um Latossolo amarelo distrófico tratado com lodo de esgoto e cultivado com cana-de-açúcar. Sci. Agric. 58(4):807-812.

Prior M, Smanhotto A, Sampaio SC, Nóbrega LHP, Uribe-Opazo MA, Dieter J (2009). Acúmulo e percolação de fósforo no solo devido a aplicação de água residuária da suinocultura na cultura do milho (Zea maysL.). Pesquisa Aplicada & Agrotecnologia. 2:89-96.

Queiroz FM, Matos AT, Pereira OG, Oliveira RA (2004). Características químicas de solo submetido ao tratamento com esterco líquido de suínos e cultivado com gramíneas forrageiras. Cienc. Rural. 34(5):1487-1492.

Rattan RK, Datta SP, Chhonkar PK, Suribabu K, Singh AK (2005). Long-term impact of irrigation with sewage effluents on heavy metal content in soils, crops and groundwater – a case study. Agriculture. Ecosyst. Environt.109(3-4):310-322.

Rusan MJM, Hinnawi S, Rousan L (2007). Long term effect of wastewater irrigation of forage crops on soil and plant quality parameters. Desalination 215:143-152.

Sampaio SC, Fiori MGS, Uribe-Opazo MA, Nóbrega LHP (2010). Comportamento das formas de nitrogênio em solo cultivado com milho irrigado com água residuária da suinocultura. Eng. Agric.30:138-149.

Scherer EE, Baldissera IT, Nesi CN (2007). Propriedades químicas de um Latossolo Vermelho sobre plantio direto e adubação com esterco de suínos. Rev. Bras. Cienc. Solo. 31:123-131.

Scherer EE, Nesi CN, Massotti Z (2010). Atributos químicos do solo influenciados por sucessivas aplicações de dejetos suínos em áreas agrícolas de Santa Catarina. Rev. Bras. Cienc. Solo. 34:1375-1383.

Singer JM, Nobre JS, Rocha FMM (2010). Análise de dados longitudinais. (Versão parcial preliminar). Departamento de estatística – USP: SP. p. 215.

Smanhotto A, Souza AP, Sampaio SC, Nóbrega LHP, Prior M (2010) Cobre e zinco no material percolado e no solo com a aplicação de água residuária de suinocultura em solo cultivado com soja. Eng. Agric. 30(2):347-357.

Toze S (2006). Reuse of effluent water-benefits and risks. Agric. Water Manage. 80(1-3):147–159.

Xu J, Xu L, Chang AC, Zhang Y (2010). Impact of long-term reclaimed wastewater irrigation on agricultural soils: A preliminary assessment. J. Hazard. Mater. 183:780-786.

24

Productivity of soil fertilised with faecal manure of cattle fed Calliandra, Gliricidia and Luecaena browse/maize silages

Habib Kato[1]*, Felix Budara Bareeba[2] and Elly Nyabombo Sabiiti[3]

[1]Department of Agriculture, Kyambogo University, P. O. Box 1, Kyambogo, Uganda.
[2]Department of Animal Science, Makerere University, P. O. Box, 7062, Kampala, Uganda.
[3]Department of Crop Science, Makerere University, P. O. Box, 7062, Kampala, Uganda.

The use of farmyard manure to improve soil productivity is a key element in mixed crop/livestock farming systems. Browse/maize silage mixtures (20% browse DM basis) of Calliandra, Gliricidia, and Leucaena and maize silage alone and their corresponding cattle faecal manure were applied to the soil to determine their effect on soil productivity. Hopi Red Dye Amaranthus (*Amaranthus cruentus*) was used as the test crop. Its dry matter (DM) yield, and crude protein and fiber content were determined. The browse/maize silages had higher total N and narrower C:N ratio than that of maize silage alone. Calliandra/maize silage mixture had higher levels of ADFN and lignin. Cattle faecal manure derived from the browse/maize silages had higher total N and ADFN content and narrower C:N ratios compared to the faecal manure from maize silage alone. Application of the browse/maize silages and the corresponding cattle faecal manures raised C, N and C:N of the soil compared to the control soil. The treated soils maintained higher levels of C and N up to the third crop but the C:N ratios were similar with the control soil. Amaranthus DM yield was highest with faecal manure treatments. Treatments with silages had no DM yield advantage over the control soil. Addition of faecal manure from maize silage alone gave highest DM yield followed by feacal manure from Gliricidia/maize and Leucaena/maize silages. Faecal manure from Calliandra/maize silage gave lower yields in spite of having similar levels of N. Much of its N was fiber bound, thus limiting availability of the N for plant growth.

Key words: Calliandra, Gliricidia, Leucaena, Amaranthus, browse.

INTRODUCTION

The use of manure to improve soil productivity is a key element in mixed agro-pastoral farming systems (Fernandez-Rivera et al., 1993; Karl et al., 1994; Twinamasiko, 2001; Kajura, 2001). Farmyard manure has greater positive effects on soil than resting periods, with crop response to farmyard manure increasing linearly with rate of application and total number of applications (Ssali, 2001). The beneficial effects of farmyard manure on the soil is due to the presence of hormones, vitamins, antibiotics and growth regulating substances such as biotin, whose stimulating effect on root growth and on the growth of micro organisms (yeast cultures) has been demonstrated (Karl et al., 1994). The nutrient release from farmyard manure is dependent on how fine it is spread, the proportion of soluble N, the C:N ratio and storage methods (Karl et al., 1994; Katuromunda et al., 2010). Also, the type of feed and passage of the feed through the ruminants` digestive tract affects the availability of nutrients in the manure. The total amount and proportion of nutrients excreted in faeces and urine vary with the lignin:neutral-detergent fiber (NDF), lignin:N and polyhenol:N ratios of the diets (Powell et al., 1994). The browses, Calliandra and Leucaena have substantial levels of tannins and lignin

---
*Corresponding author. E-mail: habibkyanda@yahoo.co.uk.

(Bareeba and Aluma, 2000) which bind protein and protect it from degradation in the rumen (Fahey et al., 1980; Navas-Camancho et al., 1993). Tannins have a binding effect and interfere with adequate utilization of browse protein by grazing animals and shift N excretion from urine to faeces and from faecal microbial to undigested feed N (Topps, 1992). The binding of protein could also subsequently affect the availability of manure N in the soil when animals are fed browse diets. In Uganda, zero-grazing farmers are encouraged to preserve forage which is usually elephant grass (*Penisetum purpureum*) by ensiling. The crude protein (CP) content of the silage could be improved to 12% DM by ensiling the elephant grass with browses such as Calliandra, Leucaena or Gliricidia (Kato et al., 2004). These browses have been identified and recommended as the most suitable species for supplementation of indigenous goats under tethering or free range grazing conditions in the sub-humid zones of Uganda (NARO, 1999; Sabiiti, 2001). The purpose of this study was to investigate the effect of applying faecal manure from cattle fed Calliandra, Gliricidia and Leucaena browse/maize silages on soil C and N and DM yield and chemical composition of Amaranthus grown on the soil.

MATERIALS AND METHODS

Experiment

The study investigated Amarathus growth response to application to the soil of Calliandra, Gliricidia and Leucaena browse/maize silages (20% browse DM basis), maize silage alone and the corresponding faecal manure from cattle fed the silages.

Preparation of the treatment materials

The silages and manure were obtained from a feeding trial of 12 weeks. The silages and manure samples were air dried, aggregated and ground in a laboratory mill to pass through a 2 mm sieve before application to the soil. The soil used in the experiment was collected from a crop field. The soil was spread out in a screen house to dry after which it was ground to pass through a 2 mm sieve and mixed thoroughly. A sample of the soil was taken for laboratory analysis.

Setting up the experiment and sampling procedures

The experiment was a pot experiment in a screen house set out in a completely randomized design (CRD) with treatments replicated four times. Four kilograms of soil were used per pot. The ground silages and manures applied at the rate of 5 g/kg of soil was equivalent to the rate of manure application of 10 t/ha for 3 years for Uganda, each hectare being equivalent to a plough share of 2,000,000 kg of soil (Anderson and Ingram, 1993; Ssali, 2001). Four successive plantings were made without changing the soil or treatments in the pots. Each planting cycle lasted four weeks. Planting done by seed broadcast in the pots and the seeds covered thinly with soil. Adequate moisture for crop growth was maintained by watering with tap water. The pot soils were sampled at the beginning and after the first, second and third harvests for chemical

analysis. Five plants from each pot were harvested at flower bud stage by cutting the plants at collar level, weighed and fresh weights recorded. Whole plant materials for the five plants from each pot were packed in paper bags and dried in the oven at 60°C for 72 h to determine dry matter (DM) content (%) and yield (kg). The dried plant materials were ground in a laboratory mill to pass through a 2 mm sieve and preserved for chemical analysis.

Chemical analyses of the soil, browse/maize silages, faecal manures and the harvested plant material

The soil samples were analyzed for soil OM and C according to Walkley and Black (1993) and N by the Kjeldal method (AOAC, 1990). Samples of the silages, faecal manures and the harvested plant material were analyzed for total N, crude fiber (CF), and ash by the AOAC (1990) procedures, non protein nitrogen (NPN) by the trichloro acetic acid method (Gaines, 1977), neutral detergent fiber (NDF), acid detergent fiber (ADF) and acid detergent lignin (ADL) by the Van Soest and Robertson (1985) procedures; neutral detergent fiber nitrogen (NDFN) and acid detergent fiber nitrogen (ADFN) were obtained by determining N in the NDF and ADF residues respectively.

Data analysis

The data obtained was subjected to statistical analysis by Genstat Release 12.2 and differences between the means were separated using the least significant difference (LSD) method at probability level of 5%.

RESULTS AND DISCUSSION

Composition of the browse/maize silages

The chemical composition of the browse/maize silages applied to the soils is shown in Table 1. The browse/maize silage mixtures had higher levels of N than maize silage alone and maize silage had a much higher C:N ratio. The browse/maize silage mixtures had much less soluble N (NPN), particularly Calliandra mixture compared to maize silage alone. Also, Calliandra mixture had higher levels of insoluble fiber (ADF), fiber bound N (ADFN) and lignin. Therefore, while application of the browse/maize silage materials would introduce more N in the soil than maize silage alone, their low levels of soluble N would limit their decomposition in the soil and availability of their nutrients for plant growth. Maize silage had less N and a wider C:N ratio, but much more soluble N, which would make it more decomposable in the soil and make its nutrients available for plant growth.

Chemical composition of the faecal manures

The chemical composition of faecal manure from feeding cattle the browse/maize silages and applied to the soil is shown in Table 2. Compared to the feed silages (Table 1), all the faecal manures had less C, higher levels of total N and narrower C:N ratio. Digestion in the animals'

Table 1. Chemical composition (% DM) of the browse/maize silages (20% browse) applied to the soil as compost.

| Composition | Browse/maize silages | | | |
	Calliandra	Gliricidia	Leuceana	Maize
Carbon	46.47	46.23	45.82	46.32
Nitrogen	1.70	1.90	1.87	1.28
C:N ratio	27.34	24.33	24.50	36.19
NPN (% Total N)	21.95	34.71	27.02	41.28
NDF	67.07	61.70	64.90	70.60
ADF	42.40	37.33	35.58	38.27
NDFN (%Total N)	34.49	25.40	31.72	21.52
ADFN (%Total N)	24.82	11.58	14.42	16.14
ADL	19.58	11.85	13.75	6.40

Table 2. Chemical composition (%DM) of the fecal from feeding cattle the browse/maize silages.

| Composition | Browse/maize silages | | | |
	Calliandra	Gliricidia	Leuceana	Maize
Carbon	41.99	40.10	40.84	41.09
Nitrogen	2.27	2.05	2.28	1.80
C:N ratio	18.50	19.56	17.91	22.83
NPN (% Total N)	19.13	18.43	14.58	12.84
ADF	46.96	46.13	46.50	46.22
ADFN (% Total N)	25.42	19.97	26.11	15.64

gut could have reduced the C content in the undigested material. The N retained in the undigested material depends on the lignin content of the feed material. The browses, Calliandra and Leucaena for example, have substantial levels of tannins and lignin (Bareeba and Aluma, 2000) which bind protein and protect it from degradation. Thus, much of the N in the browse/maize silage diets was retained compared to the non-browse maize silage diet (Fahey et al., 1980; Navas-Camancho et al., 1993). The faecal manures also had less soluble N (NPN) and higher levels of fiber-bound N (ADFN). Faecal manure from maize silage had the lowest level of total and soluble N although the maize silage feed had the highest level of soluble N compared to the browse/maize silages (Table 2). Since maize silage had a high proportion of soluble N much of its N could have been metabolized into microbial protein in the rumen and subsequently utilized by the host animal (Fahey et al., 1980; Navas-Camancho et al., 1993). Faecal manure from maize silage had the lowest level of fiber bound N compared to faecal manure from browse/maize silages. Therefore, although faecal manure from maize silage had a lower level of total N, its N would be readily available for plant growth. Faecal manure from browse/maize silages particularly Calliandra/maize silage had higher levels of fiber bound N, which would limit its OM decomposition and release of their nutrients in the soil for

plant growth. The high level of fiber bound N of the browse/maize silages, particularly Calliandra/maize silage mixture persisted in their corresponding faecal manures. This means that the faecal manure from Calliandra silage mixture would decompose slowly and its beneficial effects could last longer in the soil.

Effect of treatments on C, N and C:N in the soil

The levels of C, N and C:N ratios in the soils after the third harvest are shown in Table 3. The results at the initial stage indicated that the treated soils attained higher levels of C, N and a wider C:N ratio as a result of the materials added compared to the control soil. There were variations between treatments which could have arisen because the treatments were based on the manure kg rate of 10 t/ha rather than on the basis of N/ha to be supplied by each treatment. The values after the third harvest indicated that the treated soils maintained higher levels of C and N than the control soil. However, the C:N ratio was similar for all treatments. The fact that the treated soils maintained higher levels of C and N indicated that the treatments were effective in having the expected effects on the soil.

The C:N ratio was similar for all treatments therefore, it is possible that irrespective of rate of decomposition of

Table 3. Mean levels (%) of carbon, nitrogen and C:N ratio of soils fertilized with the browse/maize silages and corresponding fecal manure.

Composition	Control	CS	CF	GS	GF	LS	LF	MS	MF	LSD
				Soil treatments						
Initial										
Carbon	0.97	1.39	1.31	1.39	1.22	1.38	1.31	1.31	1.39	0.12
Nitrogen	0.12	0.150	0.130	0.130	0.120	0.140	0.130	0.130	0.14	0.01
C:N ratio	8.08	9.27	10.08	10.69	10.17	10.57	10.08	10.08	9.92	0.87
Mean values after three planting cycles										
Carbon	1.01	1.36	1.41	1.53	1.27	1.41	1.53	1.49	1.31	0.12
Nitrogen	0.12	0.16	0.15	0.15	0.15	0.15	0.15	0.14	0.14	0.01
C:N	10.58	8.50	9.40	10.20	8.47	9.40	10.20	10.64	9.36	0.87

LSD is at P<0.05; CS, Calliandra/maize silage; CF, fecal manure from Calliandra/maize silage; GS, Gliricidia/maize silage; GF, fecal manure from Gliricidia/Maize silage; LS, Leucaena/maize silage; LF, fecal manure from Leucaena/maize silage; MS, maize silage; MF, fecal manure from maize silage.

Table 4. Fiber and protein composition (% DM) of *Amaranthus cruentus* grown on soils fertilized with browse/maize silages and corresponding cattle fecal manure.

Composition	Control	CS	CF	GS	GF	LS	LF	MS	MF	LSD
				Soil treatments						
DM(kg)	0.23	0.26	0.36	0.20	0.40	0.23	0.43	0.22	0.46	0.08
C P	25.08	28.37	27.85	28.03	28.35	27.27	26.98	23.71	28.20	4.04
C F	9.68	10.50	10.31	9.41	11.24	11.32	12.03	9.84	10.75	2.16
NDF	32.02	32.75	25.04	32.63	34.57	34.65	36.28	33.55	42.44	7.78
ADF	14.81	14.82	13.63	15.04	17.77	15.51	18.17	13.59	17.63	2.79

LSD is at P<0.05; CS, Calliandra/maize silage; CF, fecal manure from Calliandra/maize silage; GS, Gliricidia/Maize silage; GF, fecal manure from Gliricidia/maize silage; LF, fecal manure from Leucaena/maize silage; MS, maize silage; MF, fecal manure from Maize silage.

the added organic matter to the soil, the C:N ratio subsequently settled to the constant soil ratio of about 10:1 (Russell, 1961; Brady, 1974).

Effect of treatments on DM yield

The mean DM yield (kg) of Amaranthus after treatment of soil with selected manure is as shown in Table 4. *Amaranthus cruentus* grown on soils treated with faecal manure had significantly higher DM yield than that grown on soils treated with silages and the control soil. Soil treatment with silages had no yield advantage over the control soil. Treatment of soils with the faecal manures could have given higher yields because the manures had more N, and narrower C:N ratios, which could have made more N available for plant growth. Of the manure treatments, the soil treated with fecal manure from maize silage alone gave the highest yield in spite of having lower N content and a wider C:N ratio. However, it had less fiber-bound N (Table 2) and therefore, more of its N could have been available for plant growth (Delve et al., 2001). Of the browses, treatment with fecal manure from

Gliricidia and Leucaena silage mixtures gave higher yields than treatment with fecal manure from Calliandra silage mixture. The manure from Calliandra/maize silage in spite of having a similar level of N, more of its N was fiber bound (Table 2) and therefore, may not have released as much N in the soil to support as much or higher DM yield.

Effect of treatments on the composition of the plant material

The protein and fiber composition of the Amaranthus is shown in Table 4. Treatment with either silage or manure tended to increase CP and fiber content compared to the control. Amaranthus grown on soils treated with faecal manure had significantly higher DM yield or growth. Hence, the higher fiber content of the Amaranthus. Also, the faecal manures had more N and narrower C:N ratios that could have made more N available that could have resulted into higher CP content of the Amaranthus. Also, the faecal manures had more N and narrower C:N ratios that could have made more N available which could have resulted into higher CP content of the Amaranthus. Lubis and Kumagai (2007) observed that manure application had no effect on chemical composition of maize and sorghum except for crude protein (CP) and that manure application higher than 8 ton/ha might cause greater DM yield of sorghum and maize without any increase in fiber fraction. Mpairwe et al. (2002) observed that intercropping forage legumes with cereals generally resulted in fodder with higher fodder CP concentration, lower NDF and higher DM degradability than fodder from sole cereals.

Conclusion

The results showed that faecal manure from browse/maize silage mixtures would improve soil productivity. Faecal manure from maize silage alone or forages alone would be more effective than faecal manure from browse/maize silage mixtures in improving soil productivity. Also, the faecal manure from Gliricidia/maize and Leucaena/maize silage mixtures would be more effective than the faecal manure from Calliandra/maize silage mixture. In spite of faecal manure having the ability to improve soil productivity it would not have direct bearing on crude protein and fiber content of forage DM.

RECOMMENDATIONS

The results obtained in this study are indicative results that need to be tested further under field conditions.

ACKNOWLEDGEMENTS

This study was funded by Kyambogo University.

REFERENCES

Anderson JM, Ingram JS (1993). Tropical soil biology and fertility: A hand book of methods, Second Edition. C. A. B. Int. (82):70-82.

AOAC (1990). Official methods of analysis, 15th Edition. AOAC Inc. Arlington, Virginia. p. 22201 USA.

Bareeba FB, Aluma J (2000). Chemical composition, phenolics and in vitro organic matter digestibility of some multipurpose tree species used for agroforestry in Uganda. Uganda Vet. 6:89-92

Brady NC (1974). The Nature and Properties of Soil. Macmillan Publishing Co. pp. 257-258, 268-269.

Delve RJ, Cadisch G, Tanner JC, Thope W, Thorine PJ, Giller KE (2001). Implications of livestock feeding management on soil fertility in the small holder farming systems of Sub-Saharan Africa. Agric. Ecosyst. Environ. 84:227-243.

Fahey Jr GC, At-Haydari SY, Hindis FC, Short DE (1980). Phenolic compounds in roughages and their fate in the digestive system of sheep. J. Anim. Sci.50:1165-1172.

Fernandez-Rivera S, Williams TO, Hiernaux P, Powell JM (1993). Faecal excretion by ruminants and manure availability for crop production in semi- arid West Africa. ILCA, Semi-arid Zone Programme. Niamey, Niger.

Gaines TP (1977). Determination of protein nitrogen in plants. AOAC. 60:590.

Kajura S (2001). Beef Production. In: Mukiibi, J. K. (Ed.), Agriculture in Uganda, Fountain Publishers/National Agric. Res. Organ. (NARO) Uganda 4:1-17.

Karl M, Muller–Samann, Kotschi J (1994). Sustaining growth, soil fertility management in tropical small holdings. CTA/GTZ Margraf Verlag publishers.

Kato H, Bareeba FB, Ebong C, Sabiiti E (2004). Fermentation characteristics and nutrient composition of browses ensiled with maize fodder. Afr. Crop Sci. J. 12:393-400.

Katuromunda S, Sabiiti EN, Bekunda MA (2010). Effect of method of storing cattle faeces on the physical and chemical characteristics of the resultant composted cattle manure. Uganda J. Agric. Sci. (In press).

Lubis AD, Kumagai H (2007). Comparative study on yield and chemical composition of maize (Zea mays L.) and sorghum (Sorghum bicolor Moench) using different levels of manure application. Anim. Sci. J. 78:605-612.

Mpairwe DR, Sabiiti EN, Ummuna NN, Tegegne A, Osuji P (2002). Effect of intercropping cereal crops with forage legumes and source of nutrients on cereal grain yield and fodder dry matter yields. Afr. Crop Sci. J. 10(1):81-97.

NARO (1999). NARO achievements during FY 1998/99 and plans and budgets for FY   1999/2000. National Agricultural Research Organization (NARO), Uganda.

Navas-Camacho A, Max AL, Aurora C, Hector A, Juan CL (1993). Effect of supplementation with a tree legume forage on rumen function. Livestock Research for Rural Development, 5(2).

Russell W (1961). Soil conditions and plant growth. Ninth Edition. Longmans. pp. 49-249.

Sabiiti EN (2001). Pastures and range management. In: Mukiibi JK (ed.) Agriculture in Uganda, Fountain Publishers/National Agricultural Research Organization (NARO) Uganda, 4:237-297.

Ssali H (2001). Soil fertility. In: Mukiibi JK (ed.), Agriculture in Uganda. Fountain Publishers/National Agricultural Research organization (NARO) Uganda 1:104-135.

Topps JH (1992). Potenial, composition and use of legume shrubs and trees as fodders for livestock in the tropics. J. Agric. Sci. Cambridge. (192):1-8.

Powell JM, Fernandez-Rivera S, Hoffs S (1994). Effects of sheep diet on nutrient cycling in mixed farming systems of semi-arid Agriculture d West Africa. Ecosyst. Environ. 48:262-271.

Twinamasiko NI (2001). Dairy Production In: Mukiibi JK (ed.), Agriculture in Uganda Fountain Publishers/National Agricultural Research Organization (NARO) Uganda 4:18-24.

Van Soest PJ, Robertson JB (1985). Analysis of forage and fibrous foods. A laboratory manual for Animal Science Cornell University, Ithaca. New York, USA. p. 613.

Walkley A, Black IA (1034). An examination of the detjareff method for determining soil organic matter and a proposed chromic acid titration method. Soil Sci. J. 34:29-38

25

Vertical differentiation analysis of sierozem profile characteristics in Yili-River valley, China

Ruixiang Shi[1], Xiaohuan Yang[1], Hongqi Zhang[2] and Lixin Wang[2]

[1]State Key Lab of Resources and Environmental Information System, Institute of Geographic Sciences and Natural Resources Research, Chinese Academy of Sciences, Beijing 100101, China.
[2]Institute of Geographic Sciences and Natural Resources Research, Chinese Academy of Sciences, Beijing 100101, China.

Research on the vertical differentiation of sierozem profile characteristics can provide a basis to implement soil resource assessment and develop sustainable land management system. In this study, 189 sampling data from 50 sierozem profiles along Yili River-valley in northwest of China were chosen to analyze the vertical change of edaphic characteristics and soil ionic composition – including soil organic matter (SOM), pH value, electrical conductivity (EC), total salinity (TS), contents of CO_3^{2-}, HCO_3^-, Cl^-, SO_4^{2-}, K^+, Na^+, Ca^{2+} and Mg^{2+} in soluble salt. Then sierozem characteristics in Yili-River valley were compared with those in central Asia, Ningxia and Gansu provinces to reveal the vertical change and regional difference. The results showed that vertical difference of sierozem profile characteristics in different regions had some common: SOM content decreased with soil depth, while pH value, EC, contents of TS and most ions increased with soil depth. But there was obvious regional difference in spatial variation and intensity of sierozem characteristics due to more arid climate in Yili-River valley. The results could be used for regional ecological construction and sustainable development.

Key words: Yili-River valley, sierozem, vertical differentiation.

INTRODUCTION

Currently, there is a new concern about feeding the world (UN Millennium Project, 2005) and the land needed for energy and increased animal production (UNEP, 2007). It is necessary to design and study sustainable land management system (Smyth and Dumanski, 1995). Study of soil profile characteristics can provide a basis for implementing soil resource assessment and management.

In arid or semiarid climate region, the natural environmental systems are very fragile and the ecosystems are in precarious balance. Land is believed to be a non-renewable resource at a human time-scale and some adverse effects of degradative processes on land quality are irreversible, especially in fragile environmental system (Eswaran et al., 2001). Western China is a national ecological and environmental security barrier for its special geographical position and semi-arid / arid climate. It is very important for realizing national ecological improvement and sustainable development to make western ecological construction. Studying vertical change of soil properties in Yili-River valley will provide the basis for regional ecological construction and sustainable development.

Studies on soil characteristics were mainly from descriptions of the typical soil profiles before 1990s (Xinjiang institute of comprehensive expedition and Institute of soil sciences, Chinese academy of sciences, 1965; Wu and Quan, 1985). However, it was deficient to research regional soil profile groups comprehensively. In recent years, quantitative information on soil variability has become increasingly important for soil resource assessment and management. Much research has focused on spatial distribution and variation of soil elements with the development of geo-statistics and GIS technology (Yost et al., 1982a, b; Xu and Webster, 1983; Webster and Nortcliff, 1984; Tao, 1995a, b; Haefelel and Wopereis, 2005; Liao et al., 2008). However, these studies mainly focused on single soil characteristic, such as soil nutrient, salinity or heavy metals, while less research has been done to analyze soil comprehensive elements. Research focused much more on the horizontal distribution and variation of soil characteristics than on vertical variation of soil attributes (Acosta et al., 2011; McBratney et al., 2000). Recently, although more studies have been done on soil distribution and spatial variation in Yili-River valley (AMAITT et al., 2006; Shi et al., 2009; Mamattuesun et al., 2010; Hamid et al., 2011; Zulpiya et al., 2011; Gulnar et al., 2011), it was scarce to study the soil comprehensive characteristics based on the whole soil profiles.

Sierozem is the soil developed in the dry climate and desert steppe in warm temperate zone, which is ash brown, yellow brown or brown, has low humus and weak leaching (National soil census office, 1998). There is patch or pseudohyphae calcium carbonate deposition and strong lime reaction within full sierozem profile. Most sierozem distributed on hilly loess and high terraces of river, where the underground water level is generally deep. Most parent material of sierozem is loess, while less is the material from alluvial fan and proluvial. The vegetation dominated with perennial dryland grass, shrubs and artemisia.

The distribution of sierozem is incontinuous in China, which can be divided into eastern part (including Gansu, Ningxia etc.) and western part (Yili-River valley). There is obvious difference in seasonal distribution of precipitation and composition of vegetation in two parts. The rainy season is from July to September in eastern part, while the rainfall is even, a little more in spring in western part. The dominant species are *Stipa breviflora*, *Gobi Stipa*, *sandy Stipa*, *Peganum harmala* and drought-tolerant *Artemisia* in eastern part, while *Artemisia*, *Poa annua*, *Carex*, *Kochia scoparia* and other ephemeral plants are the main species in western part. Sierozem in Yili-River valley is believed to be the extension of sierozem zone in central Asia. There are many common characteristics in soil formation between arid regions in northwest of China and central Asian countries. Study of soil profile characteristics in Yili will benefit academic exchange between China and the central Asian countries and promote the development of soil sciencs. In this study, the sierozem profile characteristics were comprehensively analyzed and compared with those from central Asia, Ningxia and Gansu to express the common and difference of vertical change of sierozem profiles, based on field survey, sampling and testing on soil organic matter (SOM), pH value, electric conductivity (EC), total salt (TS) and 8 main ions.

It is supposed that SOM content decreases with soil depth as usual; there is high pH value, EC, more soluble salt in deposition layer of sierozem profile due to the semi-arid/arid climate - strong transpiration, low rainfall and weak leaching in Yili-River valley; there is much difference in sierozem profile characteristics among different regions for that climate and vegetation are the two important factors in sierozem formation.

MATERIALS AND METHODS

Description of study area

The study area is located in Yili-River valley, surrounded by Tianshan Mountains at east, south and north, west open to Kazakhstan. It has a temperate semiarid continental climate, and is dominated by westerly winds throughout the year. The winter climate is controlled mainly by the intensity and position of the Siberia high pressure cell, and is also influenced by the north branch of the westerlies; the summer climate is in part affected by the Indian low pressure cell, when the southern branch of the westerlies shifts northwards (Li, 1991). There is the most abundant rainfalls in Yili valley of Xinjiang, because it is open on the west to humid airflow. The study area lies in the second and third terrace and tilted plain, with elevation ranging from 600 to 1100 m. It has a low annual precipitation (230 to 350 mm) and an annual potential evapotranspiration, reaching 1200 to 1900 mm. The aridity index k1 is between 2.5 to 4.2. The mean annual atmospheric temperature is 7.9 to 9.2°C, with monthly averages ranging from a minimum of -12.2°C in January to a maximum of 22.7°C in July. Natural vegetation is desert Artemisia grassland. There is always 1 to 3 cm yellow grey surface crust. The soil is mainly light-middle loam. The zonal soil of the study area is sierozem, which belongs to Aridisols (Order) – Warm Aridisols (Suborder) – Sierozem (Great Group) – Ordinary Sierozem (Subgroup) according to genetic soil classification of China (Fan and Cheng, 1986; National soil census office, 1998). The genetic soil classification was strongly based on the setting of the soil's location (Scalenghe and Ferraris, 2009). Genetic soil classification of China (GSCC) differs sharply from WRB (IUSS Working Group WRB, 2006) in its underlying understanding about the genetic process. The Aridsols (order) - sierozem (Great group) in GSCC and the Cambisols in WRB reference soil group have a 44.2% maximum referencibility (Shi et al., 2010).

Data sampling, testing and analytical methods

From late June to early July in 2008, the working group made a field survey along Yili-River valley, where 69 soil profiles were

---

[1] $k = (0.16 \sum \geq 10°C)/r$. Where $\sum 10°C$: ≥10°C accumulated temperature; r: Rainfall during the same period. K: 2.0-4.0, arid; >4: hyper arid.

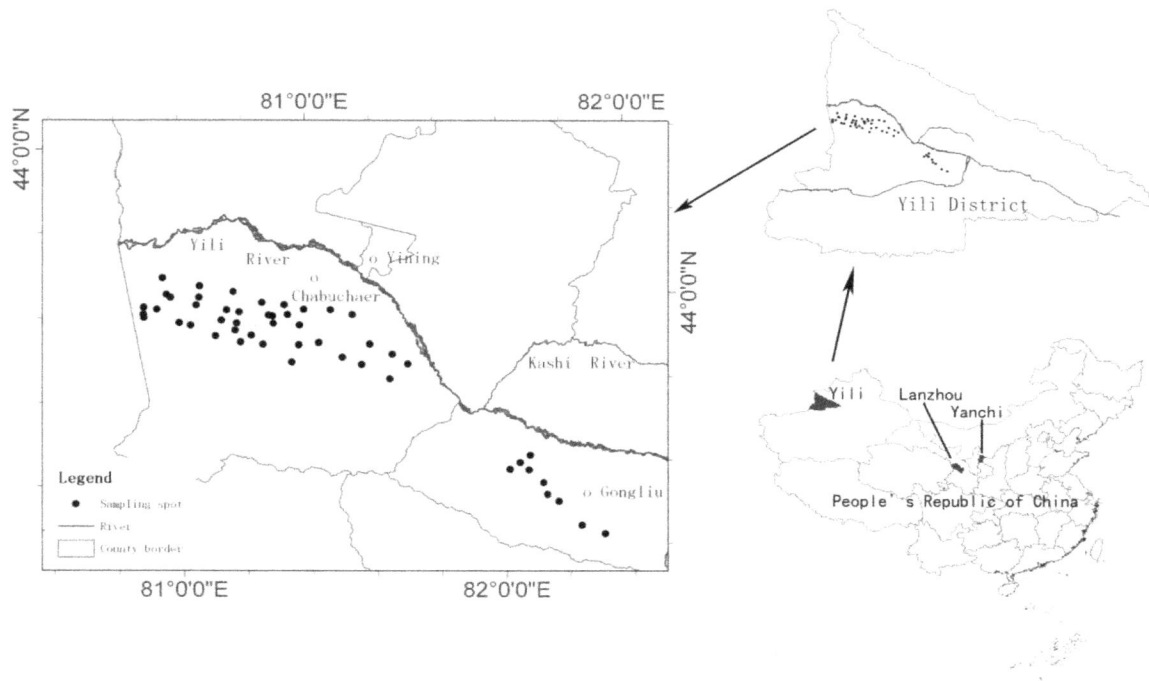

Figure 1. Position of Sierozem soil profiles in the study.

investigated and sampled. The positions of soil profiles were determined by land cover, landform, groundwater and altitude. Most soil profiles lay uncultivated land. Soil samples were collected from middle of each soil physical layer. In this study, 50 sierozem soil profiles and 189 sampling data were used. Position of sierozem soil profiles was shown in Figure 1.

The sampling data were tested in Xinjiang Institute of Ecology and Geography, Chinese Academy of Sciences. The following soil properties were measured: SOM, pH, EC and soluble salt content. SOM was analyzed using rapid dichromate oxidation techniques (Tiessen and Moir, 1993). Soil pH value was determined with pH electrode in the saturation paste, 1:5 soil: water mixture (pHS-2C, Shanghai, China). EC was measured by conductivity meter (DDS-307, Shanghai, China). TS was determined by using the distillation residue method. CO_3^{2-} and HCO_3^- were measured by means of dual indicator- neutralization titration (Bao, 2000). Cl^- was measured by using the $AgNO_3$ titration (Davey and Bembrick, 1969); SO_4^{2-} was measured by using the EDTA indirect titration. Na^+ and K^+ were measured with the flame spectrometry (Flame photometer 6410, Shanghai, China); Ca^{2+} and Mg^{2+} were measured by using the EDTA compleximetry (Working group on analytical chemistry in department of chemistry, Hangzhou University, 1997). Then sample data were analyzed to get the vertical distribution of soil elements using Matlab7.0 software programming (Figures 2 and 3). Data from each physical layers in soil profiles were identified in the middle of the layer.

In order to get data in every 25 cm below the surface, soil sample data were recalculated by weighted average. Then mean and standard deviation was separately calculated using spss 16.0 software in every 25 cm below the surface (Table 1). Statistical regression analysis was used to model the change of SOM content with soil depth using SPSS 16.0 software. Firstly, specified SOM content as dependent variable, soil depth as independent variable. Then take natural logarithm of dependent variable, and make linear regression using independent variable and natural logarithm of

dependent variable. Finaly, calculate the dependent variable. Regional comparison of sierozem profile properties between Yili-River valley and Central Asia, Gansu and Ningxia was used to learn the common and difference of vertical change of sierozem soil profiles.

RESULTS

Characteristic and vertical differentiation of SOM, pH value, EC and TS

Topsoil SOM content, between 7.0 and 19.0 g kg^{-1}, was low in sierozem soil profiles in Yili-River valley. SOM content decreased with soil depth in sierozem profiles (Figure 2 and Table 1). Averaged SOM content was 11.18 g kg^{-1} at 0 to 25 cm depth, while 5.92 g kg^{-1} at 50 to 75 cm depth, 3.64 g kg^{-1} at 100 to 125 cm depth. The distribution of SOM content with soil depth was simulated to exponential function,

$$y = \exp\left(\frac{x + 109.85}{36.957}\right) \qquad r^2 = 0.4696$$

Where y was the SOM content (g kg^{-1}); x was the soil depth (cm); r: Correlation coefficient.

The quantity and vertical distribution of sierozem SOM content were tightly related to the remnant body of plant and decomposition, mineralization and leaching of humus. In the study area, zonal vegetation was desert

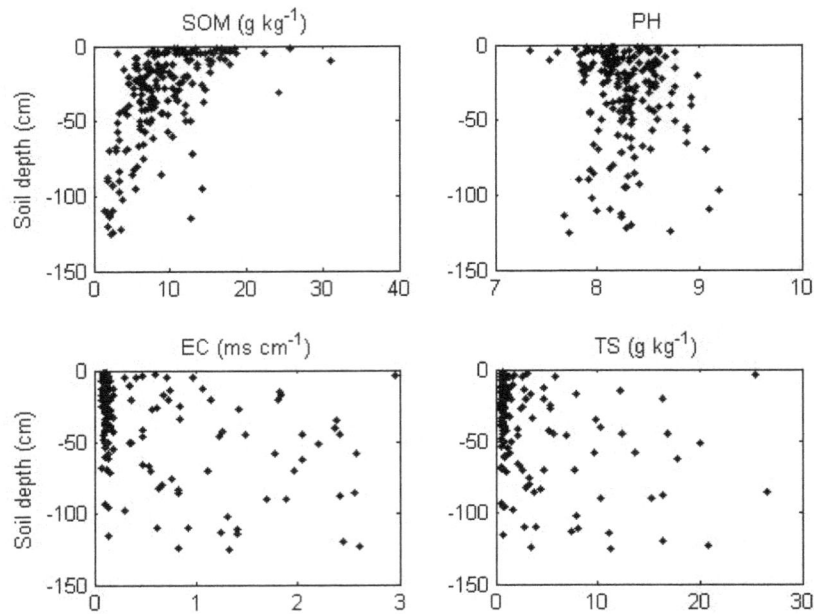

Figure 2. Vertical distribution of SOM, pH value, EC and TS in Sierozem soil profiles in Yili-River valley.

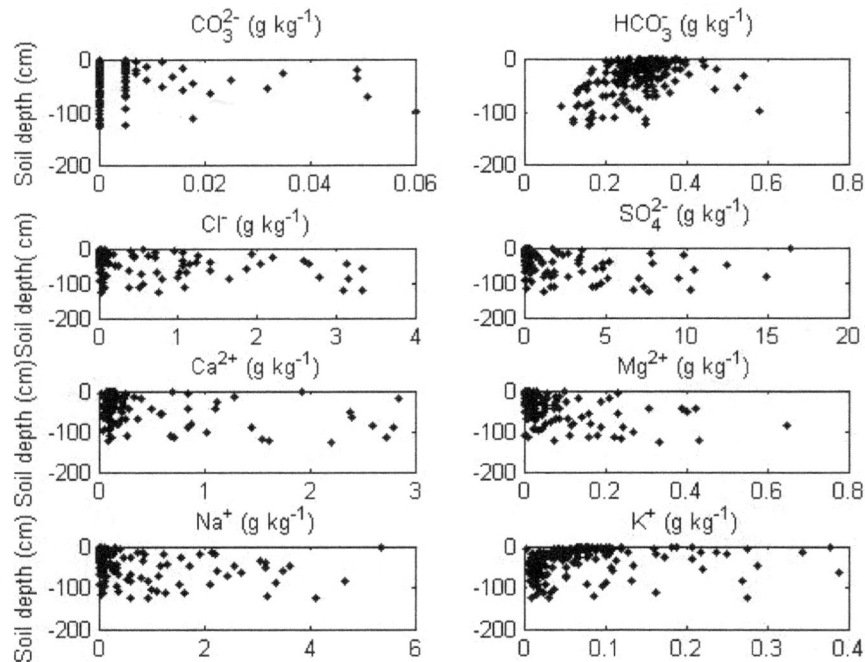

Figure 3. Vertical distribution of main cations and anions in Sierozem profiles in Yili.

steppe, whose cover degree was about 20 to 35% (Wu and Quan, 1985). Main species were *Stipa breviflora*, *Artemisia* and *Chenopodiaceae* pod. The vegetation structure was simple and had low production. In arid climate, decomposition and leaching of humus was slow,

which had low contribution on sierozem SOM content and made SOM mainly accumulate in surface soil.

The pH value in Sierozem topsoil mainly ranged from 7.8 to 8.7, which showed soil was alkaline in Yili-River valley. pH value fluctuantly increased with soil depth in

Table 1. Statistics of soil characteristics every 25 cm depth in sierozem profiles in Yili-River valley.

Soil depth (cm)	SN	SOM (g kg^{-1})		pH		EC (ms cm^{-1})		TS (g kg^{-1})	
		Mean	St.Dev.	Mean	St.Dev.	Mean	St.Dev.	Mean	St.Dev.
0-25	42	11.18	4.39	8.24	0.24	0.32	0.44	1.90	3.05
25-50	28	8.07	3.69	8.36	0.23	0.70	0.81	4.02	5.20
50-75	20	5.92	2.81	8.40	0.31	0.95	0.87	5.91	6.79
75-100	16	4.82	2.98	8.29	0.37	1.02	0.79	6.69	7.23
100-125	10	3.64	3.45	8.34	0.50	1.18	0.82	8.08	7.00

Soil depth (cm)	SN	CO_3^{2-} (g kg^{-1})		HCO_3^- (g kg^{-1})		Cl^- (g kg^{-1})		SO_4^{2-} (g kg^{-1})	
		Mean	St.Dev.	Mean	St.Dev.	Mean	St.Dev.	Mean	St.Dev.
0-25	42	0.00	0.01	0.31	0.05	0.16	0.32	0.79	1.94
25-50	28	0.01	0.01	0.28	0.07	0.62	0.90	1.74	2.85
50-75	20	0.01	0.01	0.26	0.10	0.75	0.95	2.89	3.88
75-100	16	0.01	0.02	0.24	0.11	0.73	1.00	3.49	4.08
100-125	10	0.01	0.02	0.23	0.14	0.83	1.26	4.29	3.80

Soil depth (cm)	SN	Ca^{2+} (g kg^{-1})		Mg^{2+} (g kg^{-1})		Na^+ (g kg^{-1})		K^+ (g kg^{-1})	
		Mean	St.Dev.	Mean	St.Dev.	Mean	St.Dev.	Mean	St.Dev.
0-25	42	0.20	0.35	0.03	0.04	0.27	0.59	0.08	0.07
25-50	28	0.30	0.48	0.08	0.11	0.81	1.15	0.06	0.08
50-75	20	0.50	0.74	0.11	0.13	1.14	1.33	0.08	0.10
75-100	16	0.66	0.87	0.15	0.15	1.17	1.32	0.06	0.07
100-125	10	0.96	1.02	0.17	0.15	1.22	1.40	0.08	0.08

SN, Sample number.

sierozem profiles (Figure 2 and Table 1). pH values were discrete at the bottom of soil profiles, which showed much difference among parent materials beneath soil profiles. Soil pH value was related to soil parent material, bio-climate and agricultural activities (Xiong and Li, 1987). Sierozem parent material was alkaline, which was composed of loess and loess-like material with high calcium. Calcium was usually leached to the lower layer slowly and deposited to calcium carbonate, which made pH value high in the deposition layer. Besides, there was weak humification in the surface soil, which could neutralize part alkalescence. So pH value increased with depth in sierozem profile and had a peak value in the deposition layer.

Soil EC and TS are indicators of water-soluble salt content. Figure 2 showed that most sierozem soil profiles depth less than 60 cm had low EC and TS, which were respectively less than 0.2 ms cm^{-1} and 2.0 g kg^{-1}. In general, EC and TS increased with soil depth in sierozem profiles (Table 1). The standard deviations of EC and TS were very high, which indicated that there was much difference among different soil profiles. Soil soluble salt content was related to the climate, groundwater level and composition, parent material, vegetation and human activities (National soil census office, 1998). In the study area, underground water level was low (Wu and Quan, 1985), which had little effect on soil salt. In the parent

material of sierozem soil, quartz and feldspar contents were more. But silicate and aluminum silicate were mostly insoluble. So soluble salt content was less and EC was low in the sierozem soil. Due to the arid climate, evaporation was much greater than precipitation. Soluble salt could partly leach in soil profile with low rainfall.

Characteristic and vertical differentiation of main ions

Among 4 anions, SO_4^{2-} content was the most, while CO_3^{2-} content was the least. Among 4 cations, Na^+ content was the most, followed by Ca^{2+}, K^+ and Mg^{2+} contents were less (Table 1). So there were more SO_4^{2-}, Na^+ and Ca^{2+} contents in sierozem profiles in Yili-River valley. In four anions, only HCO_3^- content decreased with soil depth, while other anions contents increased with soil depth. Averaged SO_4^{2-} content was 0.79 g kg^{-1} at 0 to 25 cm.depth, while 2.89 g kg^{-1} at 50 to 75 cm depth, 4.29 g kg^{-1} at 100 to 125 cm depth. Averaged Cl^- content was 0.16 g kg^{-1} at 0 to 25 cm.depth, while 0.75 g kg^{-1} at 50 to 75 cm depth, 0.83 g kg^{-1} at 100 to 125 cm depth. Averaged HCO_3^- content was 0.31 g kg^{-1} at 0 to 25cm.depth, while 0.26 g kg^{-1} at 50 to 75cm depth, 0.23 g kg^{-1} at 100 to 125cm depth.

In sierozem profiles calcium carbonate lay from top to

bottom. In the surface soil, there was more CO_2 resulted from respiration of plant roots. Carbonate could be easily hydrolyzed to bicarbonate under the effect of humus. Thus, in the sierozem profiles, HCO_3^- content was more in the surface soil, less and less with the profile down.

In four cations, only K^+ content fluctuates with soil depth, while other cations' contents increased with soil depth. Averaged Na^+ content was 0.27 g kg^{-1} at 0 to 25 cm.depth, while 1.14 g kg^{-1} at 50 to 75 cm depth, 1.22 g kg^{-1} at 100 to 125 cm depth. Averaged Ca^{2+} content was 0.20 g kg^{-1} at 0 to 25 cm.depth, while 0.50 g kg^{-1} at 50 to 75 cm depth, 0.96 g kg^{-1} at 100 to 125 cm depth.

DISCUSSION

The result showed that our assumption was partly right. SOM content really decreased with soil depth as usual and there is high pH value in the deposition layer of sierozem profile. But EC, contents of TS and 8 ions was different from our assumption. Above analysis showed that other ions and TS' contents and EC increased with soil depth, except HCO_3^- and K^+.

Comparison of sierozem SOM in Yili-River valley and in other regions

Sierozem in Yili-River valley is believed to be the extension of sierozem zone in central Asia. The climate in Yili-River valley is similar to the climate of northern Sierozem zone in central Asia. Annual temperature was 8°C and annual rainfall was 150 to 400 mm in northern sierozem zone in central Asia (Luozan, 1958). Northern sierozem in central Asia was divided into light northern sierozem and ordinary northern sierozem by A.H. Luozan. Light northern sierozem was mainly located in the ground and Alluvial slope, where humus layer was thin (30 to 60 cm), humus content was 0.8 to 1.7%; ordinary northern sierozem mainly lay hillside, where humus layer was thick (50 to 80 cm), humus content was higher, about 1.5 to 3.0% (Luozan, 1958). Compared with them, the sierozem humus layer was thinner, only 8 to 15cm in Yili-River valley (Xinjiang institute of comprehensive expedition and Institute of soil sciences, Chinese academy of sciences, 1965; Soil cencus office in Chabucha'er, 1984); SOM content was also less, only between 0.7 and 1.9% in the topsoil. This situation may be related to the arid climate in Yili-River valley.

The sierozem humus layer was 20 to 40 cm and average SOM content was 0.89%, which ranged from 0.45 to 1.33% in Ningxia hui autonomous region (Ningxia Institute of agriculture exploration and design, 1990). Meanwhile, the sierozem humus layer was 20 to 30 cm and averaged SOM content was 1.19%, which ranged from 0.64 to 1.74% in Gansu province (Gansu soil census office, 1993). According to our research,

averaged sierozem SOM content was 1.20% at 0 to 12 cm depth in Yili-River valley. So the sierozem SOM content in Humus layer in Yili-River valley was close to that in Gansu province, a little more than that in Ningxia hui autonomous region. But the thickness of sierozem humus layer was thinner in Yili-River valley than that in Gansu and Ningxia. The differences were related to the different climate, vegetation composition and growing season in three places. Compared with typical Sierozem soil profiles in Lanzhou, Gansu and Yanchi, Ningxia, the Sierozem SOM content decreased faster with soil depth in Yili-River valley (Figure 4). The situation was related to both arid summer and light-might loam in Yili-River valley, which led to weak infiltration of humus.

Above sierozem SOM characteristic showed that vegetation had a little contribution for soil SOM and the vegetation - soil system was not quite stable in Yili-River valley. The sierozem could only be moderately reclaimed, with the ecological protection in the area.

Comparison of sierozem pH value, TS and 8 ions in Yili-River valley and in other regions

In Gansu province, average soil pH value was 8.45 in the topsoil of sierozem (22 cm thickness), 8.5 in the middle of soil profiles (41.2 cm thickness), 8.61 at the bottom (50.1 cm thickness) (Gansu soil census office., 1993). Meanwhile, averaged sierozem pH value was 8.24 at 0 to 25 cm depth, while 8.40 at 50 to 75 cm depth, 8.34 at 100 to 125 cm depth in Yili-River valley. Obviously, sierozem alkaline was stronger in Gansu than that in Yili-River valley. Both pH value increased with soil depth and had a peak value in the deposition layer. Figure 5 showed that sierozem pH value also increased with soil depth and had a peak value in the deposition layer in Yanchi, Ningxia.

In Ningxia hui autonomous region, averaged sierozem TS was 0.04% in Humus layer, 0.13% in sabach, 0.07% in parent material horizon (Ningxia Institute of agriculture exploration and design, 1990). While averaged Sierozem TS was 0.19% at 0 to 25 cm depth, 0.59% at 50 to 75 cm depth, 0.81% at 100 to 125 cm depth in Yili-River valley. It was obvious that both sierozem TS contents increased with soil depth, but sierozem TS contents were many more in Yili-River valley than that in Ningxia. Figure 5 showed that sierozem TS content also increased with soil depth in Lanzhou, Gansu province. Vertical differentiation of sierozem TS content was related to local arid climate. Soluble salt could partly leach in soil profile with low rainfall. TS contents' characteristics showed that drainage system should be set up if the sierozem was reclaimed. Otherwise, second salinization of soil would easily appear.

In the central profile about 75 cm, which was located transition of deposition layer and parent rock horizon, sierozem TS contents had an obvious peak in Lanzhou and Yili-River valley. The reason remain to further

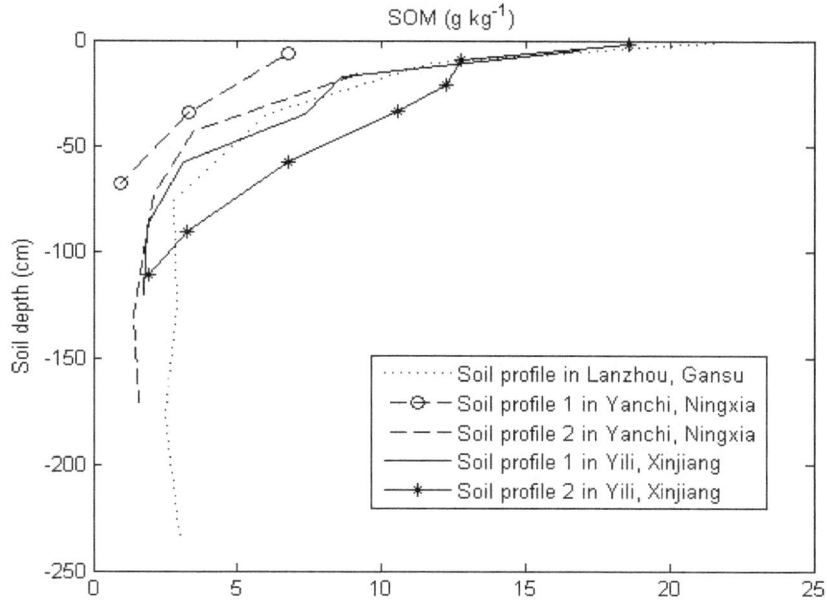

Figure 4. Comparison of SOM contents in typical Sierozem soil profiles (Data in Yili-River valley was from the field survey; Data in Yanchi, Ningxia was from reference (Hu et al., 1991); Data in Lanzhou, Gansu was from reference (Xiong and Li, 1987).

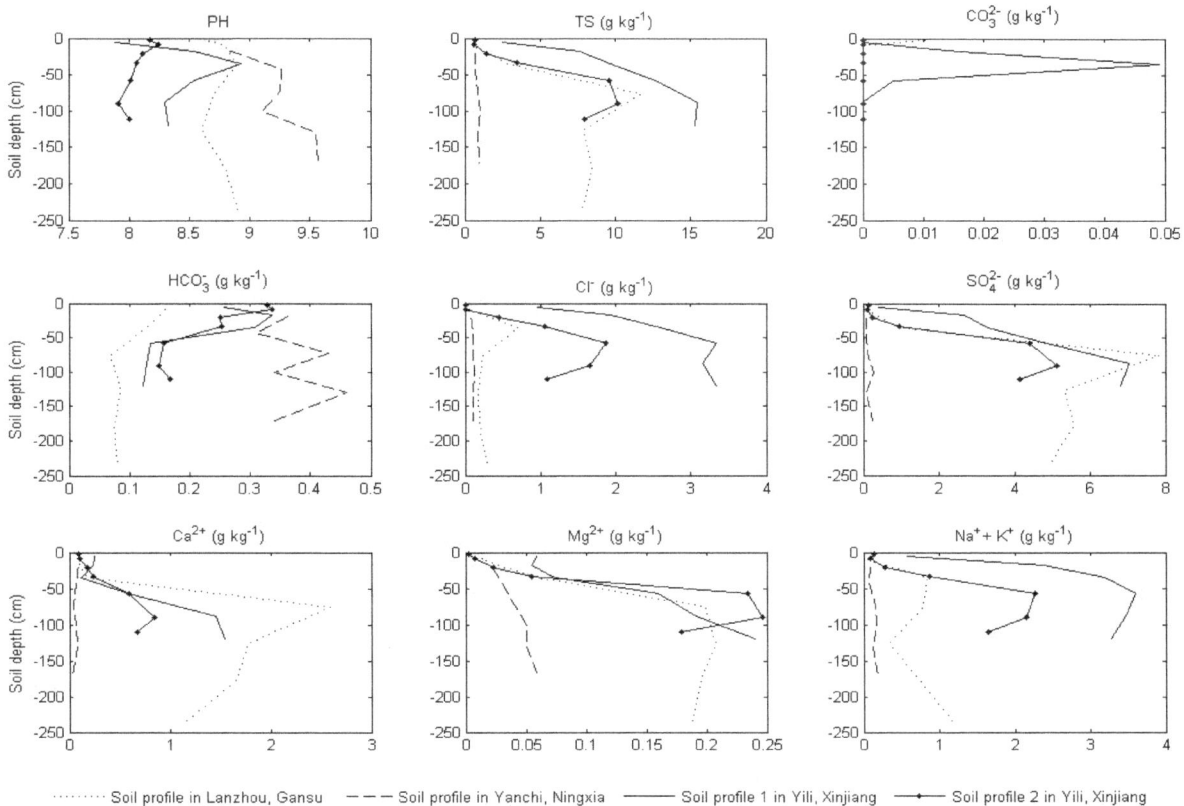

Figure 5. Comparison of Sierozem pH value, TS and 8 ions in Yili-River valley, Yanchi, Ningxia and Lanzhou, Gansu (Data in Yili-River valley was from the field survey; Data in Yanchi, Ningxia was from reference (Hu et al., 1991); Data in Lanzhou, Gansu was from reference (Xiong and Li, 1987).

research, which was perhaps related to the above $CaCO_3$ deposited (Ningxia Institute of agriculture exploration and design, 1990). Among the anions, sierozem CO_3^{2-} content was the least in the three regions. Sierozem SO_4^{2-} content was more in Yili-River valley and Lanzhou, Gansu. However, HCO_3^- content was more in Yanchi, Ningxia (Figure 5). Among the cations, sierozem Na^++K^+ and Mg^{2+} contents were many more in Yili-River valley. While Ca^{2+} and Mg^{2+} contents were many more in Lanzhou, Gansu. The composition of the soil salt was related to that of parent material and water table fluctuations (Agriculture Bureau in xinjiang uygur autonomous region, 1996; Riha et al., 1986; Johnston et al., 2001).

Figure 5 showed that sierozem HCO_3^- content decreased with soil depth in Yili-River valley and Lanzhou, Gansu. However, sierozem HCO_3^- content increased with soil depth in Yanchi, Ningxia. Sierozem SO_4^{2-} and Cl^- contents increased with soil depth in Yili-River valley, Yanchi and Lanzhou.

Sierozem Ca^{2+} content increased with soil depth in Yili-River valley and Lanzhou. However, sierozem Ca^{2+} content decreased with soil depth in Yanchi. Sierozem Mg^{2+}, Na^++K^+ contents increased with soil depth in Yili-River valley, Yanchi and Lanzhou.

Conclusions

Sierozem was a particular soil type which was formed between desert and grassland in arid region. Vertical difference of sierozem profile characteristics in China different regions had some common: SOM content decreased with soil depth, while pH value, EC, contents of TS and most ions increased with soil depth. But there was obvious regional difference in spatial variation and intensity of sierozem characteristics due to more arid climate in Yili-River valley. Above vertical change of sierozem properties could provide important theoretical basis for implementing regional ecological construction and sustainable development in Yili-River valley.

ACKNOWLEDGEMENTS

The authors wish to thank Yuancun Shen, Tao Liang and Hongyan Cai from IGSNRR, CAS for their fruitful discussions and comments. Also we would like to thank Hanqing Ma, Yang Yang and Ying Zhang from IGSNRR, CAS for their sampling of soil profiles.

REFERENCES

Acosta JA, Fazb A, Jansen B (2011). Martínez-Martínez S. Assessment of salinity status in intensively cultivated soils under semiarid climate, Murcia, SE Spain. J. Arid Environ. 75:1056-1066.

Agriculture Bureau in xinjiang uygur autonomous region, Soil census office in Xinjiang uygur autonomous region (1996). Soil in Xinjiang. Science press: Beijing. (In Chinese), pp. 168-169.

AMAITT N, Xu H, Li J (2006). The Soil characteristic of the light Sierozem of the land of the forestry ecology in Yili river basin. Arid Land Geogr. (In Chinese) 29:867-871.

Bao SD (2000). Soil chemical analysis of agriculture (Fourth Edition). Beijing: China Agriculture Press, (In Chinese), pp. 12–21.

Davey BG, Bembrick MJ (1969). The potentiometric estimation of chloride in water extracts of soils. Soil Sci. Soc. Am. Proc. 33:385-387.

Eswaran H, Lal R, Reich PF (2001). Land degradation: an overview. In: Bridges, E.M., et al. (eds.) Responses to Land Degradation. Proc. 2nd. International Conference on Land Degradation and Desertification, Khon Kaen, Thailand. Oxford Press, New Delhi, India.

Fan ZL, Cheng XJ (1986). Land types and its reasonable use in Yili, Xinjiang. In: 1:1000000 land type map committee (Eds.), Land type in China. Science press: Beijing. (In Chinese), pp. 216-228.

Gansu soil census office (1993). Soil in Gansu. Agriculture press: Beijing, pp. 150-157. (In Chinese)

Gulnar T, Hamid Y, Mihrigul M (2011). Study on spatial variability and patterns of soil salinity in the Yili river valley. Agric. Res. Arid Areas, (In Chinese) 29:151-158.

Haefelel SM, Wopereis MCS (2005). Spatial variability of indigenous supplies for N, P and K and its impact on fertilizer strategies for irrigated rice in West Africa. Plant Soil 270:57-72.

Hamid Y, Mihrigul M, Gulnar T (2011). Spatial variability of soil salinity ions in the saline area of the Yili river basin. Agric. Res. Arid Areas. (In Chinese). 29:64-69.

Hu SX., Shen YC., Liu ZC (1991). Relationship between characters of heilu soil and Sierozem and forestry – Yanchi county as an example. In: Wang Y.M., et al. (Eds.).Remote sensing application essays on renewable resources in Gan-Qing-Ning region of the "Three-North Shelter Forest Programme". Science press: Beijing (In Chinese), pp. 130-138.

IUSS Working Group WRB (2006). World reference base for soil resources 2006. World Soil Resources Reports No. 103. FAO, Rome.

Johnston CA, Bridgham SD, Schubauer-Berigan JP (2001). Nutrient dynamics in relation to geomorphology of riverine wetlands. Soil Sci. Soc. Am. J. 65(2):557-577.

Li JF (1991). Climate in Xinjiang. China Meteorological Press, Beijing (in Chinese).

Liao FQ, Zhou SL, Zhang HF (2008). Spatial distribution and changes of heavy metals of agricultural lands in typical pregrading coast in Dongtai city, Jiangsu province, China. Chin. Geogr. Sci. 18(3):276-283.

Luozan AH, (Wen ZW, Xu Q translated) (1958). Sierozem in central Asia. Science press: Beijing (In Chinese), pp. 277-281.

Mamattuesun E, Hamid Y, Anwar M (2010). Characteristics of soil salinity in Yili river valley, western Tianshan Mountains. Res. Environ. Sci. (In Chinese) 23:774-781.

McBratney AB, Bishop TFA, Teliatnikov IS (2000). Two soil profile reconstruction techniques. Geoderma 97:209–221.

National soil census office (1998). Soil in China; China agriculture press: Beijing. (In Chinese), pp. 56-434.

Ningxia Institute of agriculture exploration and design (1990). Soil in Ningxia. Ningxia people press: Yinchuan (In Chinese), pp. 99-102.

Riha SJ, James BR, Senesac GP (1986). Spatial variability of soil pH and organic matter in forest plantations. Soil Sci. Soc. Am. J. 50(5):1347-1352.

Scalenghe R, Ferraris S (2009). The first forty years of a Technosol. Pedosphere 19(1):40–52.

Shi RX, Yang XH, Wang LX (2009). Soil nutrient properties and land development in the new reclamation area of Yili, Xinjiang, China. Resour. Sci. (In Chinese), 31:2016-2023.

Shi XZ, Yu DS, Xu SX, Warner ED. Wang HJ. Sun WX, Zhao YC, Gong ZT (2010). Cross-reference for relating Genetic Soil Classification of China with WRB at different scales. Geoderma 155:344–350.

Smyth AJ, Dumanski J (1995). A framework for evaluating sustainable land management. Canadian J. Soil Sci. 75:401–406.

Soil cencus office in Chabucha'er (1984). Soil census report in Chabucha'er, Yili. (In Chinese), pp. 35-52.

Tao S (1995a). Kringing and mapping of copper, lead, and mercury contents in surface soil in Shenzhen area. Water Air Soil Poll. 83:161-172.

Tao S (1995b). Spatial structure of Copper, Lead and mercury contents in surface soil in Shenzhen area. Water Air Soil Poll. 82:583-591.

Tiessen H, Moir JO (1993). Total and organic carbon. In: Carter, M.R. (Ed.), Soil Sampling and Methods of Analysis. Lewis Publishers, Boca Raton, FL, pp. 187–199.

UN Millennium Project (2005). Halving hunger: it can be done. Task force on hunger. Earthscan, London.

UNEP (2007). Global Environmental Outlook GEO4 — Environment for Development. United Nations Environment Programme, Nairobi.

Webster R, Nortcliff S (1984). Nortcliff S. Improved estimation of micronutrients in hectare plots of the Sonning series. J. Soil Sci. 35:667-672.

Working group on analytical chemistry in department of chemistry, Hangzhou University (1997). The Handbook of Analytical Chemistry (2nd edition), vol. 2, Chemical Industry Press, Beijing, (In Chinese), pp. 558-559.

Wu R, Quan Z (1985). Soil in Yili region, Yili soil cencus office: Yili, Xinjiang (In Chinese), 5:118-139.

Xinjiang institute of comprehensive expedition, Institute of soil sciences, Chinese academy of sciences (1965). Xinjiang soil geography. Science press: Beijing. (In Chinese), pp. 130-141.

Xiong Y, Li Q (1987). Soil in China (2nd edtion); Science press: Beijing; (In Chinese) 2-11:143.

Xu J, Webster R (1983). Optimal estimation of soil survey data by geostatistical method – semivariogram and block kriging estimation of topsoil Nitrogen of Zhangwu county. Acta Pedologiga Sinica. (In Chinese) 20:419-430.

Yost RS, Uehara G, Fox RL (1982a). Geostatistical analysis of soil chemical properties of large land areas: i. Semivariograms. Soil Sci. Soc. Am. J. 46:1028-1032.

Yost RS, Uehara G., Fox RL (1982b). Geostatistical analysis of soil chemical properties of large land areas: ii. Kriging. Soil Sci. Soc. Am. J. 46:1033-1037.

Zulpiya M, Haimiti Y, Mamattursun A (2011). The distribution of salinity of soil and groundwater in Yili river valley. Agric. Res. Arid Areas (In Chinese), pp. 29:58-63.

26

Short-term effects of sustainable agricultural practices for spring maize (*Zea mays* L.) production on soil organic carbon characteristics

Zhang Jinjing[1], Gao Qiang[1], Wang Qinghe[1], Dong Peibo[1], Feng Guozhong[1], Li Cuilan[1] and Wang Lichun[2]

[1]College of Resource and Environmental Science, Jilin Agricultural University, Changchun 130118, China.
[2]Agricultural Environments and Resources Research Centre, Northeast Agricultural Research Centre of China, Changchun 130124, China.

The effects of the integrated adoption of several recommended sustainable agricultural practices (SAP) on maize grain yield and the quantitative and qualitative characteristics of soil organic carbon (SOC) were studied in 2 years, spring maize monoculture field plot experiment. The recommended agricultural practices included conservation tillage, combined application of organic manure and chemical fertilizers, and crop residue return. Compared with the conventional agricultural practices (CAP), the maize grain yield increased in the SAP treatment, and the difference between the two treatments was statistically significant (P<0.05) in the second season of the experiment (2010). The content of soil total organic carbon (TOC) and organic C fractions (that is, water soluble organic C, easily oxidizable organic C, particulate organic C, humus C and black C) were higher in the SAP than in the CAP treatment, although the differences between the two treatments were not significant. The relative intensities of O-alkyl C and carbonyl C and the aliphatic C / aromatic C ratio were higher, while the relative intensities of alkyl C and aromatic C and the ratios of alkyl C / O-alkyl C and hydrophobic C / hydrophilic C were lower in solid-state ^{13}C CPMAS NMR spectra of HF-treated soils in the SAP than in the CAP treatment. The recommended sustainable agricultural practices were beneficial for the increase of maize grain yield and the improvement of the quantity and quality of SOC during a short-term period.

Key words: Soil organic carbon, organic carbon fraction, agricultural practice, spring maize, ^{13}C CPMAS NMR.

INTRODUCTION

Anthropogenic activities have led to an increase in atmospheric concentration of carbon dioxide (CO_2) from 280 ppm in the pre-industrial era to almost 400 ppm at present, and is increasing at the rate of about 2.2 ppm per year (Lal, 2011). The sequestration of atmospheric CO_2 into terrestrial soils is a vital solution for mitigating climate change. The soil organic carbon (SOC) storage in the global agro-ecosystem nearly accounts for 10% of the total terrestrial SOC storage (Tang et al., 2010), and thus agricultural soils play an important role in the global carbon (C) cycle. At the same time, SOC in agro-ecosystem also is also vital for soil quality, crop productivity, and the sustainability of farming systems (Rees et al., 2001). Agricultural practices can have profound effects on the organic C contents of agricultural soils. Sustainable agricultural practices such as conservation tillage (Chen et al., 2009), manure application (Gong et al., 2009), crop residue return (Bakht et al., 2009) and crop rotation (Kelley et al., 2003) are effective in increasing SOC levels. However, the rate of soil C sequestration through the adoption of sustainable agricultural practices differs among eco-regions and is dependent on soil texture and structure, rainfall, temperature, farming systems and soil management (Lal, 2004).

Maize (*Zea mays* L.) is the second most important cereal crop after rice in China. Jilin province, located in the Songnen Plain of northeast China, is one of the major maize-growing regions of China and is also one of the major contributors to the maize belt of the world's temperate zone (Zhang et al., 2005). Nevertheless some conventional agricultural practices used by local farmers in this region are not correct for achieving the increase of SOC and thus increasing/stabilizing maize yield. These conventional practices included: (1) conventional tillage, (2) single application of chemical fertilizers, and (3) crop residues burning or removal. Analogous problems also exist in other major agricultural regions of China (Wang et al., 2008) and in many developing counties (Chivenge et al., 2007; Bakht et al., 2009; Moloto, 2009). Thus, the introduction of sustainable agricultural practices is necessary to increase SOC. A number of researches have been carried out on the effects of a single agricultural practice on SOC characteristics. However, the effect of the integrated and contemporary introduction of some sustainable agricultural practices on SOC characteristics, especially on the qualitative properties of SOC, was not yet fully investigated. The aim of this study is to assess the effects of the introduction of some sustainable agricultural practices on both quantitative and qualitative characteristics of SOC in 2 years, spring maize monoculture field plot experiment.

MATERIALS AND METHODS

Site description

The field experiment was started in April 2009 on a farm field located in Sikeshu Township, Lishu County, Jilin Province, northeast China (124°03′ E, 43°20′ N). The site has a semi-humid monsoon continental climate of the temperate zone. The average annual temperature is 6.5°C during the study period. The annual effective accumulated temperature above 10°C ranges from 3 000 to 3 200°C. The average annual precipitation is between 525 and 550 mm, with 60% occurring between June and August. The average amount of sunshine each year is about 2 500 h, and the frost-free period is between 127 and 148 days.

The soil in the study site was classified as an Alluvic Primosols, according to the Chinese Soil Taxonomy (CRGCST, 2001), a Fluvent according to the USDA Soil Taxonomy (Soil Survey Staff, 1998), and a Fluvisol according to the World Reference Base (FAO-ISRIC-ISSS, 1998). Alluvic Primosols was one common soil type used for maize production in Jilin province. Before starting the experiment in 2009, the content of organic C, total N, hydrolysable

N, available P, available K, and pH in the 0 to 20 cm soil layer were 7.08, 1.04, 92.0, 29.1, 52.0 and 5.15 mg kg^{-1}, respectively.

Experimental design and sampling

Two agricultural strategies, namely conventional agricultural practice (CAP) and sustainable agricultural practice (SAP), were adopted in the present study. Continuous spring maize monoculture is common in the region. The maize cultivar used was Xianyu 335, which was the dominant hybrid used in Chinese agriculture with planting area above 5 000 000 ha between 1980 and 2009 (Wu et al., 2011). The seeds were coated prior to sowing. The weeds were controlled by atrazine after sowing. The size of each plot was 120 m^2 (20 × 6 m). The experimental design was a completely randomized block with three replications for each of the two treatments.

CAP

Before sowing in spring, rotary tillage and maize stubble breaking were conducted in the 10 to 15 cm soil layer. The main purposes of stubble breaking were to ensure the quality of sowing and facilitate the decomposition of stubble. The sowing date was selected according to the usual practice of the local farmers (around 20 April of every year). The seeding rate was chosen to guarantee the final number of maize seedlings was 55 000 plants ha^{-1}. During the growth period of maize, intertillage was not conducted while N fertilizer was top dressed at the jointing stage. Only chemical fertilizers were applied for this treatment. The application rates of fertilizers were 225 kg N ha^{-1}, 100 kg P$_2$O$_5$ ha^{-1} and 60 kg k$_2$O ha^{-1}. All of the P$_2$O$_5$ and K$_2$O and half of the N were applied as basal fertilizers before sowing, and the remaining N was applied as jointing fertilizer. After harvesting, the maize stubble was retained in the field, while the maize stalk was either removed from the field or burnt *in situ*.

SAP

There was no tillage operation prior to sowing in spring for this treatment. Sowing was carried out when the temperature of soil layer within 5 cm was stabilized at 5 to 6°C, and the air temperature was stabilized at 7 to 9°C (around 1 May of every year). The seed was sown manually in the plough furrow using disseminator by precision dibbling method. The sowing rate was chosen to guarantee the final number of maize seedlings was between 70 000 and 80 000 plants ha^{-1}. The maize roborant, which could confer enhanced resistance to stalk lodging, was sprayed at jointing stage. Both chemical and organic fertilizers were applied. The application doses of fertilizers were 240 kg N ha^{-1} as urea, 100 kg P$_2$O$_5$ ha^{-1} as superphosphate, 100 kg k$_2$O ha^{-1} as potassium sulfate, 75 kg ha^{-1} of ZnSO$_4$ and 2.0 to 2.5 t ha^{-1} of organic fertilizer. All of the P$_2$O$_5$ and organic fertilizer and 45% of the N and 80% of the K$_2$O were applied as basal fertilizers, ZnSO$_4$ and 35% of the N was applied as seed fertilizer, and the remaining N and K$_2$O were applied as ear fertilizer. After harvesting, the standing stubble of about 30 cm in height was retained with all maize stalk left as a mulch cover. Maize was cultivated in a three-year fallow rotation.

The soil samples were collected to a depth of 0 to 20 cm in October 2010 from five locations in each plot, and then thoroughly mixed into a composite sample. The collected soil samples were air-dried, milled and sieved through a 2 mm sieve. Prior to solid-state ^{13}C NMR analysis, all soil samples were pre-treated with 10% hydrofluoric acid (HF) as recommended by Schmidt et al. (1997) to remove magnetic materials, concentrate the organic matter, and increase the signal-to-noise (S/N) ratio of the resultant NMR

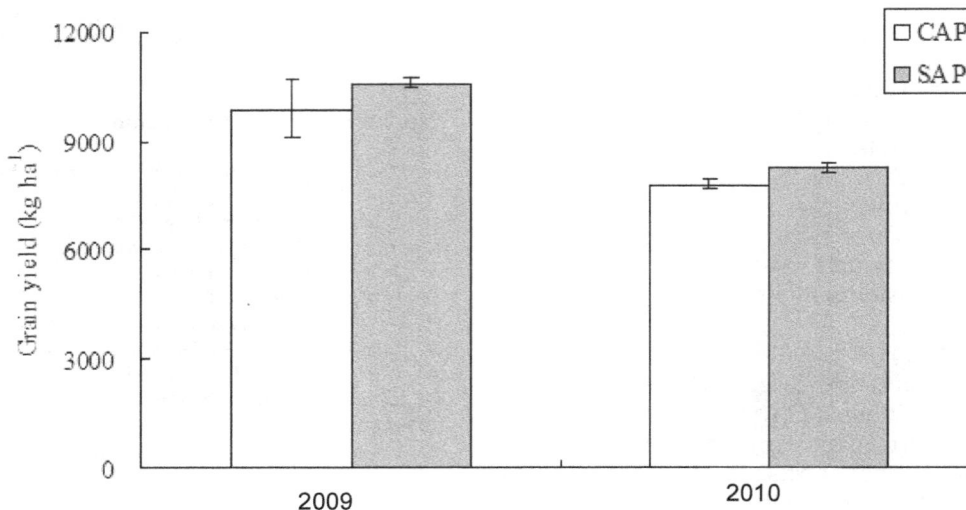

Figure 1. Maize grain yield under conventional agricultural practices (CAP) and sustainable agricultural practices (SAP) in 2009 and 2010.

spectra.

Soil analysis

Soil total organic carbon (TOC) was determined by $K_2Cr_2O_7$ oxidation and total N by semi-micro Kjeldahl method (Lao, 1988). The water soluble organic carbon (WSOC) and humus carbon (HC) in each sample were successively analyzed according to the method described by Zhang et al. (2010). Briefly, the soil samples were first suspended in distilled water at 70±1°C for 60 min. The supernatant was referred to as the water soluble fraction (WSF). After centrifugation, the remaining soil was further extracted using a solution of 0.1 mol l^{-1} NaOH and 0.1 mol l^{-1} $Na_4P_2O_7$ at 70±1°C for 60 min. The dark brown alkaline supernatant solution, corresponding to the total alkali-soluble humic extract (HE), was separated into the acid-insoluble humic acid (HA) and the acid-soluble fulvic acid (FA) fractions by acidifying the alkaline supernatant to pH 1.0. The residue remaining after extraction was referred to as the humin (HM) fraction. The carbon contents of WSF (WSOC), HE (HEC) and HA (HAC) were directly determined, while that of HM (HMC) was calculated by subtraction. Easily oxidizable organic carbon (EOC) was determined as described by Blair et al. (1995). Soil samples containing 15 mg of organic carbon were reacted with 333 mmol l^{-1} $KMnO_4$ solution for 60 min, and the amount of EOC was spectrophotometrically determined from the amount of $KMnO_4$ reduced. Particulate organic carbon (POC) was measured following Cambardella and Elliott (1992). Soil samples were dispersed in 100 ml of 5 gL^{-1} $(NaPO_3)_6$ solution and shaken at 90 r min^{-1} for 18 h. The suspension was passed through a 53 μm screen and the retained coarse fraction was rinsed with distilled water, dried at 65°C, weighed and ground for determination of organic C.

Black carbon (BC) was analyzed by the method given by Aiken et al. (1985). Soil samples were reacted with 25 ml of 0.1 mol L^{-1} $K_2Cr_2O_7$+2 mol L^{-1} H_2SO_4 solution at 55±1°C for 60 h, and the oxidized organic C was determined by titration using 0.2 mol L^{-1} $FeSO_4$ solution. The content of BC was calculated by subtracting the oxidized organic carbon from the TOC.

The solid-state ^{13}C CPMAS NMR experiment was performed in the National Analytical Research Center of Electrochemistry and Spectroscopy, Changchun Institute of Applied Chemistry, Chinese Academy of Sciences, Changchun, China. The spectra were obtained on a Bruker AVANCE III 400 WB spectrometer (Fällanden, Switzerland) operating at 100.62 MHz, equipped with a 4 mm probe head. The dried and finely powdered soil samples (<0.1 mm) were packed in the ZrO_2 rotor closed with Kel-F cap. The conditions used were: spinning rate 5 KHz, contact time 2 ms, recycle delay 6 s, line broadening 50 Hz, and 8000 total scans. Chemical shifts were referenced to the resonances of Adamantane standard (δ=29.5). Spectra were divided into four main chemical shift regions (Mathers et al., 2003), namely alkyl C (0 - 50 ppm), O-alkyl C (50 - 110 ppm), aromatic C (110 - 160 ppm) and carbonyl C (160 to 210 ppm). The relative intensity for each chemical shift region was obtained with the integration routine of the spectrometer. The ratios of alkyl C to O-alkyl C (alkyl C / O-alkyl C), of aliphatic C to aromatic C (aliphatic C / aromatic C), and of hydrophobic C to hydrophilic C (hydrophobic C / hydrophilic C) (Zhang et al., 2009) were calculated.

Statistical analysis

Data were analyzed statistically by analysis of variance (ANOVA) procedure. Least significant difference (LSD) was employed to assess differences between treatment means at 5% significance level. Standard deviations were calculated for means values of all the determination. All statistical analyses were performed using the SPSS 16.0 for Windows statistical software package (SPSS, Chicago, IL, USA).

RESULTS

Maize yield

Maize grain yields under CAP and SAP treatments are shown in Figure 1. In the first experimental season (2009), the difference between maize grain yields obtained by different agricultural practices was not statistically significant, but a higher value could be observed in the SAP treatment than in CAP. The grain yield was significantly higher in the second experimental

Table 1. The contents of total organic carbon (TOC), water soluble organic carbon (WSOC), easily oxidizable organic carbon (EOC), particulate organic carbon (POC), total alkali-soluble humic extract carbon (HEC), humic acid carbon (HAC), humin carbon (HMC) and black carbon (BC) in soils under conventional agricultural practices (CAP) and sustainable agricultural practices (SAP).

Agricultural practices	TOC (g kg^{-1})	WSOC (g kg^{-1})	EOC (g kg^{-1})	POC (g kg^{-1})	HEC (g kg^{-1})	HAC (g kg^{-1})	HMC (g kg^{-1})	BC (g kg^{-1})
CAP	7.30±0.65	0.23±0.12	2.05±0.63	1.17±0.24	2.91±0.54	2.45±0.22	4.40±0.12	3.29±0.61
SAP	8.07±0.80	0.24±0.10	3.66±1.01	1.53±0.13	3.29±0.38	2.50±0.30	4.78±0.42	3.49±1.75
$F(P)$	1.58 (>0.05)	0.02 (>0.05)	5.48 (>0.05)	5.26 (>0.05)	1.00 (>0.05)	0.06 (>0.05)	2.34 (>0.05)	0.03 (>0.05)

Mean values ± standard error of three replicates are presented. F, variance ratio; P, significance level.

season (2010) for the SAP treatment (8.24 t ha^{-1}) than for the CAP treatment (7.78 t ha^{-1}). Moreover, maize grain yield was significantly lower in 2010 than in 2009.

Soil total organic carbon and organic carbon fractions

The contents of soil total organic C and organic C fractions are shown in Table 1. The contents of TOC, WSOC, EOC, POC, HEC, HAC, HMC, BC were all higher in the SAP than in the CAP treatment, although the differences between the two treatments were not statistically significant. The increase amplitudes were larger for the EOC and POC (78.5 and 30.8%, respectively) than for the HEC, HAC, HMC and BC (13.1, 2.04, 8.64 and 6.08%, respectively).

Solid-state ^{13}C CPMAS NMR spectra of HF-treated soils

The solid-state ^{13}C CPMAS NMR spectra of HF-treated soils under CAP and SAP treatments are shown in Figure 2. In the alkyl C region, three major peaks at 20, 25 and 43 ppm could be identified. The peaks were assigned as -CH$_3$, -CH$_2$-, and branched aliphatic C, respectively. In the region for O-alkyl C, the signals at 54 - 57, 71 - 73 and 101 - 104 ppm were generally ascribed to methoxyl C, carbohydrate C and di-O-alkyl C, respectively. In the aromatic C region, the signal at 122 to 129 ppm was due to aryl C. The small peaks at 151 to 156 ppm were the signal of phenolic C. The signals in the carbonyl C region were concentrated between 170 and 188 ppm, indicating that there was carbonyl C of carboxylic acids, esters and amides (Kögel-Knabner, 1997; Mathers et al., 2003; Zhang et al., 2009).

The relative intensity of carbon functional groups of HF-treated soils is shown in Table 2. Although there was no significant difference in the relative intensity of alkyl C, O-alkyl C, aromatic C and carbonyl C regions between the CAP and SAP treatments, an increase in the O-alkyl C and carbonyl C and a decrease in the alkyl C and aromatic C were observed in the SAP treatment. The ratios of alkyl C / O-alkyl C and hydrophobic C /

hydrophilic C were lower, whereas the aliphatic C / aromatic C ratio was higher in the SAP than in the CAP treatment.

DISCUSSION

Previous studies showed that the maize grain yield was higher under the combined application of organic manure and chemical fertilizers than under the application of chemical fertilizers only (Fan et al., 2005), under rational sowing date and planting density than under conventional ones (Andrade, 1995; Otegui et al., 1996), and under crop residue return than under residue removal (Sharma et al., 2011). However, the results on the effect of tillage practices on maize grain yield were not consistent among studied. Some studies showed that maize grain yield was higher under reduced/no tillage than under conventional tillage (Wang et al., 2007; Sharma et al., 2011), other studies indicated that reduced/minimum/no tillage (Atreya et al., 2008; Sharma et al., 2011) could give similar or even lower maize grain yield as compared to conventional tillage. In the present study, we found that maize grain yield increased by the integrated application of the recommended sustainable agricultural practices with respect to the conventional agricultural practices (Figure 1). Thus, the recommended agricultural practice was suitable for the local maize production. The significantly lower maize grain yield during the second year than during first could be due to lower precipitation from early June to early July in 2010 than in 2009.

In previous studies, higher soil organic C contents under no tillage with residue return than under conventional tillage with residue removal (Razafimbelo et al., 2008), under reduced tillage than under conventional tillage (Šimanský et al., 2008), under tillage with organic manure than under tillage without organic manure (Agbede and Ojeniyi, 2009), and under the combined application of organic manure and chemical fertilizers than under the application of organic manure or chemical fertilizers (Cai and Qin, 2006; Zhu et al., 2007), have been reported. These could explain our present result that the content of SOC was higher in the recommended than in the conventional agricultural practices (Table 1). By the application of some different sustainable

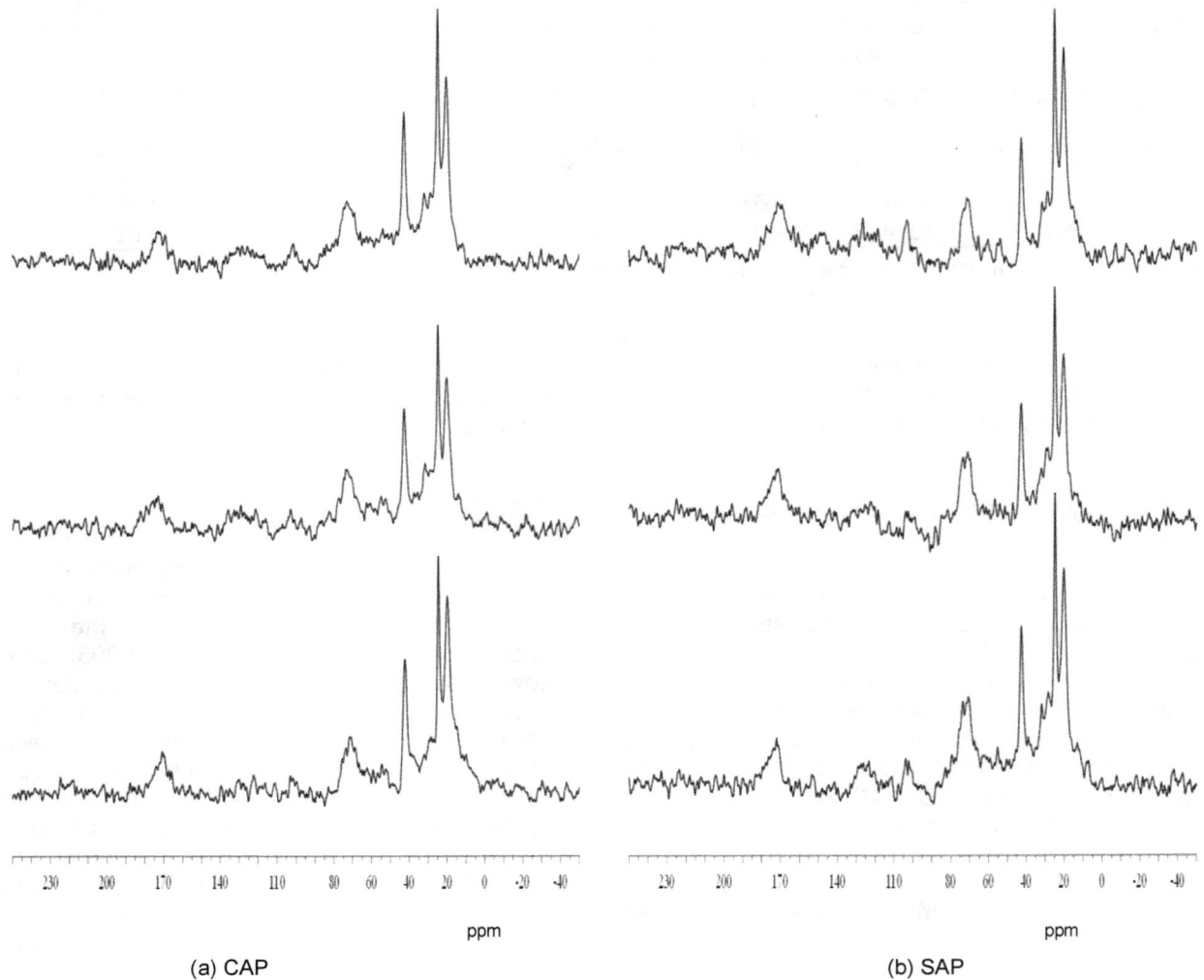

Figure 2. ^{13}C CPMAS NMR spectra of HF-treated soils under (a) conventional agricultural practices (CAP) and (b) sustainable agricultural practices (SAP).

Table 2. Distribution of different carbon types from ^{13}C CPMAS NMR spectra of HF-treated soils under conventional agricultural practices (CAP) and sustainable agricultural practices (SAP) (%).

Agricultural practices	Alkyl C	O-alkyl C	Aromatic C	Carbonyl C	Alkyl C/ O-alkyl C	Aliphatic C/ Aromatic C	Hydrophobic C/ Hydrophilic C
CAP	58.2±6.73	24.6±2.71	8.46±2.27	8.37±1.89	2.41±0.56	10.3±2.84	2.07±0.47
SAP	56.0±7.15	27.0±1.09	7.37±3.05	9.12±3.15	2.08±0.33	13.1±6.92	1.78±0.30
F(P)	0.14 (>0.05)	2.09 (>0.05)	0.24 (>0.05)	0.12 (>0.05)	0.78 (>0.05)	0.44 (>0.05)	0.82 (>0.05)

Mean values ± standard error of three replicates are presented. F: variance ratio; P: significance level.

agricultural techniques, some researchers (Jhamtani, 2007; Moloto, 2009) found that farms that were managed under sustainable agricultural practices generally contain higher soil organic C content than farms that were managed under conventional agricultural practices, in accordance with our results. The larger increase in the

active organic carbon fractions (EOC and POC) than in the resistant organic carbon fractions (HEC, HAC, HMC and BC) (Table 1) implied that the two active organic carbon fractions could be a more sensitive index for the effects of agricultural practices.

The ratios of alkyl C / O-alkyl C, aliphatic C / aromatic

C, and hydrophobic C / hydrophilic C have been used as indicators of the degree of decomposition or humificationaliphaticity or aromaticity, and hydrophobicity of SOC, respectively (Webster et al., 2001; Chen and Chiu, 2003; Mathers et al., 2003; Ussiri and Johnson, 2003; Chen et al., 2004; Zhang et al., 2009). A larger value of the ratios indicates that SOC was more decomposed, aliphatic and hydrophobic. Thus, our results (Table 2) implied that the degree of decomposition and hydrophobicity of SOC was lower while that of aliphaticity was higher in the SAP than in the CAP treatment. Moreover, the alkyl C: O-alkyl C ratio has been used as an indicator of the quality of SOC (Chen et al., 2004). The lower alkyl C/O-alkyl C ratio in our study in soil under SAP than under CAP treatment indicated that less accumulation of relative recalcitrant carbon components and thus the quality of SOC was better in the SAP treatment. The increase of aliphatic C / aromatic C ratio and the decrease of hydrophobic C / hydrophilic C ratio were identical with the decrease of alkyl C / O-alkyl C ratio under SAP with respect to the CAP treatment. The improved SOC quality under SAP could also be ascribed to the use of conservation tillage, combined organic and chemical fertilizers, and crop residue return. In the previous studies, it was found that the carbohydrate was higher and aromatic C was lower in no tillage than in conventional tillage soils (Arshad et al., 1990).

In conclusion, the recommended sustainable agricultural practice could increase maize grain yield and improve the quantitative and qualitative characteristics of SOC with respect to the conventional agricultural practice during a short-term period. Further studies are necessary to assess long-term effects of sustainable agricultural practice on SOC dynamics.

ACKNOWLEDGEMENTS

This work was supported by the National Basic Research Program of China (grant no. 2009CB118600), the National Agricultural Department Public Benefit Research Foundation (grant no. 201103030), the Postdoctoral Project of Northeast Agricultural Research Centre of China (grant no. 00225), and the Postdoctoral Project of Jilin Province (grant no. 01912). We wish to thank Zhiyong Liu for help in the collection of soil samples. We also wish to thank Dr. Zijiang Jiang for his technical support in solid-state ^{13}C CPMAS NMR spectroscopy. Moreover, we would like to express our great respect for the editors and anonymous reviewers.

REFERENCES

Aiken GR, Mcknight DM, Wershaw RL (1985). Humic Substances in Soils, Sediment and Water: Geochemistry, Isolation and Characterization. New York: John Wiley & Sons.

Agbede TM, Ojeniyi SO (2009). Tillage and poultry manure effects on soil fertility and sorghum yield in southwestern Nigeria. Soil Tillage Res. 104:74-81.

Andrade FH (1995). Analysis of growth and yield of maize, sunflower and soybean grown at Balcarce, Argentina. Field Crops Res. 41:1-12.

Arshad MA, Schnitzer M, Angers DA, Ripmeester JA (1990). Effects of till vs no-till on the quality of soil organic matter. Soil Biol. Biochem. 22:595-599.

Atreya K, Sharma S, Bajracharya RM, Rajbhandari NP (2008). Developing a sustainable agro-system for central Nepal using reduced tillage and straw mulching. J. Environ. Manage. 88:547–555.

Bakht J, Shafi M, Jan MT, Shah Z (2009). Influence of crop residue management, cropping system and N fertilizer on soil N and C dynamics and sustainable wheat (Triticum aestivum L.) production. Soil Till. Res. 104:233-240.

Blair GJ, Lefroy RDB, Lisle L (1995). Soil carbon fractions based on their degree of oxidation, and the development of a carbon management index for agricultural systems. Aust. J. Agric. Res. 46:1459-1466.

Cai ZC, Qin SW (2006). Dynamics of crop yields and soil organic carbon in a long-term fertilization experiment in the Huang-Huai-Hai Plain of China. Geoderma 136:708–715.

Cambardella CA, Elliott ET (1992). Particulate soil organic matter changes across a grassland cultivation sequence. Soil Sci. Soc. Am. J. 56:777–783.

Chen CR, Xu ZH, Mathers NJ (2004). Soil carbon pools in adjacent natural and plantation forests of subtropical Australia. Soil Sci. Soc. Am. J. 68:282–291.

Chen H, Hou R, Gong Y, Li H, Fan M, Kuzyakov Y (2009). Effects of 11 years of conservation tillage on soil organic matter fractions in wheat monoculture in Loess Plateau of China. Soil Till. Res. 106:85-94.

Chen JS, Chiu CY (2003). Characterization of soil organic matter in different particle-size fractions in humid subalpine soils by CP/MAS ^{13}C NMR. Geoderma, 117:129–141.

Chivenge PP, Murwira HK, Giller KE, Mapfumo P, Six J (2007). Long-term impact of reduced tillage and residue management on soil carbon stabilization: Implications for conservation agriculture on contrasting soils. Soil Till. Res. 94:328-337.

Cooperative Research Group on Chinese Soil Taxonomy (CRGCST) (2001). Chinese Soil Taxonomy. Beijing & New York: Science Press.

Fan T, Wang S, Tang X, Luo J, Stewart BA, Gao Y (2005). Grain yield and water use in a long-term fertilization trial in Northwest China. Agric. Water Manage. 76:36–52.

FAO-ISRIC-ISSS (1998). World Reference Base for Soil Resources. Rome: World Soil Resources Report 84, FAO.

Gong W, Yan X, Wang J, Hu T, Gong Y (2009). Long-term manure and fertilizer effects on soil organic matter fractions and microbes under a wheat-maize cropping system in northern China. Geoderma 149:318-324.

Jhamtani H (2007). Putting Farmers First in Sustainable Agriculture Practices. Macalister: Third World Network.

Kelley KW, Long Jr JH, Todd TC (2003). Long-term crop rotations affect soybean yield, seed weight, and soil chemical properties. Field Crops Res. 83:41-50.

Kögel-Knabner I (1997). ^{13}C and ^{15}N NMR spectroscopy as a tool in soil organic matter studies. Geoderma 80:243-270.

Lal R (2004). Soil carbon sequestration impacts on global climate change and food security. Science 304:1623-1627.

Lal R (2011). Sequestering carbon in soils of agro-ecosystems. Food Policy, 36:533-539.

Lao JC (1988). Handbook of Soil Agro-Chemistry Analysis. Beijing: China Agriculture Press.

Mathers NJ, Mendham DS, O'Connell AM, Grove TS, Xu Z, Saffigna PG (2003). How does residue management impact soil organic matter composition and quality under Eucalyptus globulus plantations in southwestern Australia. Forset Ecol. Manag. 179:253-267.

Moloto KP (2009). The potential of sustainable agricultural practices to enhance soil carbon sequestration and improve soil quality. Thesis of Masters of Philosophy, The University of Stellenbosch, Matieland, South Africa.

Otegui M, Ruiz RA, Petruzzi D (1996). Modeling hybrid and sowing date effects on potential grain yield of maize in a humid temperate region. Field Crops Res. 47:167-174.

Razafimbelo TM, Albrecht A, Oliver R, Chevallier T, Chapuis-Lardy L, Feller C (2008). Aggregate associated-C and physical protection in a tropical clayey soil under Malagasy conventional and no-tillage systems. Soil Till. Res. 98:140-149.

Rees RM, Ball BC, Campbell CD, Watson CA (2001). Sustainable Management of Soil Organic Matter. UK: CABI Publishing.

Schmidt MWI, Knicker H, Hatcher PG, Kögel-Knabner I (1997). Improvement of ^{13}C and ^{15}N CPMAS NMR spectra of bulk soils, particle size fractions and organic material by treatment with 10% hydrofluoric acid. Eur. J. Soil Sci. 48:319-328.

Sharma P, Abrol V, Sharma RK (2011). Impact of tillage and mulch management on economics, energy requirement and crop performance in maize-wheat rotation in rainfed subhumid inceptisols, India. Eur. J. Agron. 34:46–51.

Šimanský V, Tobiašová E, Chlpík J (2008). Soil tillage and fertilization of Orthic Luvisol and their influence on chemical properties, soil structure stability and carbon distribution in water-stable macro-aggregates. Soil Till. Res. 100:125–132.

Soil Survey Staff (1998). Keys to Soil Taxonomy. 8th edition. USDA Natural Resources Conservation Service, Washington.

Tang HJ, Qiu JJ, Wang LG, Li H, Li CS, van Ranst E (2010). Modeling soil organic carbon storage and its dynamics in croplands of China. Agr. Sci. China 9:704-712.

Ussiri DAN, Johnson CE (2003). Characterization of organic matter in a northern hardwood forest soil by ^{13}C NMR spectroscopy and chemical methods. Geoderma 111:123–149.

Wang L, Qiu J, Tang H, Li H, Li C, Van Ranst E (2008). Modelling soil organic carbon dynamics in the major agricultural regions of China. Geoderma 147:47-55.

Wang XB, Cai DX, Hoogmoed, WB, Oenema O, Perdok UD (2007). Developments in conservation tillage in rainfed regions of North China. Soil Till. Res. 93:239–250.

Webster EA, Hopkins DW, Chudek JA, Haslam SFI, Šimek M, Pîcek T (2001). The relationship between microbial carbon and the resource quality of soil carbon. J. Environ. Qual. 30:147–150.

Wu QP, Chen FJ, Chen YL, Yuan LX, Zhang FS, Mi GH (2011). Root growth in response to nitrogen supply in Chinese maize hybrids released between 1973 and 2009. Sci. China Life Sci. 54:642–650.

Zhang JJ, Dou S, Song XY (2009). Effect of long-term combined nitrogen and phosphorus fertilizer application on ^{13}C CPMAS NMR spectra of humin in a Typic Hapludoll of Northeast China. Eur. J. Soil Sci. 60:966-973.

Zhang J, Hayakawa S, Zhou D, Zhang H (2005). Risk assessment and regionalization of agro-meteorological hazards in Jilin province, China. J. Agric. Meteorol. 60:921-924.

Zhang JJ, Wang LB, Li CL (2010). Humus characteristics after maize residues degradation in soil amended with different copper concentrations. Plant Soil Environ. 56:120-124.

Zhu P, Ren J, Wang L, Zhang X, Yang X, MacTavish D (2007). Long-term fertilization impacts on corn yields and soil organic matter on a clay-loam soil in Northeast China. J. Plant Nutr. Soil Sci. 170:219–223.

Permissions

List of Contributors

Ramadan AGAMY
Department of Agricultural Botany, Faculty of Agriculture, Fayoum University, Fayoum, Egypt

Mohamed HASHEM
Botany Department, Faculty of Science, Assiut University, Assiut, Egypt, 71516
Biology Department, Faculty of Science, King Khalid University, P. O. Box 10255, Abha 61321, Saudi Arabia

Saad ALAMRI
Biology Department, Faculty of Science, King Khalid University, P. O. Box 10255, Abha 61321, Saudi Arabia

Roberto Botelho Ferraz Branco
Agency Paulista Agribusiness Technology, APTA, 14030-670 Ribeirao Preto, Sao Paulo (SP) - Brazil

Denizart Bolonhezi
Agency Paulista Agribusiness Technology, APTA, 14030-670 Ribeirao Preto, Sao Paulo (SP) - Brazil

Fernando André Salles
Agency Paulista Agribusiness Technology, APTA, 14030-670 Ribeirao Preto, Sao Paulo (SP) - Brazil

Geraldo Balieiro
Agency Paulista Agribusiness Technology, APTA, 14030-670 Ribeirao Preto, Sao Paulo (SP) - Brazil

Eduardo Suguino
Agency Paulista Agribusiness Technology, APTA, 14030-670 Ribeirao Preto, Sao Paulo (SP) - Brazil

Walter Seiti Minami
Moura Lacerda University Center, 14085-420 Ribeirão Preto, SP - Brazil

Ely Nahas
Department of Microbiology, Universidade Estadual Paulista (UNESP), 14884-900 Jaboticabal, SP - Brazil

B. Ndukwu
Department of Soil Science and Technology, Federal University of Technology Owerri, Imo State, Nigeria

S. U. Onwudike
Department of Soil Science and Technology, Federal University of Technology Owerri, Imo State, Nigeria

M. C. Idigbor
Department of Soil Science and Technology, Federal University of Technology Owerri, Imo State, Nigeria

C. E. Ihejirika
Department of Environmental Technology, Federal University of Technology Owerri, Imo State, Nigeria

K. S. Ewe
Department of Soil Science and Technology, Federal University of Technology Owerri, Imo State, Nigeria

Bridget B. Umar
Department of International Environment and Development Studies, Norwegian University of Life Sciences, Box 5003, 1432. Aas, Norway
Geography and Environmental Studies Department, University of Zambia, Box 32379. Lusaka, Zambia

Jens. B. Aune
Department of International Environment and Development Studies, Norwegian University of Life Sciences, Box 5003, 1432. Aas, Norway

Obed. I. Lungu
Soil Science Department, University of Zambia, Box 32397, Lusaka, Zambia

Johan VAN TOL
Department of Agronomy, University of Fort Hare, Alice, 5700, South Africa

Johan BARNARD
Department of Soil, Crop and Climate Sciences, University of the Free State, Bloemfontein, 9300, South Africa

Leon VAN RENSBURG
Department of Soil, Crop and Climate Sciences, University of the Free State, Bloemfontein, 9300, South Africa

Pieter LE ROUX
Department of Soil, Crop and Climate Sciences, University of the Free State, Bloemfontein, 9300, South Africa

N. N. Buthelezi
Soil Science, School of Agricultural and Environmental Sciences, University of Limpopo, Sovenga, South Africa

J. C. Hughes
Crop Science, School of Agricultural Sciences and Agribusiness, University of KwaZulu-Natal, Pietermaritzburg, South Africa

A. T. Modi
Crop Science, School of Agricultural Sciences and Agribusiness, University of KwaZulu-Natal, Pietermaritzburg, South Africa

H. Lili-Sahira
Natural Product Division, Forest Research Institute Malaysia, 52109, Kepong, Selangor, Malaysia

K. Getha
Natural Product Division, Forest Research Institute Malaysia, 52109, Kepong, Selangor, Malaysia

A. Mohd Ilham
Malaysian Institute of Pharmaceutical and Nutraceutical, Bukit Gambir 11700, Penang, Malaysia

I. Norhayati
Natural Product Division, Forest Research Institute Malaysia, 52109, Kepong, Selangor, Malaysia

M. M. Siti-Syarifah
Natural Product Division, Forest Research Institute Malaysia, 52109, Kepong, Selangor, Malaysia

A. Muhd Syamil
Natural Product Division, Forest Research Institute Malaysia, 52109, Kepong, Selangor, Malaysia

J. Muhd Haffiz
Natural Product Division, Forest Research Institute Malaysia, 52109, Kepong, Selangor, Malaysia

G. Hema-Thopla
Natural Product Division, Forest Research Institute Malaysia, 52109, Kepong, Selangor, Malaysia

Denton Oluwabunmi Aderonke
Institute of Agricultural Research and Training, Obafemi Awolowo University, Moor plantation Ibadan, Nigeria

Ganiyu Adeniyi Gbadegesin
Department of Geography, University of Ibadan, Nigeria

Daljit Singh Karam
Department of Forest Production, Faculty of Forestry, Institute of Tropical Forestry and Forest Products, Universiti Putra Malaysia, 43400 UPM Serdang, Selangor, Malaysia

A. Arifin
Department of Forest Production, Faculty of Forestry, Institute of Tropical Forestry and Forest Products, Universiti Putra Malaysia, 43400 UPM Serdang, Selangor, Malaysia
Laboratory of Sustainable Bioresource Management, Institute of Tropical Forestry and Forest Products, Universiti Putra Malaysia, 43400 UPM Serdang, Selangor, Malaysia

O. Radziah
Department of Land Management, Faculty of Agriculture, Universiti Putra Malaysia, 43400 UPM Serdang, Selangor, Malaysia
Laboratory of Food Crops and Floriculture, Institute of Tropical Agriculture, Universiti Putra Malaysia, 43400 UPM Serdang, Selangor, Malaysia

J. Shamshuddin
Department of Land Management, Faculty of Agriculture, Universiti Putra Malaysia, 43400 UPM Serdang, Selangor, Malaysia

Hazandy Abdul-Hamid
Department of Forest Production, Faculty of Forestry, Institute of Tropical Forestry and Forest Products, Universiti Putra Malaysia, 43400 UPM Serdang, Selangor, Malaysia
Laboratory of Sustainable Bioresource Management, Institute of Tropical Forestry and Forest Products, Universiti Putra Malaysia, 43400 UPM Serdang, Selangor, Malaysia

Nik M. Majid
Department of Forest Production, Faculty of Forestry, Institute of Tropical Forestry and Forest Products, Universiti Putra Malaysia, 43400 UPM Serdang, Selangor, Malaysia

I. Zahari
Forestry Department Peninsular Malaysia, Universiti Putra Malaysia, 43400 UPM Serdang, Selangor, Malaysia

Nor Halizah Ab. Halim
Forestry Department Peninsular Malaysia, Universiti Putra Malaysia, 43400 UPM Serdang, Selangor, Malaysia

Cheng Kah Yen
Department of Forest Production, Faculty of Forestry, Institute of Tropical Forestry and Forest Products, Universiti Putra Malaysia, 43400 UPM Serdang, Selangor, Malaysia

Hermann Désiré Mbouobda
Department of Biology, Higher Teachers' Training College (HTTC), University of Bamenda, P.O. Box 39, Bamenda, Cameroon
Laboratory of Plant Biology, Department of Biological Sciences, Ecole Normale Supérieure (ENS), University of Yaoundé 1, P.O. Box 47, Yaoundé, Cameroon

Fotso
Department of Biology, Higher Teachers' Training College (HTTC), University of Bamenda, P.O. Box 39, Bamenda, Cameroon
Laboratory of Plant Biology, Department of Biological Sciences, Ecole Normale Supérieure (ENS), University of Yaoundé 1, P.O. Box 47, Yaoundé, Cameroon

Carole Astride Djeuani
Laboratory of Plant Biology, Department of Biological Sciences, École Normale Supérieure (ENS), University of Yaoundé 1, P.O. Box 47, Yaoundé, Cameroon

Kilovis Fai
Department of Biology, Higher Teachers' Training College (HTTC), University of Bamenda, P.O. Box 39, Bamenda, Cameroon

Ndoumou Denis Omokolo
Laboratory of Plant Biology, Department of Biological Sciences, École Normale Supérieure (ENS), University of Yaoundé 1, P.O. Box 47, Yaoundé, Cameroon

G. Cucci
Department of Agricultural and Environmental Science, University of Bari Via G. Amendola 165/A 70126 Bari, Italy

G. Lacolla
Department of Agricultural and Environmental Science, University of Bari Via G. Amendola 165/A 70126 Bari, Italy

P. Rubino
Department of Agricultural and Environmental Science, University of Bari Via G. Amendola 165/A 70126 Bari, Italy

Ademir Sérgio Ferreira Araújo
Federal University of Piauí, Agricultural Science Center, Soil Quality Laboratory, Campus da Socopo, CEP 64000-000, Teresina, PI, Brazil

Maria Dorotéia Marçal Silva
Federal University of Piauí, Agricultural Science Center, Soil Quality Laboratory, Campus da Socopo, CEP 64000-000, Teresina, PI, Brazil

Luiz Fernando Carvalho Leite
Embrapa Mid-North, Av. Duque de Caxias, Teresina, PI, Brazil

Fabio Fernando de Araujo
UNOESTE, Campus II, Presidente Prudente, SP, Brazil

Nildo da Silva Dias
Federal University of Semi-Arid, UFERSA, Mossoró, RN, Brazil

Zahida Rashid
Department of Agronomy, Punjab Agricultural University, Ludhiana, 141 004, Punjab, India

Mudasir Rashid
Department of Agronomy, Punjab Agricultural University, Ludhiana, 141 004, Punjab, India

Suhail Inamullah
Department of Agronomy, Punjab Agricultural University, Ludhiana, 141 004, Punjab, India

Souliha Rasool
Department of Agronomy, Punjab Agricultural University, Ludhiana, 141 004, Punjab, India

Fayaz Ah. Bahar
Department of Agronomy, Punjab Agricultural University, Ludhiana, 141 004, Punjab, India

Abebe Zerihun
Bako Agricultural Research Center, P. O. Box 03, Bako West Shoa, Ethiopia

J. J. Sharma
Haramaya University, P. O. Box, 138, Dire Dawa, Ethiopia

Dechasa Nigussie
Haramaya University, P. O. Box, 138, Dire Dawa, Ethiopia

Kanampiu Fred
International Maize and Wheat Improvement Centre (CIMMYT), P.O. Box 1041-00621, Nairobi, Kenya

Tanziman Ara
Plant Breeding and Gene Engineering Laboratory, Department of Botany, University of Rajshahi, Rajshahi - 6205, Bangladesh

M. Rezaul Karim
Plant Breeding and Gene Engineering Laboratory, Department of Botany, University of Rajshahi, Rajshahi - 6205, Bangladesh

M. Abdul Aziz
Plant Breeding and Gene Engineering Laboratory, Department of Botany, University of Rajshahi, Rajshahi - 6205, Bangladesh

Rezaul Karim
Plant Breeding and Gene Engineering Laboratory, Department of Botany, University of Rajshahi, Rajshahi - 6205, Bangladesh

Rafiul Islam
Plant Breeding and Gene Engineering Laboratory, Department of Botany, University of Rajshahi, Rajshahi - 6205, Bangladesh

Monzur Hossain
Plant Breeding and Gene Engineering Laboratory, Department of Botany, University of Rajshahi, Rajshahi - 6205, Bangladesh

C. L. Kizza
College of Agricultural and Environmental Sciences, Makerere University, Uganda

J. G. M. Majaliwa
College of Agricultural and Environmental Sciences, Makerere University, Uganda

B. Nakileza
College of Agricultural and Environmental Sciences, Makerere University, Uganda

G. Eilu
College of Agricultural and Environmental Sciences, Makerere University, Uganda

I Bahat
College of Agricultural and Environmental Sciences, Makerere University, Uganda

F. Kansiime
College of Agricultural and Environmental Sciences, Makerere University, Uganda

J. Wilson
Centre for Ecology and Hydrology, Bush Estate, Penicuik, Midlothian, EH26 0QB, U.K.

Pramod Kumar
Department of Soil Science, College of Agriculture S.V.P.U.A.T., Meerut 250 110(U.P.), India

Ashok Kumar
Department of Soil Science, College of Agriculture S.V.P.U.A.T., Meerut 250 110(U.P.), India

B. P. Dhyani
Department of Soil Science, College of Agriculture S.V.P.U.A.T., Meerut 250 110(U.P.), India

Pardeep Kumar
Department of Soil Science, College of Agriculture S.V.P.U.A.T., Meerut 250 110(U.P.), India

U. P. Shahi
Department of Soil Science, College of Agriculture S.V.P.U.A.T., Meerut 250 110(U.P.), India

S. P. Singh
Department of Soil Science, College of Agriculture S.V.P.U.A.T., Meerut 250 110(U.P.), India

Ravindra Kumar
Department of Soil Science, College of Agriculture S.V.P.U.A.T., Meerut 250 110(U.P.), India

Yogesh Kumar
Department of Soil Science, College of Agriculture S.V.P.U.A.T., Meerut 250 110(U.P.), India

Amit Kumar
Department of Soil Science, College of Agriculture S.V.P.U.A.T., Meerut 250 110(U.P.), India

Sumit Raizada
Department of Soil Science, College of Agriculture S.V.P.U.A.T., Meerut 250 110(U.P.), India

Mohd Hadi-Akbar Basri
Department of Forest Management, Faculty of Forestry, Universiti Putra Malaysia, 43400 Serdang, Selangor, Malaysia

Nasima Junejo
Department of Forest Production, Faculty of Forestry, Universiti Putra Malaysia, 43400 Serdang, Selangor, Malaysia
Institute of Tropical Forestry and Forest Products, Universiti Putra Malaysia, 43400 Serdang, Selangor, Malaysia

Arifin Abdu
Department of Forest Management, Faculty of Forestry, Universiti Putra Malaysia, 43400 Serdang, Selangor, Malaysia
Institute of Tropical Forestry and Forest Products, Universiti Putra Malaysia, 43400 Serdang, Selangor, Malaysia

Hazandy Abdul Hamid
Department of Forest Production, Faculty of Forestry, Universiti Putra Malaysia, 43400 Serdang, Selangor, Malaysia
Institute of Tropical Forestry and Forest Products, Universiti Putra Malaysia, 43400 Serdang, Selangor, Malaysia

Mohd Ashadie Kusno
Department of Forest Production, Faculty of Forestry, Universiti Putra Malaysia, 43400 Serdang, Selangor, Malaysia
Institute of Tropical Forestry and Forest Products, Universiti Putra Malaysia, 43400 Serdang, Selangor, Malaysia

I. M. Hilmi
Department of Crop Science, Faculty of Agriculture and Food Science, University Putra Malaysia Bintulu Sarawak Campus, 97008 Bintulu, Sarawak, Malaysia

K. Susilawati
Department of Crop Science, Faculty of Agriculture and Food Science, University Putra Malaysia Bintulu Sarawak Campus, 97008 Bintulu, Sarawak, Malaysia

O. H. Ahmed
Department of Crop Science, Faculty of Agriculture and Food Science, University Putra Malaysia Bintulu Sarawak Campus, 97008 Bintulu, Sarawak, Malaysia

Nik M. Majid
Institute of Tropical Forestry and Forest Products (INTROP), Universiti Putra Malaysia 43400 Serdang, Selangor, Malaysia

Zelalem Bekeko
Department of Plant Sciences, Haramaya University Chiro Campus, P. O. Box 335, Chiro, Ethiopia

P. Arunachalam
Dryland Agricultural Research Station, Chettinad-630 102, Tamil Nadu Agricultural University, Tamil Nadu, India

P. Kannan
Dryland Agricultural Research Station, Chettinad-630 102, Tamil Nadu Agricultural University, Tamil Nadu, India

G. Prabukumar
Dryland Agricultural Research Station, Chettinad-630 102, Tamil Nadu Agricultural University, Tamil Nadu, India

M. Govindaraj
Centre for Plant Breeding and Genetics, Coimbatore-641 003, Tamil Nadu Agricultural University, Tamil Nadu, India

Trust Manyiwa
Department of Environmental Science, University of Botswana, Private Bag 00704, Gaborone, Botswana

Oagile Dikinya
Department of Environmental Science, University of Botswana, Private Bag 00704, Gaborone, Botswana

Shaiane D.M. Lucas
Department of Water Resources and Environmental Sanitation of the Western Paraná State University – University Street, 2069, 85819-110, Cascavel - PR, Brazil

Silvio C. Sampaio
Department of Water Resources and Environmental Sanitation of the Western Paraná State University – University Street, 2069, 85819-110, Cascavel - PR, Brazil

Miguel A. Uribe-Opazo
Department of Water Resources and Environmental Sanitation of the Western Paraná State University – University Street, 2069, 85819-110, Cascavel - PR, Brazil

Simone D. Gomes
Department of Water Resources and Environmental Sanitation of the Western Paraná State University – University Street, 2069, 85819-110, Cascavel - PR, Brazil

Nathalie C. H. Kessler
Department of Water Resources and Environmental Sanitation of the Western Paraná State University – University Street, 2069, 85819-110, Cascavel - PR, Brazil

Naimara V. Prado
Department of Water Resources and Environmental Sanitation of the Western Paraná State University – University Street, 2069, 85819-110, Cascavel - PR, Brazil

Habib Kato
Department of Agriculture, Kyambogo University, P. O. Box 1, Kyambogo, Uganda

Felix Budara Bareeba
Department of Animal Science, Makerere University, P. O. Box, 7062, Kampala, Uganda

Elly Nyabombo Sabiiti
Department of Crop Science, Makerere University, P. O. Box, 7062, Kampala, Uganda

Ruixiang Shi
State Key Lab of Resources and Environmental Information System, Institute of Geographic Sciences and Natural Resources Research, Chinese Academy of Sciences, Beijing 100101, China

Xiaohuan Yang
State Key Lab of Resources and Environmental Information System, Institute of Geographic Sciences and Natural Resources Research, Chinese Academy of Sciences, Beijing 100101, China

Hongqi Zhang
Institute of Geographic Sciences and Natural Resources Research, Chinese Academy of Sciences, Beijing 100101, China

Lixin Wang
Institute of Geographic Sciences and Natural Resources Research, Chinese Academy of Sciences, Beijing 100101, China

Zhang Jinjing
College of Resource and Environmental Science, Jilin Agricultural University, Changchun 130118, China

Gao Qiang
College of Resource and Environmental Science, Jilin Agricultural University, Changchun 130118, China

Wang Qinghe
College of Resource and Environmental Science, Jilin Agricultural University, Changchun 130118, China

Dong Peibo
College of Resource and Environmental Science, Jilin Agricultural University, Changchun 130118, China

Feng Guozhong
College of Resource and Environmental Science, Jilin Agricultural University, Changchun 130118, China

Li Cuilan
College of Resource and Environmental Science, Jilin Agricultural University, Changchun 130118, China

Wang Lichun
Agricultural Environments and Resources Research Centre, Northeast Agricultural Research Centre of China, Changchun 130124, China

www.ingramcontent.com/pod-product-compliance
Lightning Source LLC
Chambersburg PA
CBHW080644200326
41458CB00013B/4732